D1159996

Cell Interactions

and

Receptor Antibodies

in

Immune Responses

Proceedings of the Third
Sigrid Jusélius Symposium

Cell Interactions
and
Receptor Antibodies
in
Immune Responses

Proceedings of the Third
Sigrid Jusélius Symposium

Edited by

O. MÄKELÄ, ANNE CROSS
and
T. U. KOSUNEN

Department of Serology and Bacteriology
University of Helsinki, Helsinki, Finland

1971

LONDON AND NEW YORK

ACADEMIC PRESS INC. (LONDON) LTD
Berkeley Square House
Berkeley Square,
London, W1X 6BA

U.S. Edition published by
ACADEMIC PRESS INC.
111 Fifth Avenue,
New York, New York 10003

PRINTED IN GREAT BRITAIN BY
The Whitefriars Press Ltd., London and Tonbridge

List of Participants

*The address of all organizers is Department of Serology and Bacteriology,
University of Helsinki, Finland*

AHLQUIST, J., *Helsinki, Finland*
AIRD, J., *organizer*
AKAAN-PENTTILÄ, E., *Helsinki, Finland*
ANDERSSON, B., *Stockholm, Sweden*
ARVILOMMI, H., *Turku, Finland*
ASKONAS, B., *London, England*
AUERBACH, R., *Madison, U.S.A.*
BACH, J.-F., *Paris, France*
BRAUN, W., *New Brunswick, U.S.A.*
BRITTON, S., *Stockholm, Sweden*
BUSSARD, A., *Paris, France*
CAMPBELL, D., *Uppsala, Sweden*
CELADA, F., *Stockholm, Sweden*
CLAMAN, H. N., *Denver, U.S.A.*
CROSS, A. M., *organizer*
DAVID, J. R., *Boston, U.S.A.*
DAVIES, A. J. S., *London, England*
DENMAN, A. M., *Stockholm, Sweden*
EHNHOLM, C., *organizer*
FELDMAN, M., *Rehovot, Israel*
FICHTELIUS, K. E., *Uppsala, Sweden*
FISCHER, H., *Freiburg, Germany*
GOOD, R. A., *Minneapolis, U.S.A.*
GOWANS, J. L., *Oxford, England*
GREAVES, M. F., *London, England*
HAIMOVICH, J., *St. Louis, U.S.A.*
HÄKKINEN, I., *Turku, Finland*
HARTMANN, K.-U., *Tübingen, Germany*
HASKILL, J. S., *Kingston, Ontario, Canada*
HATCHER, V. B., *organizer*

v

HÄYRY, P., *Helsinki, Finland*
HUMPHREY, J. H., *London, England*
IVERSON, G. M., *London, England*
JOKIPII, A., *organizer*
JOKIPII, L., *Helsinki, Finland*
JORMALAINEN, S., *organizer*
KLEIN, E., *Stockholm, Sweden*
KONTIAINEN, S., *organizer*
KOSKIMIES, S., *organizer*
KOSTIALA, A., *organizer*
KOSUNEN, T. U., *organizer*
KOUVALAINEN, K., *Turku, Finland*
LEHTINEN, M., *Tampere, Finland*
LESKOWITZ, S., *Boston, U.S.A.*
LINDER, E., *organizer*
MÄKELÄ, O., *organizer*
MITCHISON, N. A., *London, England*
MÖLLER, E., *Stockholm, Sweden*
MÖLLER, G., *Stockholm, Sweden*
MOORE, M. A. S., *Oxford, England*
NATVIG, J., *Oslo, Norway*
NORDMAN, C. T., *Helsinki, Finland*
NOSSAL, G. J. V., *Melbourne, Australia*
NUSSENZWEIG, V., *New York, U.S.A.*
OKER-BLOM, N., *Helsinki, Finland*
PASANEN, V., *organizer*
PENTTINEN, K., *Helsinki, Finland*
PERLMANN, P., *Stockholm, Sweden*
PETTAY, O., *Helsinki, Finland*
PLOTZ, P., *London, England*
RAFF, M. C., *London, England*
RAJEWSKY, K., *Cologne, Germany*
RÄSÄNEN, J., *Helsinki, Finland*
RAUNIO, V., *Oulu, Finland*
RENKONEN, K. O., *organizer*
RICHTER, W., *Uppsala, Sweden*
SAKSELA, E., *Helsinki, Finland*
SALMI, A., *Turku, Finland*
SARVAS, H., *organizer*
SAXÉN, L., *Helsinki, Finland*
SEPPÄLÄ, I., *organizer*
SCHIRRMACHER, V., *Cologne, Germany*

SERCARZ, E., *Los Angeles, U.S.A.*
SIMONSEN, M., *Copenhagen, Denmark*
SINGHAL, S. K., *London, Ontario, Canada*
SJÖBERG, O., *Stockholm, Sweden*
SMITH, R. T., *Gainesville, U.S.A.*
SOOTHILL, J. F., *London, England*
ŠTERZL, J., *Prague, Czechoslovakia*
TALLBERG, T., *Helsinki, Finland*
TALWAR, P. J., *New Delhi, India*
TAYLOR, R. B., *Bristol, England*
TIILIKAINEN, A., *Helsinki, Finland*
VIROLAINEN, M., *Helsinki, Finland*
VISAKORPI, R. S., *organizer*
WAKSMAN, B. H., *New Haven, U.S.A.*
WEIGLE, W. O., *La Jolla, U.S.A.*
WIGZELL, H., *Stockholm, Sweden*
WILSON, D. B., *Philadelphia, U. S. A.*

Preface

The third Sigrid Jusélius Symposium was held at the Savings Banks Institute near Helsinki in June, 1970. The proceedings of the symposium are published in this book. Attempting prompt publication we set an early dead-line for manuscripts. Unfortunately this meant omission of some lectures from the book. On the other hand, formal papers were invited covering some points that were central to the topic and were actively discussed at the meeting even though they were not formally presented as a lecture. A great portion of the discussion was omitted from the book. With some exceptions only those contributions are included that were given to the editors written.

The Sigrid Jusélius Foundation is the most important non-governmental supporter of medical research in Finland. It entrusted the scientific responsibility for this Symposium to a group of scientists from the Department of Serology and Bacteriology, Helsinki University. They are identified in the list of participants. This group benefited from discussions with Professor Nils Oker-Blom, scientific secretary to the Foundation who kindly opened the Symposium. Mr Helge von Knorring, until recently Director of the Foundation was responsible for the practical arrangements for the Symposium. His competent assistants were Mrs Barbara Hatcher, Miss Tuula Hindikka and Mrs Hilkka Kontiopää. The work of organizing the symposium was greatly facilitated by the excellent facilities provided by the Savings Banks Institute.

April 1971. Olli Mäkelä, Anne M. Cross, Timo U. Kosunen

Introduction

This is a book about lymphocytes, their classification, characteristics and interactions with other cells. It is greatly concerned with two classes of lymphocytes which were originally distinguished by the work of Szenberg and Warner (*Nature, Lond.* 1962, *194*, 146), Cooper, Peterson and Good (*Nature, Lond.* 1965, *205*, 143), Claman, Chaperon and Triplett (*J. Immunol.* 1966, *97*, 828), Davies *et al.* (*Transplantation* 1967, *5*, 222), Mitchell and Miller (*J. exp. Med.* 1968, *128*, 821) and others. One of the classes was determined by its approximate absence in neonatally thymectomized mice and in normal mouse bone marrow (it may reside in rabbit bone marrow). Another important characteristic of this thymus-dependent lymphocyte is that it cannot transform into a plasma cell. The other class is characterized by its absence from the thymus, presence in neonatally thymectomized mice, relative abundance in the bone marrow and differentiation under the influence of antigen into plasma cells. Both experimental work and observation of human patients (Good and Finstad; Soothill, Kay and Batchelor in this book) suggested that the latter class is responsible for humoral antibody production, while the former class is greatly involved in homograft rejection, and delayed hypersensitivity reactions. An additional function of the latter class is to help in the antigen-induced activation of plasma cell precursors. This introduction will summarize some of the topics that are discussed in the book. Selection of topics was haphazard.

NOMENCLATURE AND CHARACTERISTICS OF THE LYMPHOCYTE CLASSES

Table I gives some of the synonyms used for the two lymphocyte classes. The editors decided against making the nomenclature uniform, the names used by the authors are generally maintained. The only exception is that the term "thymus-derived cell" was changed to "thymus-dependent cell". The term "antigen sensitive cell" in this book seems to mean small lymphocytes of both classes.

Both classes of small lymphocytes seem to carry antibody-like receptors for antigens on their surface. In the case of B cells these receptors generally have the same specificity as the antibodies that are produced if the cell is triggered. This is suggested for instance by the papers of Humphrey, Roelants and Willcox;

TABLE I

Nomenclature and characteristics of lymphocyte classes

	Thymus-dependent lymphocytes	Thymus-independent lymphocytes
Other names	Thymus-derived lymphocyte. Antigen-sensitive cell. T cell.	(Bone) marrow derived cell. Antibody-forming cell precursor (AFCP). Plasma cell precursor (PCP). B cell.
Receptors for antigens on cell surface	Demonstrable but probably scarce	Demonstrable
Antigenic markers of classical Ig classes on cell surface	Absent with possible exception of IgM	Demonstrable
Organ-specific antigens theta, Ly^a Ly^b	Present	Absent
Receptors for the third component of complement	Absent?	Present?
Organ-specific antigen PC-1	Absent at least in thymus cells	Present in plasma cells

Mitchison; E. Möller and Greaves; and Wigzell, Andersson, Mäkelä and Walters. B cells may have more receptor antibodies than T cells; this is suggested by the fact that radioactive antigen destroyed plasma cell precursors of the corresponding specificity, but not helper activity (T cells) of the same specificity. This argument is weakened by Miller's finding that irradiated (dead) T cells co-operate *in vitro*. Another finding suggesting scarcity of receptor antibodies in T cells was that Wigzell *et al.* could not deplete helper activity by passing a cell population through antigen columns, although antibody forming activity was arrested.

An important finding suggesting scarcity of antibodies on the T cell surface was made by Raff. Using fluorescent antibodies he could not demonstrate markers of conventional immunoglobulins on cells carrying the T cell marker theta, although they could be demonstrated on theta-negative lymphocytes. On the other hand, using different techniques, Greaves and Hogg as well as Humphrey, Roelants and Willcox produced evidence for IgM and L chain markers on theta-positive cells. Similar findings were made by Ada (*Developmental aspects of Antibody Formation and Structure,* Academic Press, 1970, *2,* 503).

A way to force most of these findings into conformity might be to assume that T cell receptors belong to a new Ig-class related to IgM. Some anti-IgM sera cross react with it (Ada; Greaves and Hogg; Humphrey, Roelants and Willcox) while others do not (Raff).

There are antigenic markers (Theta, Ly^a, Ly^b) on T cells that cannot be demonstrated on B cells (Raff; E. Möller and Greaves; Greaves and Hogg; Takahashi, Old and Boyse, *J. exp. Med.* 1970, *131*, 1325). Possible markers of B cells are more recent and less firmly documented. They include the complement receptors described by Nussenzweig, Bianco and Dukor. Another possible marker for B cells is the PC-1 antigen, which has not yet been demonstrated in lymphocytes but has been demonstrated in plasma cells (Takahashi, Old and Boyse, *J. exp. Med.*, 1970, *131*, 1325). The third marker reported in September by Raff and Owen (*Proceedings of the 3rd International Congress of Lymphatic Tissue and Germinal Centers in Immune Reactions*) looks very useful. Its antiserum was obtained by immunizing rabbits with bone marrow cells of mice. The donors were thymectomized, lethally irradiated and reconstituted with fetal liver cells.

PREPARATION OF PURIFIED LYMPHOCYTE CLASSES

Progress is less impressive in preparation of purified T cells or purified B cells. This might be desired for example by students of tumour immunity. A promising possibility is selective complement lysis of T cells from a population using anti-theta, or B cells using anti-Ig or the Raff-Owen antiserum. Nussenzweig, Bianco and Dukor produced lymphocyte populations depleted of complement reactive cells by density gradient centrifugation of rosettes; these may have been B cell depleted populations. Andersson showed that mature T cells can be enriched in the thymus at the expense of immature cells by prior cortisone treatment of the animal. Finally, Miller; Mitchison; and Hartmann prepared "educated thymus cells", which are probably reasonably free of B cells and contain an increased proportion of T cells primed with antigen.

TRIGGERING OF LYMPHOCYTES

Lymphocytes can respond to signals in two ways, by proliferating (immune induction), or by committing the equivalent of suicide (tolerance). While the difference between the two signals still remains unknown important data and new ideas were put forward at the meeting. One of them was demonstration by Iverson of a tolerogen which is not immunogenic. It is C3H myeloma protein 5563. This protein injected into syngeneic mice is not immunogenic even in Freund's adjuvant, but coupled with DNP or injected across an allotype barrier, it immunizes mice against an idiotype. Native protein can paralyse cells that are capable of producing the anti-idiotype.

Other important data about triggering were presented by Mitchison. They suggest that T cells are triggered by lower antigen concentrations than B cells (both for immune induction and paralysis induction). This phenomenon could be explained by one of two hypotheses. According to one, the differential affinity hypothesis, T cells have higher affinity receptors than B cells. This has been sponsored by Taylor and Iverson (*Proc. Roy. Soc. Series B*, 1971, *176*, 393). The other, differential triggering hypothesis, is favoured by G. Möller and by Mitchison, mainly because of its power to explain the high number of T lymphocytes reacting to transplantation antigens. The power of their argument is somewhat reduced but not depleted by a finding of Wilson and Nowell, which is supported by findings of Lafferty and Jones (*Austr. J. exp. Biol. Med. Sci.* 1969, *47*, 17). These people found that the high number of reacting T cells is only valid for allogeneic interactions: not many T cells react to xenogeneic transplantation antigens. On the other hand the finding that T cells but not B cells are induced by phytohaemagglutinin (Greaves and Roitt, *Clin. exp. Immunol.* 1968, *3*, 393, Doenhoff, Davies, Leuchars and Wallis, *Proc. Roy. Soc. Lond. B* 1970, *176*, 69) might suggest that the former are more easily triggered than the latter.

Weigle, Chiller and Benjamin demonstrated that both collaborating cell populations, thymus cells and bone marrow cells, exhibit specific tolerance to human gamma globulin. They presented suggestive evidence that tolerance may break down earlier in the B cell population than in T cell population. If the kinetics of paralysis were different in the T and B cell populations this might explain why Taylor (*Transplant. Rev.* 1969, *1*, 114) failed to find specific tolerance in bone marrow cells of paralysed donor mice.

Weigle and his collaborators made rabbits unresponsive to BSA. When they were immunized with serum albumins of other species they produced anti-BSA whose quantity and quality were similar to those of anti-BSA in control rabbits. The authors suggest that the tolerant rabbits lacked only BSA-specific T (helper) cells and not BSA-specific B cells. When human serum albumin, for instance, provided new determinants for the helper function plasma cell precursors against the determinants shared by HSA and BSA were triggered.

ARE POST-PCP CLONES RESTRICTED TO ONE IG CLASS?

While clones deriving from a plasma cell precursor are restricted to one allotype of a locus (Bosma and Weiler, *J. Immunol.* 1970, *104*, 203) and to a maximum of a few antigenic specificities (Wigzell, Andersson, Mäkelä and Walters) they may not be restricted to one class. Most convincing demonstrations that a clone deriving from a plasma cell precursor can make both IgM and IgG are the following: (i) The demonstration by Nossal, Lewis and Warner of single cells synthesizing both IgM and IgG anti-sheep erythrocyte antibodies. (ii) The demonstration of Wang *et al.* (*Proc. Nat. Acad. Sci.* 1970, *66*, 657) that two para-proteins in one patient, one IgM and the other IgG, had identical (probably H chain) idiotypes and identical N-terminal (27 residues) H chain sequences. (iii) The finding of Oudin and Michel (*J. exp. Med.* 1969, *130*, 619) that IgM and IgG antibodies of individual rabbits share idiotypic determinants. It would be attractive to deduce from these findings that the switch from early IgM to late IgG production which occurs at least in anti-erythrocyte responses is caused by a switch within clones and not by selection between clones. The data of Šterzl and Nordin support this deduction. They found a considerable number of spleen colonies which synthesized both IgM and IgG antibody. If it can be demonstrated that the number of B cells and not T cells was limiting in the formation of these colonies this is a very strong argument. Against this deduction is the finding of Cosenza and Nordin (*J. Immunol.* 1970, *104*, 976) that the number of double IgM + IgG cells does not increase at the time of the IgM → IgG switch and is low (0.79% of all antibody producers). Also against this deduction are papers of Wigzell, Andersson, Mäkelä and Walters; Schirrmacher; Mäkelä, Pasanen and Sarvas. They argue that at least a majority of secreted antibody of a given class is produced by cells whose PCP ancestor had receptor antibody of this class, and whose whole progeny mainly produced this Ig class.

Whether specialized IgM and IgG clones are responsible for most immuno-globulin production cannot be decided at present. If the answer is no, a "mobile" gene for the variable region of the H· chains might be the cause of doubly active cells or doubly active clones. It might fuse to constant γ and μ chain genes in some cells or float from one the μ gene to the γ gene in another.

Cell Interactions in the Induction of a Humoral Response

An important contribution to the meeting was the bringing together by Raff of two phenomena, the Claman-Miller interaction of two cell types in humoral responses and the Overay-Benacerraf interaction of several different determinants in a humoral anti-hapten response (carrier-specificity). Mitchison, Rajewsky and Taylor (*Developmental Aspects of Antibody Formation and Structure*, ed. J. Šterzl and J. Říha, Academic Press, 1970, *2*, 547) had demonstrated that the basis of the Ovary-Benecerraf phenomenon is a need for hapten-specific cells and carrier-specific helper cells to collaborate in anti-hapten production. They also produced suggestive evidence that the carrier-specific helper cells are the thymus-dependent partners of the Claman-Miller interaction. This was now demonstrated by Raff who showed that the carrier-specific helper cells carry the T cell marker theta.

Collaboration between T cells and B cells was shown to enhance *in vitro* responses to sheep erythrocytes (Hartmann). Both Mitchison and Weigle, Chiller and Benjamin presented evidence that animals with high-zone tolerance lack both B and T cell activities for extended periods.

Several authors produced evidence that this collaboration is not necessary in all humoral responses. Good and Finstad pointed to the high Ig-levels and/or reasonable humoral responses when thymus function is eliminated. Andersson and Blomgren demonstrated fully thymus-independent antigens. Common to them seems to be a small number of different determinants, but a high multiplicity of one kind. Schirrmacher argued that B cells with high affinity receptors are not in need of anti-carrier help and Kontiainen produced evidence that B cells with IgM receptors are less in need of anti-carrier help than B cells with IgG receptors.

Contents

LYMPHOCYTE–LYMPHOCYTE INTERACTIONS IN IMMUNE RESPONSES

Interactions in vivo

Interactions in vivo

Interactions via Humoral Factors

EFFECTOR MECHANISMS SECONDARY TO LYMPHOCYTE ACTIVATION

Co-operative Antibody: A Concentrating Device

GUNTHER DENNERT, HELMUT POHLIT and
KLAUS RAJEWSKY

Institut für Genetik der Universität Köln
D-5 Köln 41, Weyertal 121, W. Germany

Introduction

Antigen receptors are heterogeneous. This applies to both their specificity for antigen and their function in the regulation of the immune response. The final step in the induction process is the activation of antibody forming cell precursors, which is mediated by receptors on the surface of these cells. However, the regulation of the immune response starts before this. Antigens are selected by receptors on helper cells and by humoral antibodies for presentation to the final receptor on the antibody forming cell precursor. Our experiments are concerned with the co-operative function of 19S antibodies, which have been shown by Henry and Jerne (1968) to be capable of stimulating the response to sheep red cells (SRBC). We are going to present evidence that the mechanism of stimulation by 19S antibodies is based on a specific accumulation of antigen in the spleen.

Materials and Methods

The mice were outbred NMRI strain, Ivanovas, Kisslegg, W. Germany.

1. 19S anti-SRBC antibodies. Mice were bled four days after intravenous injection of a large dose of SRBC (Behringwerke, Marburg, W. Germany). 19S antibodies were purified by acid precipitation, filtration through Sephadex G-200 (Pharmacia, Uppsala, Sweden) and ultracentrifugation. The purification procedure is described in detail elsewhere (Dennert and Rajewsky, 1971). The 19S antibody preparations were still heavily contaminated with other serum proteins, but were free of any detectable 7S antibody.

4×10^5 SRBC were either injected at various times up to 2 hr after injection of the 19S preparation, or mixed with the antibodies immediately before

3

injection. The same result was obtained in both cases. All injections (0.1-0.2 ml/mouse) were intravenous.

2. Assay of immune response. 19S anti-SRBC antibodies were assayed by the usual haemolytic assay. Reaction mixtures consisted of 0.025 ml diluted antibody, 0.025 ml of a suspension containing 2×10^8 SRBC/ml, and 0.025 ml 1 : 10 diluted guinea pig serum (absorbed with SRBC). Antibody concentration is expressed by the lytic titre. The microtitre equipment was obtained from Cooke Engineering Company, Alexandria, Va., U.S.A.

Direct plaque-forming cells (PFC) were determined as described by Jerne and Nordin (1963) with the modifications of Mishell and Dutton (1967).

3. Determination of the uptake of radioactively labelled red blood cells into the spleen. 10^8 red blood cells were suspended in 1 ml balanced salt solution and incubated with $100 \mu C$ ^{51}Cr (sodium chromate) for 30 min at $37°C$. The suspension was then cooled to $0\text{-}4°C$ and the cells were washed twice with balanced salt solution.

2×10^6 labelled red blood cells were injected intravenously either alone or together with 19S antibody. After various times the spleen was excised, weighed and counted directly in a Packard gamma counter.

RESULTS

1. Stimulation of SRBC uptake into the spleen. The kinetics of SRBC uptake into the spleen and liver in the presence and absence of 19S anti-SRBC antibodies are shown in Fig. 1. In the spleen the radioactivity was highest 10 min after injection, and later decreased. The uptake was considerably enhanced when the SRBC were injected together with 19S antibodies. In contrast, uptake into the liver was not affected by 19S antibodies, and the label in the liver remained constant over a time period of at least 2 hr. The enhancement of splenic uptake

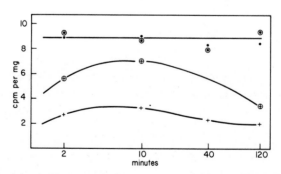

Fig. 1. Kinetics of SRBC uptake into spleen and liver. + SRBC alone, spleen; ⊕ SRBC + 19S anti-SRBC, spleen; ● SRBC alone, liver; ◉ SRBC + 19S anti-SRBC, liver. The 19S preparation had a lytic titre of 2^5.

of SRBC by 19S anti-SRBC antibodies appears to be specific, since the same antibody preparation did not increase the uptake of mouse red blood cells by the spleen (Table I).

TABLE I

Specificity of enhanced erythrocyte uptake induced by 19S antibody

Injection of	Uptake (cpm/mg spleen)[b]
2×10^6 SRBC	0.91
2×10^6 SRBC + 19S anti-SRBC[a]	2.64
2×10^6 MRC	1.11
2×10^6 MRC + 19S anti-SRBC[a]	0.97

[a] 0.1 ml of a preparation having a lytic titre of 2^6.
[b] determined 10 min after injection.
MRC = mouse red blood cells.

In one experiment, of the total injected ^{51}Cr-activity 30-35% was detected in the liver, 1.5% (3% with 19S antibody, titre 2^6) in the spleen, and c. 2% in blood after 10 min.

2. *Antigen uptake and stimulation of the immune response.* Figure 2 shows the relationship between dose of 19S antibody and stimulation of the immune response, and also the dependence of antigen uptake on the antibody dose. The two curves appear to be correlated perfectly.

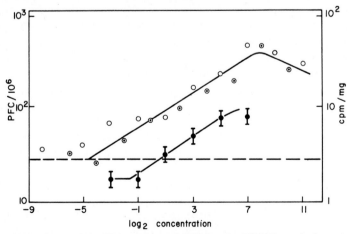

Fig. 2. Dependence of the PFC response to SRBC and of SRBC uptake into the spleen on the dose of 19S anti-SRBC. Open circles represent mean PFC responses on day 6 of groups of 5-6 mice; (2 separate experiments) 4×10^5 SRBC injected per mouse. Closed circles represent means of radioactivity in the spleens of five mice 10 min after injection of 2×10^6 ^{51}Cr-labelled SRBC. Vertical bars are standard error.

DISCUSSION

Our 19S anti-SRBC antibody preparations stimulate the response to SRBC at surprisingly small doses, much smaller than the doses reported in the original publication of Henry and Jerne (1968). This discrepancy could be due to different degrees of purity of the 19S preparations. On the assumption that approximately one IgM molecule is required to lyse one erythrocyte (Humphrey and Dourmashkin, 1965), we calculate that at the lowest detectable stimulation, the IgM dose amounted to approximately 10^6 molecules of haemolytic antibody. In view of the 10^{14}-10^{15} IgM molecules present in the serum of a normal mouse (Barth et al., 1965), one would have to assume that lytic 19S anti-SRBC antibody has a frequency of not more than 10^{-8} to 10^{-9} in "normal" IgM. The problem of this surprisingly low value seems to us worth further consideration.

The rapid concentration of antigen in the spleen seems to be the primary co-operative effect of specific IgM. Secondary effects, but still occurring rather early in the immune response, might be indicated by the reduction in the enhanced radioactivity concentration in the second hour after antigen injection. The specificity of the stimulatory effect in the system (Henry and Jerne, 1968) is underlined by the finding that the uptake into the spleen is not enhanced for non-cross reacting erythrocytes. We want to mention in this context that the dose response relationship recorded in Fig. 2 holds fairly well for a number of 19S preparations, including preparations from antisera to cross reacting antigens and from normal sera (Dennert and Rajewsky, 1971).

Comparison of uptake data for spleen and liver demonstrates the relevance of the stimulatory effect to the lymphatic system and therefore the immune response. That the observed effect might be of general importance for the induction of immunity is suggested by the finding of a similar stimulation in the response to keyhole limpet haemocyanin (Dennert and Rajewsky, 1971). The antibody-mediated co-operation of erythrocyte isoantigens described by McBride and Schierman (1970) may be due to the concentrating mechanism described here.

The relevance of this mechanism for thymus-marrow co-operation cannot be established by our data, and must be borne out by further investigations.

ACKNOWLEDGEMENTS

This work was supported by the Sonderforschungsbereich 74 and the Stiftung Volkswagenwerk.

We are grateful to Miss G. v. Hesberg for excellent assistance.

Helmut Pohlit is on leave from the Basle Institute of Immunology.

SUMMARY

The immune response of mice to low doses of SRBC is enhanced by exceedingly low doses of anti-SRBC IgM. The co-operative function of the IgM antibodies is correlated with their capacity to concentrate antigen in the spleen.

REFERENCES

Barth, W. F., McLaughlin, C. L. and Fahey, J. L. (1965). *J. Immun. 95,* 781-790.
Dennert, G. and Rajewsky, K. (1971). In preparation.
Henry, C. and Jerne, N. K. (1968). *J. exp. Med. 128,* 133-155.
Humphrey, J. H. and Dourmashkin, R. R. (1965). Ciba Fdn. Symp. "Complement" (G. E. W. Wolstenholme and J. Knight, eds), p. 175.
Jerne, N. K. and Nordin, A. A. (1963). *Science, N.Y. 140,* 405.
McBride, R. A. and Schierman, L. W. (1970). *J. exp. Med. 131,* 377-390.
Mishell, R. I. and Dutton, R. W. (1967). *J. exp. Med. 126,* 423-442.

DISCUSSION

LESKOWITZ. If you give small amounts of IgG antibody, do you get similar localization of SRBC to the spleen? The effect here is one of suppression rather than stimulation. How do you account for this difference?

RAJEWSKY. The uptake of SRBC into the spleen seems also to be enhanced by IgG antibody, although much higher doses of antibody are required than in the case of IgM. I imagine that within the spleen, antibodies of different classes may direct the antigen to different pathways.

In vitro Immune Response to Sheep Erythrocytes with Different Cell Populations

M. VIROLAINEN, V. PASANEN, E. AKAAN-PENTTILÄ
and P. HÄYRY

Laboratory of Immunobiology, 3rd Dept. of Pathology
University of Helsinki, Helsinki 29
and
State Serum Institute, Helsinki, Finland

The primary immune response to sheep red blood cells (SRBC) in dissociated mouse spleen cell cultures closely parallels the *in vivo* response (Mishell and Dutton, 1967; Marbrook, 1967), and offers a model system for studying cells involved in immune responses. An interaction of two or more cell types appears to be required for the induction of the primary response *in vitro* (Mosier, 1967; Mosier and Coppleson, 1968; Haskill *et al.*, 1970). Two populations of mouse spleen cells, both of which are required for the response, can be separated on the basis of their adherence to plastic. One is an adherent cell population, possibly macrophages, the other is a non-adherent lymphoid cell population (Mosier, 1967; Pierce, 1969).

In order to characterize the cell types involved in the response in greater detail, we have replaced the adherent and non-adherent spleen cell populations with cells of more exactly known type and origin. In addition we have preliminary results of the effect of macrophage-specific antiserum on the adherent cell population.

CBA mice spleen cells were cultured according to the method of Mishell and Dutton (1967), and the method of Mosier (1967) was used to separate adherent and non-adherent spleen cell populations. The number of plaque forming cells (PFC) was assayed on day 4.

In 19 unseparated control spleen cell cultures the number of PFC was 26 ± 16 (mean $\pm$ SD) per 10^6 recovered cells, and in cultures stimulated with 3×10^6 SRBC the number was 127 ± 58. In each experiment the number of PFC was significantly higher in the stimulated cultures than in the corresponding controls.

As found by Mosier (1967) and by Pierce (1969) adherent or non-adherent spleen cells alone gave little or no response. Combining these two populations reconstituted the response.

Unseparated lymph node cell cultures stimulated with SRBC gave little or no response (Table I). However, when lymph node cells were combined with spleen adherent cells a good response was obtained in all experiments (Table 1).

TABLE I

Cultured cells	SRBC	PFC per culture		
LN	−	9	0	14
LN	+	154	2	49
LN + MR	−	37	23	116
LN + MR	+	548	284	629
MR	+	N.D.	1	13

Primary immune response to SRBC in lymph node cell cultures (LN). MR = adherent spleen cells. N.D. = not done. Three different experiments.

Buffy coat cells alone, or combined with spleen adherent cells, did not produce any significant response (Table II). In the experiments reported in Table II, the mice were injected with supernatant fluid from *B. pertussis* cultures to increase the yield of buffy coat cells (Morse and Bray, 1969). Spleen cell cultures from these mice gave a normal primary response. Buffy coat cells from non-pertussis-treated mice were also tested both alone and in combination with spleen adherent cells. They were found to be equally unresponsive.

The spleen adherent cell population was replaced with pure macrophages, with freshly collected peritoneal cells, and with bone marrow cells. Pure

TABLE II

Cultured cells	SRBC	PFC per culture		
BC	−	N.D.	22	45
BC	+	12	8	33
BC + MR	−	N.D.	N.D.	38
BC + MR	+	46	0	33
MR	+	0	2	4

Lack of primary immune response to SRBC in buffy coat cultures (BC). Mice injected i.v. with 0.2 ml of supernatant fluid from *B. pertussis* culture three days before sacrifice. MR = spleen adherent cells. Three different experiments.

populations of mouse macrophages were obtained by culturing mouse peritoneal macrophages in the presence of conditioned medium from mouse fibroblast cultures (Virolainen and Defendi, 1967). None of these cell populations was able to replace the adherent cells of splenic origin, and restore the response of non-adherent spleen cells. Moreover, the response of unseparated spleen cell cultures was even inhibited by adding peritoneal exudate cells or bone marrow

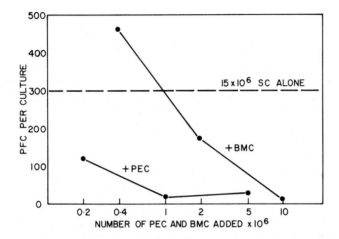

Fig. 1. Effect of freshly collected peritoneal exudate cells (PEC) or bone marrow cells (BMC) on the primary immune response to SRBC (3×10^6) in spleen cell (SC) cultures.

cells to the cultures. Higher numbers of bone marrow cells than of peritoneal exudate cells were required for the inhibitory effect (Fig. 1). A similar inhibitory effect of peritoneal exudate cells on the *in vitro* response has been reported by Diener *et al.* (1970).

The effect of antigen dose on the inhibitory effect of peritoneal exudate and bone marrow cells was tested by adding different numbers of SRBC to the cultures. With higher amounts of antigen the inhibition was partially reversed (Table III). This excludes the possibility that the inhibition is due to some inhibitory substances produced by the added cells.

The effect of anti-macrophage serum (AMS) on the adherent spleen cell population was studied by treating the cells with different concentrations of unabsorbed or lymph node cell-absorbed AMS in the presence of guinea pig complement. The AMS was produced by immunizing rabbits with macrophages cultured for two weeks in the presence of conditioned medium. At this time all lymphocytes had disappeared from the cultures, and the number of macrophages had increased to approximately 15 times the initial number (Virolainen and Defendi, 1967; Virolainen and Lahti, 1970). In double diffusion and indirect

TABLE III

Number of SRBC per culture	PFC per culture				
	Exp. 47		Exp. 52		
	SC	SC + PEC	SC	SC + PEC	SC + BMC
0	145	56	15	6	40
3×10^4	406	38			
3×10^5	555	66	376	13	212
3×10^6	482	75	296	18	172
3×10^7	200	255	158	91	280

Effect of antigen dose on primary immune response to SRBC in cultures containing 15×10^6 spleen cells (SC) and in cultures containing in addition 1×10^6 peritoneal exudate cells (PEC) or 2×10^6 bone marrow cells (BMC). Two different experiments.

immunofluorescence the unabsorbed AMS reacted only with macrophages, but the cytotoxic titre for macrophages was 1 : 160 and for lymph node lymphocytes 1 : 40. After absorption with lymph node cells the cytotoxicity to lymphocytes disappeared (lowest tested dilution 1 : 5) and that to macrophages decreased to 1 : 80 (Virolainen and Lahti, 1970). The adherent cell layer was treated with different dilutions of unabsorbed or absorbed AMS and complement for 30 min, washed, and either lymph node cells or non-adherent spleen cells were added to the cultures (Table IV). Unabsorbed AMS inhibited the

TABLE IV

No. of exp.	Cells cultured	PFC per culture					
		Control	Dilution of AMS				
			1 : 160	1 : 80	1 : 40	1 : 20	1 : 10
38	LN	49					
	LN + MR (unabs. AMS)	466	259	152	103		
	LN + MR (LN-abs. AMS)	466		408	264	350	
52	LR	72					
	LR + MR (unabs. AMS)	385	53	62			
	LR + MR (LN-abs. AMS)	385	250	223	270	172	35

Primary immune response to SRBC in cultures of spleen adherent cells (MR) combined with spleen non-adherent cells (LR) or lymph node cells (LN). The MR population was treated with different dilutions of either unabsorbed or lymph node cell absorbed AMS. Controls MR treated with complement only. All cultures were stimulated with SRBC.

reconstituting effect of adherent spleen cells even at high dilutions, whereas absorbed macrophage-specific AMS was effective only at a considerably lower dilution.

The interpretation of these preliminary data is not unequivocal, and more experiments are needed to draw any firm conclusions. Apparently, since the non-adherent spleen cell population can be replaced by lymph node cells (consisting of nearly 100% lymphocytes), the non-adherent cell required for the response shares the characteristics of a lymphocyte. It has been suggested by Mosier and Coppleson (1968) that two types of non-adherent cells are required (a thymus-dependent and a bone marrow-derived cell?) for the response. Both of these cells are apparently present in the lymph node, but surprisingly not in the buffy coat.

The adherent cell population cannot be replaced by peritoneal or bone marrow-derived macrophages. Moreover, these cells inhibited the response of unseparated spleen cells. Furthermore, since the cytotoxic titres of unabsorbed and absorbed AMS for peritoneal macrophages differed by only one dilution step, while the effects on the adherent cell population differed by up to four dilution steps, we would be inclined to suggest that the inhibitory effect with non-absorbed AMS is non-specific, and that the adherent cell required for the primary *in vitro* response is not necessarily a macrophage in the sense that it would have the same antigenic properties as peritoneal macrophages. This observation would not be surprising since Mosier and Coppleson (1968) suggested that only one out of 1000-10,000 adherent cells is required for the response, and Haskill *et al.* (1970) found that the required adherent cells sediment separately from the phagocytic spleen cells in bovine serum albumin gradients.

ACKNOWLEDGEMENTS

The work has been supported by the Sigrid Jusélius Foundation, Helsinki, and the National Research Council for Medical Sciences, Finland.

SUMMARY

In order to characterize the cell types involved in primary immune response of mouse spleen cells to sheep red cells *in vitro,* we replaced the adherent or non-adherent spleen cell populations with cells from other sources. The non-adherent spleen cell population could be replaced by lymph node cells but not by buffy coat cells. The adherent cell population could not be replaced by peritoneal macrophages or by bone marrow cells. Moreover, these cells had an inhibitory effect. Treatment of the spleen adherent cell population with

C.I.—2

14   M. VIROLAINEN, V. PASANEN, E. AKAAN-PENTTILÄ AND P. HÄYRY

macrophage-specific antiserum at concentrations cytotoxic to peritoneal macrophages did not abolish the response. The results indicate that the two types of non-adherent spleen cells, suggested to be necessary for the response, are also present in the lymph node but not in the buffy coat, and that the adherent cell required for the response is not necessarily the macrophage.

REFERENCES

Diener, E., Shortman, K. and Russell, P. (1970). *Nature, Lond. 225*, 731-732.
Haskill, J. S., Byrt, P. and Marbrook, J. (1970). *J. exp. Med. 131*, 57-76.
Marbrook, J. (1967). *Lancet ii*, 1279-1281.
Mishell, R. I. and Dutton, R. W. (1967). *J. exp. Med. 126*, 423-442.
Morse, S. I. and Bray, K. K. (1969). *J. exp. Med. 129*, 523-549.
Mosier, D. E. (1967). *Science, N.Y. 158*, 1573-1575.
Mosier, D. E. and Coppleson, L. W. (1968). *Proc. natn. Acad. Sci. U.S.A. 61*, 542-547.
Pierce, C. W. (1969). *J. exp. Med. 130*, 345-364.
Virolainen, M. and Defendi, V. (1967). *In* "Growth Regulating Substances for Animal Cells in Culture" (V. Defendi and M. Stoker, eds), pp. 67-85. Wistar Institute Press.
Virolainen, M. and Lahti, A. (1970). Manuscript in preparation.

Microcinematographic Analysis of the Co-operation Between Lymphocytes and Dendritic Macrophages Following Primary and Secondary Stimulation with Soluble Antigen

MARIE-LUISE MATTHES, WOLFGANG AX
and HERBERT FISCHER

Max-Planck-Institut für Immunbiologie, Freiburg, West Germany

INTRODUCTION

Many functional and morphological observations indicate that the induction of antibody synthesis requires the interaction of several cell types (see this Symposium). However, such intercellular co-operation has not previously been visualized *in vitro* under conditions approaching those existing *in vivo*. We have therefore used time-lapse cinematography to follow the general behaviour and co-operative interactions of omental macrophages and lymphocytes, following primary and secondary stimulation with a soluble antigen. The mouse omentum is a favourable model for such studies because its lymphocytes and macrophages, lying in a mesothelial framework, are easily and continuously visible.

We have attempted to answer the following questions:

1. Do primary and secondary antigenic stimuli significantly increase the number of encounters between lymphocytes and antigen-bearing macrophages?

2. Can activated lymphocytes be recognized by their motility?

3. Does antigenic stimulus induce a proliferative response in which direct intercellular contacts are involved?

We will only describe macrophage-lymphocyte interactions, although lymphocyte-lymphocyte contacts were also observed frequently.

MATERIALS AND METHODS

We used 25 g female NMRI mice (Zentralinstitut für Versuchstierzucht, Hannover). The antigen was horseradish peroxidase (Boehringer, Mannheim, RZ 0.6). Primary immunization was 2 mg i.p., secondary was 1 mg i.p. 15 days later. Omenta were removed under sterile conditions 24 hr and 56 hr after immunization, washed in Eagle's medium (Dulbecco modification) and suspended in Eagle's medium containing 10% calf serum. They were then mounted on mica in a diffusion chamber located in a 37°C thermostatically controlled case (Ax *et al.,* 1966). Observations were via an inverted Zeiss Phase Contrast microscope with an Apochromat 40 objective and a C 5 X or microprojection ocular, using an Askania Z camera and Kodak Double X, 35 mm. Time-lapse analyses were with a Kodak LW Motion Analyzer. During the observation period the culture medium was changed every 2 hr. Anti-peroxidase was located enzymatically by the method of Leduc *et al.* (1968). To localize macrophages in final preparations, animals were given i.p. 0.5 ml of a 15% colloidal carbon solution (Pelikan-Spezialtusche 11/1331a); the omenta were removed 6 hr later, dried in air and stained with May-Grünwald-Giemsa.

Omenta for control experiments were taken from animals injected with physiologic saline, i.p.

RESULTS

1. Omental macrophages. Six hours after intraperitoneal injection of colloidal carbon, two types of cells had ingested carbon particles: round cells and cells with extensively branched, "dendritic" processes (Fig. 1).

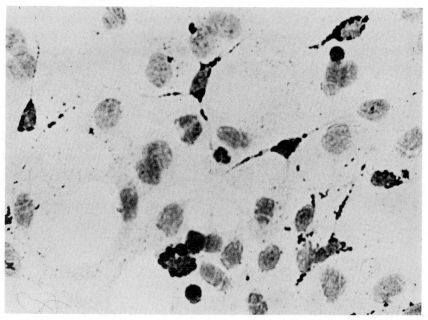

Fig. 1. Round and dendritic macrophages 6 hr after injection of 0.5 ml of 15% suspension of colloidal carbon. May-Grünwald-Giemsa x250.

Both types of cell had increased in number 48 hr after primary immunization (Table I). The increase in round cells was due to immigration of macrophages from the peritoneal fluid.

TABLE I

Round and dendritic macrophages before and after antigenic stimulus

	n	Controls	48 hr after stimulus
Round macrophages	10	16.2 ± 3.3	27 ± 3.9
Dendritic macrophages	10	83.2 ± 3.7	112. 4 ± 6

n = number of areas examined.
Means and standard deviations. Numbers of both cells were significantly different after immunization (P < 0.001).

2. Time-lapse cinematographic analysis of omental lymphocytes before and after antigenic stimulation. Without antigenic stimulation few lymphocytes were seen except in the vicinity of "milky spots". To analyse the motile behaviour of a given lymphocyte, the position of the cell in consecutive frames was recorded, defining the frames with the Kodak LW Motion Analyzer. Only definitely moving lymphocytes were analysed, but this included practically all lymphocytes. Wandering lymphocytes follow a non-erratic, smooth course of migration during the period of observation. We measured migration velocities of lymphocytes using the most advanced point relative to the overall migration course in each frame. We followed the wanderings of 50 lymphocytes from different control omenta, each for 6 min. The migration velocities fell into two groups with means of 1.18 mm/hr and 1.87 mm/hr. Without antigenic stimulation most lymphocytes moved at the slow speed.

Fifty-six hours after primary immunization the regions free of milky spots were richly populated with lymphocytes and lymphoblasts. Moreover a time-lapse analysis of the migration of 50 individual lymphocytes showed that most of these lymphocytes moved at the fast speed (Table II).

In the stimulated omentum, unlike the controls, lymphocytes had extensive encounters with antigen-containing, dendritic macrophages (Figs 1, 2, 3). We have determined the frequency of lymphocyte-macrophage contacts in 10 fields from different omenta, each observed for 1 hr. Lymphocyte-macrophage contact became more common 56 hr. after primary antigenic stimulation (Table III). After secondary immunization the change was already detectable at 12 hr.

3. Lymphocyte proliferation after antigenic stimulation. Most lymphocytes leave the macrophage after contact, but a few remain, round up and divide

M-L. MATTHES, W. AX AND H. FISCHER

TABLE II

Motility of lymphocytes before and after antigenic stimulus

	Control		56 hr after stimulus	
	Number	Mean motility	Number	Mean motility
Fast lymphocytes	9	1.86 mm ± 0.16[a]/hr	29	1.90 mm ± 0.15/hr
Slow lymphocytes	41	1.19 mm ± 0.10/hr	21	1.15 mm ± 0.11/hr

[a] Standard deviation.

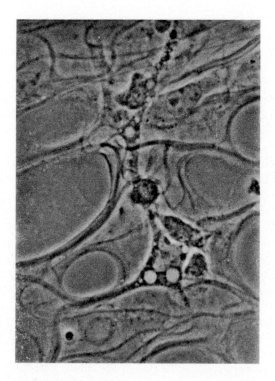

Fig. 2. Omentum 24 hr after secondary stimulus. Four lymphocytes in contact with a dendritic macrophage. ×400, Phase Contrast.

TABLE III

Lymphocytes and lymphocyte-macrophage-contacts before and after antigenic stimulus

	Control	24 hr after primary stimulus	56 hr after primary stimulus	24 hr after booster
No. of lymphocytes/field Mean ± S.D.	5.2 ± 0.9	5.4 ± 0.9	38 ± 2.6	35.6 ± 2.5
No. of lymphocytes/field touching macrophages Mean ± S.D.	2.2 ± 0.6	2.4 ± 0.8	30.8 ± 2.4	30.8 ± 1.9

Each field was observed for 1 hr.
Ten fields from different omenta were examined in each situation.

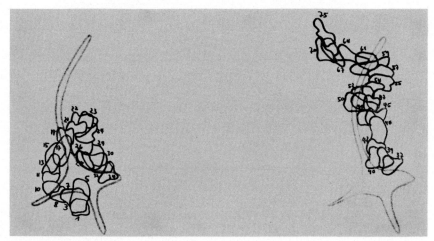

Fig. 3. Time-lapse analysis 24 hr after secondary stimulus showing the same lymphocyte migrating over the surface of a dendritic macrophage.

(Fig. 4). After primary stimulation mitoses were only seen in cells in contact with dendritic macrophages. Mitotic figures were more frequent after secondary stimulation (Table IV) and a few of them were seen at some distance from the macrophages. Because of the time required for these observations, Table IV only shows data from four omenta and areas of 1.2 mm² in each. Therefore only absolute values for each area are given.

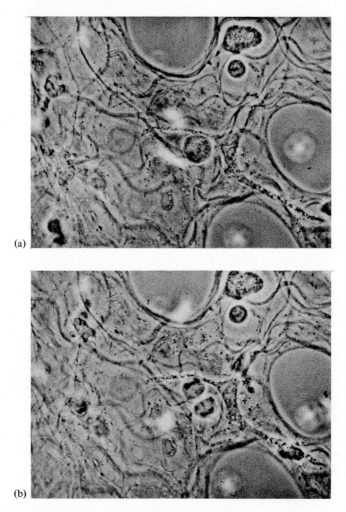

(a)

(b)

Fig. 4. Omentum 24 hr after secondary antigenic stimulus: a dividing lymphocyte in contact with a macrophage process. ×400, Phase Contrast.

4. Synthesis of specific antibody in new cell clusters. Three to five days after antigenic stimulation clusters of lymphoid or plasma cells accumulate near or at some distance from dendritic macrophages (Malchow *et al.*, 1969). (Prior studies of ^{3}H-thymidine incorporation showed that these cell populations arise by proliferation.) We tested omenta for antibodies against peroxidase five days after the booster injection, and found that many of the cells in the new cell clusters were synthesizing specific antibody. Moreover all dendritic macrophages adsorb anti-peroxidase on their surfaces.

TABLE IV

Lymphocyte division before and after antigenic stimulus

Controls	24 hr after primary stimulus	56 hr after primary stimulus	24 hr after secondary stimulus
0	0	1	3
0	0	1	4
0	0	0	3
0	0	1	2

Number of divisions in areas of 1.2 mm^2, each area observed for 5 hr. Four different omenta.

DISCUSSION

White was the first to show that highly specialized non-phagocytic dendritic reticular cells in germinal centres play a special role in antigen fixation and presentation. His observations were confirmed and extended by Nossal *et al.* (1968), Sordat *et al.* (1970) and by others. It is quite clear, that the main parts of the bodies of the dendritic cells on the omentum are phagocytic. In the electron microscope, however, their long dendrites resemble, in morphology and behaviour, the dendritic cells in the germinal centres (Freund-Mölbert, unpublished observations). Whether the cells found on the omentum and in developing milky spots can transform into specialized dendritic reticular cells is still open and deserves further consideration.

Since macrophages divide infrequently (Forbes and Mackaness, 1963), we suspect that the increased number of dendritic cells after antigenic stimulation is primarily due to conversion of round peritoneal macrophages. We are currently testing this point by a time-lapse analysis.

Our data show that antigenic stimulation increases the proportion of lymphocytes with high motility, but we cannot say whether these are sensitized cells, whether they have a unique origin or function, or whether they proliferate. However, it is clear that 56 hr after primary stimulation and 12 hr after secondary stimulation these cells migrate towards the dendritic macrophages that are scattered throughout the omentum. We are not certain whether this is due to non-specific chemotaxis or to the action of antigen-antibody complexes on the macrophage membrane.

The intimate contact between lymphocytes and macrophage-processes is reminiscent of the relationship between lymphocytes and "target" cells. In both cases "not-self" is recognized; in both cases there are extensive membrane appositions (Ax *et al.,* 1968).

After primary immunization mitoses were only seen close to macrophages, but after a booster injection mitoses were also observed remote from macrophages. We suggest that competent cells making their first contact with antigen on a macrophage surface, maintain this contact throughout their subsequent proliferation. This morphologic observation is well in accord with the results of Uhr and Finkelstein (1963) and Hanna *et al.* (1969). They found that primary induction of antibody formation requires the presence of antigen during the process of cell activation and differentiation. However, memory cells might be stimulated to proliferate even by transient contact with antigen-bearing macrophages. We realize that our experiments do not distinguish between non-specific proliferative stimuli and mechanisms leading to synthesis of specific antibody. However, analogous studies (Holub *et al.*, 1970; Fischer *et al.*, 1969) reveal antigen in dendritic macrophages and specific antibody in the lymphocyte clusters formed by proliferation.

ACKNOWLEDGEMENTS

This work was supported by a grant from the Deutsche Forschungs-gemeinschaft and a donation from the Volkswagenstiftung.

We wish to thank Mrs Ute Brehmer for excellent technical assistance, and Dr Donald F. Hoelzl Wallach for his helpful criticism.

SUMMARY

We have examined the behaviour and co-operative interactions of omental lymphocytes before and after primary and secondary stimulation with horse-radish peroxidase, using time-lapse cinematography. The omentum contains both round and dendritic macrophages: both increase in number after antigenic stimulation. Omental lymphocytes fall into two classes with distinct migration velocities: 1.18 mm/hr and 1.87 mm/hr. The slow species predominates in unstimulated omenta, and the fast after immunization. Antigenic stimulation promotes migration of lymphocytes towards dendritic macrophages leading to intimate contact between the surfaces of the two cell types. After primary antigenic stimulus, lymphocytes proliferate predominantly in contact with dendritic macrophages. After a booster mitoses are also seen remote from macrophages. Histochemical studies show synthesis of specific antibodies in the lymphocyte clusters formed by proliferation.

REFERENCES

Ax, W., Kaboth, U. and Fischer, H. (1966). *Z. Naturf. 21b,* 782.
Ax, W., Malchow, H., Zeiss, I. and Fischer, H. (1968). *Expl Cell Res. 53,* 108-116.

Fischer, H., Ax, W., Freund-Mölbert, E., Holub, M., Krüsmann, W. F. and Matthes, M. L. (1969). *In* "Conference on Mononuclear Phagocytes". In press.

Forbes, I. J. and Mackaness, G. B. (1963). *Lancet ii,* 1203-1204.

Hanna, M. G., Jr., Nettesheim, P. and Francis, M. W. (1969). *J. exp. Med. 129,* 953-971.

Holub, M., Ax, W., Fischer, H., Freund-Mölbert, E., Krüsmann, W. F., Matthes, M. L., Ríha, I., Šulc, J. and Tlaskalová, H. (1970). *In* "Developmental Aspects of Antibody Formation and Structure" (J. Šterzl and I. Ríha, eds). In press.

Leduc, E. H., Avramaes, S. and Bouteille, M. (1968). *J. exp. Med. 127,* 109-118.

Malchow, H., Ax, W. and Fischer, H. (1969). *Z. Naturf. 24b,* 61-66.

Nossal, G. J. V., Abbot, A., Mitchell, J. and Lummus, Z. (1968). *J. exp. Med. 127,* 277-290.

Sordat, B., Sordat, M., Hess, M. W., Stoner, R. D. and Cottier, H. (1970). *J. exp. Med. 131,* 77-92.

Uhr, J. W. and Finkelstein, M. S. (1963). *J. exp. Med. 117,* 457-477.

LYMPHOCYTE CLASSES

Experimental and Clinical Models of Immune Deficiency and Reconstitution of Immunologic Capacity

ROBERT A. GOOD and JOANNE FINSTAD

*Department of Pathology and the Pediatric Research Laboratory
of the Variety Club Heart Hospital
University of Minnesota, Minneapolis, Minnesota 55455, U.S.A.*

Over the past six or seven years we in Minneapolis and numerous investigators throughout the world have been considering experimental models and clinical human immunodeficiencies in terms of a division of labor within the lymphoid apparatus (Warner *et al.*, 1962; Cooper *et al.*, 1965, 1966c) (Fig. 1). The thymus

THYMUS SYSTEM DEVELOPMENT

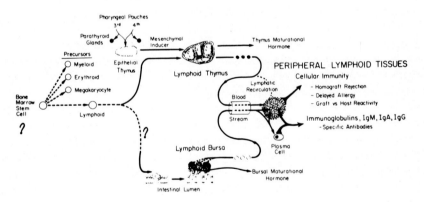

BURSAL SYSTEM DEVELOPMENT

Fig. 1.

has been shown to be essential for differentiation of the lymphocytes responsible for the development of the cell-mediated immunities (Good *et al.*, 1962; Miller, 1961). We include among cell-mediated immunities prototype delayed allergy, the capacity to recognize and eliminate solid tissue allografts, to initiate graft-*versus*-host reactions and to provide mammals and birds with effective defense against certain viruses, fungi, facultative intracellular bacterial pyogenic pathogens and certain forms of malignant cells. An excellent measure of this thymic differentiative influence in man, rats and chickens are the *in vitro* PHA responses, and *in vitro* responses to allogeneic cells. Sharply to be distinguished from this system of lymphocytes subserving cell-mediated immunities is a system of relatively sessile lymphoid cells, which develops independently of thymic influence. Both systems of cells are marrow-derived. In birds, the latter population of lymphoid cells has been shown to develop under the differentiative influence of the bursa of Fabricius (Glick *et al.*, 1956; Wolf *et al.*, 1963; Warner and Szenberg, 1964). In fully differentiated form, the latter population represents the sessile lymphocytes and plasma cells responsible for production and secretion of the conventional immunoglobulin molecules (Turk, 1969). In man evidence indicates that immunoglobulin molecules are extraordinarily heterogeneous including at least five major classes. Subclasses of each seem to be highly specialized in domain of influence. Viewed biologically each potential restriction has been developed by genetic interactions during phylogeny, and represents precise differentiative events experienced during ontogeny. As such, each cell system and subsystem represents a major bulwark of the bodily defense against certain organisms and agents capable of threatening or restricting the reproductive potential of the individual. The production and secretion into the lymph, blood and bodily fluids of the different types of circulating antibodies is the major business of the thymus-independent system of cells. The old argument that we have all enjoyed so much through the years "Is it lymphocytes or plasma cells that produce antibody?" no longer seems germane. Both lymphocytes and plasma cells are known to produce and secrete antibodies, but when they do they reflect a differentiative influence quite separate from that which subserves differentiation of the thymus-dependent cells responsible for cell-mediated immunities (Good *et al.*, 1968). For mammals we do not yet know the anatomic location of the crucial differentiative site or sites equivalent to the bursa of Fabricius of birds. It seems, however, that we have now learned enough about the characteristics of the bursa itself to be able to define this equivalent site within a year or two (Cooper *et al.*, 1970). We in Minneapolis have been excited by similarities between certain gut epithelium-associated lymphoid follicles located especially in the appendix, Peyer's patches and Sacculus rotundus in rabbits (Archer *et al.*, 1963; Sutherland *et al.*, 1970; Cooper *et al.*, 1966a; Perey *et al.*, 1968a, 1970) and the bursa of Fabricius of chickens (Perey *et al.*, 1968a; Cooper *et al.*, 1968a). We do know from experimental, genetic and differentiative perturbations that in man and mammals the bursa equivalent site is not the thymus (Perey *et al.*, 1968b; Cooper *et al.*, 1968b; Good and Finstad, 1968). It is located separately and represents a separate crucial differentiative influence just as is the case with birds. Thus, we are pleased with currently popular terminology which sharply distinguishes T and B type lymphocytes. It must be realized that, although the dichotomy of the two cell types is a prevailing view, the extraordinary heterogeneity that exists among the B cells also exists to some degree among populations of T cells. For instance, B cells can be separated into those synthesizing IgM, IgG, IgA, IgD and IgE type immunoglobulins. Among those producing IgA type immunoglobulins separate subdivisions seem to exist for those contributing primarily to the circulating pool of immunoglobulins (Hong *et al.,* 1969) and those contributing to the secretory pool. Indeed the distribution of IgA subtypes seems different in the secretory IgA plasma cell system and the circulating pool of immunoglobulins. Similarly exquisite specialization of function seems to characterize even certain subclasses of the IgG immunoglobulins (Frommel and Good, 1970). Further, some evidence suggests that for T cells as for B cells specialization in the form of a secretory system exists. Differences between B cells and T cells might indeed represent differentiation to secretory capacity in the latter which is lacking in the former. Thus the heterogeneity of T cells might be every bit as great as the heterogeneity of B cells.

Underlying the development of knowledge of the complexity of the immunologically competent cells has been an extremely useful interplay between clinic and laboratory (Good *et al.,* 1968; Gabrielson *et al.,* 1969; Stutman *et al.,* 1969a). It has, indeed, become clear that certain models of immunodeficiency can offer useful questions to those concerned with potential cellular interactions in immunity. In turn, these models, both clinical and experimental, can provide crucial tests of postulates developed from study of *in vitro* behavior of cells.

What we consider to be crucial models may be experimental models where the specific deficiencies are most sharply defined. Alternatively, clinical models based on genetically determined perturbations of development of the immunological systems may be the source of the crucial tests of our developing concepts of cellular relationships and interactions.

Among the clinical models of immunodeficiency we consider most helpful for analysis of questions concerning interactions of cells in immunity are the following:

a. Rare patients with infantile sex-linked agammaglobulinemia who have normal thymus and normal thymus-dependent lymphoid system, while having gross deficiency of ability to produce immunoglobulins, antibodies, plasma cells and germinal centers.

b. Complete form of DiGeorge syndrome in which thymus and thymus-dependent lymphoid system and functions are lacking.

c. Autosomal or sex-linked lymphopenic agammaglobulinemia with dual system immunodeficiency and lack of significant development in both lymphoid systems.

d. De Vaal syndrome of reticular dysgenesis.

e. Patients with complete or virtually complete absence of one immuno-globulin system, who have normal development of all the others.

Experimental models considered in our laboratory to be particularly useful are the following:

1. Neonatally thymectomized mice that received 500 R total body irradiation as neonates.
2. Chickens bursectomized at hatching and subjected to near lethal irradiation (more than 700 R).
3. Chickens completely bursectomized at 15 days of embryonation which grow up lacking both IgM and IgG immunoglobulins (Cain et al., 1969).
4. Chickens possessing IgM but lacking IgG immunoglobulins. These are prepared by complete bursectomy on the 18th or 19th day of embryonation (Cooper et al., 1969; Van Alten et al., 1968).
5. Mice lethally irradiated as adults and given a marrow transplant from a neonatally thymectomized-irradiated syngeneic donor.

Less completely evaluated models that may be equally clean and useful, but which we think need more study include:

a. Chickens given, at a critical time, sufficiently large doses of testosterone or progestational hormones, e.g. 19 nortestosterone.
b. Chickens given at hatching crucial doses of cyclophosphamide.
c. Combinations of a and b.
d. Immunological prevention of immunoglobulin development using anti-μ and complement.
e. Nude mice, which like patients with DiGeorge syndrome, are born lacking the thymus.

Certain strains of mice have been especially useful for studying the role of thymic-dependent lymphocytes in the body economy. These include C3HBi, CBA and Balb/c. At birth these mice have differentiated and disseminated only very little of the thymic-dependent lymphoid system of cells. In contrast, mice of C57B1/1 and 6 and Swiss albino mice have already, by the time of birth, achieved a substantial population of thymus-dependent cells in the periphery. With all strains mice lacking both thymus and thymus-dependent cell lines can be produced by combining neonatal thymectomy with near lethal total body irradiation. Stutman et al. (1969a) have also noted that neonatally thymectomized mice of certain strains, which already at birth have a peripheralized thymus-dependent population of cells, lose their post thymic population with time. Thus after 35-40 days in a conventional environment neonatally thymectomized C3HBi mice seem to have lost much, if not all, of the small population of post thymic cells with which they were born (Stutman and

Good, 1969). The loss of such cells is accelerated in the neonatally thymectomized mouse by injection of Freund's complete adjuvant. The difference between neonatally thymectomized mice deprived of this small population of post-thymus lymphocytes and neonatally thymectomized mice possessing a peripheralized post-thymic population of thymus-dependent lymphocytes is seen in experiments where cell-mediated immune functions are improved by treatment with functional carcinogen-induced thymic stromal epitheliomas (Stutman *et al.*, 1967, 1968, 1969b). These epitheliomas transplanted either within or without cell impenetrable (0.1 and 0.2 μ)millipore chambers are able to reconstitute cell-mediated immunologic functions when the thymectomized mice possess even a relatively small population of post-thymic cells. If the neonatally thymectomized mice on the other hand do not possess a population of post-thymic cells the functional stromal epitheliomas cannot achieve immunologic restoration. Influences similar to those attributable to the carcinogen-induced functional thymic stromal epitheliomas can be achieved, somewhat less consistently in our hands, by normal thymus grafts in millipore chambers (Stutman *et al.*, 1969a).

The current enthusiasm for cell-cell interactions originates from the earliest studies of the influence of neonatal thymectomy on development of immune capacity. When we first reported that the thymus plays a crucial role in development of immunologic capacity we were, in the very first experiments, working with capacity to produce circulating antibody (Archer and Pierce, 1961; Good *et al.*, 1962). Indeed in rabbits, mice and rats many workers found that neonatal thymectomy interferes with the development of antibody producing capacity to certain antigens (Archer *et al.*, 1964; Miller, 1964; Arnason *et al.*, 1964). To be sure it soon became apparent that the cell-mediated immunities were even more consistently depressed by neonatal thymectomy than were antibody responses. Parrott and East (1964) then pointed out that production of antibody to some antigens is relatively independent of neonatal thymectomy. Many investigators found that neonatal thymectomy produced no significant depression of immunoglobulin levels (Fahey and Robinson, 1963; Mosser *et al.*, 1970). Development of antibody responses to sheep red blood cells and to some simple soluble protein antigens, however, were always especially dependent upon a vigorous post-thymic population of lymphoid cells. Claman *et al.* (1966) reported exciting results indicating synergism between thymus cells and marrow cells in reconstituting the antibody-forming capacity of lethally irradiated mice. This was followed by a spate of papers showing that even though it was thymus cells, or thymus-dependent cells, that were sensitive to antigen, it was the marrow-derived cells that were responsible for antibody production to sheep red blood cells (Davies *et al.*, 1967; Miller and Mitchell, 1968; Mitchell and Miller, 1968; Nossal *et al.*, 1968). All of these findings appeared after Warner *et al.* (1962); Warner and Szenberg (1964) had shown that in chickens the ability to produce antibody developed independently of the thymus. The chicken model was all the more telling when Cooper et al. (1965, 1966b, c) showed that agammaglobulinemia and unresponsiveness in antibody formation to many antigens is produced by bursectomy combined with irradiation in the newly-hatched birds. One of the most important contributions of the Cooper work was that it established that the thymus of chickens subserves exactly the same functions as it serves in the mouse, rabbit and rat. It is essential to development of the cell-mediated immunities. Davies et al. (1967), Mitchell and Miller (1968) and Nossal et al. (1968) then proved that the antibody produced against sheep red blood cells was not produced by thymus-dependent or thymus-derived cells, but by thymus-independent or so-called bone marrow-derived cells. From these studies the concept became popular that marrow-derived cells, themselves insensitive to antigen, were responsible for antibody synthesis and secretion. The concept that there existed one antigen sensitive cell and another separate antibody producing cell was stated and restated.

To us in Minneapolis, the latter concept made no sense at all. Repeatedly we spoke out against this view. We pointed out that in the absence of the thymus and demonstrable thymus-derived cells, our chickens, mice and rabbits made normal amounts of immunoglobulins. Who should argue at this late date that at least the bulk of these immunoglobulins are not antibodies against something? Further, the thymus-dependence of antibody synthesis seemed in large degree to be a special case for sheep erythrocytes and certain protein antigens. Irradiated, neonatally-thymectomized mice were capable of making absolutely normal responses to Salmonella H antigens and very good responses to Brucella antigens (Mosser et al., 1970). In contrast, neonatal thymectomy combined with irradiation in mice and chickens completely eliminated expression of cell-mediated immunities. Finally, in what we considered to be most crucial experiments, the thymus-dependence of antibody responses to sheep red blood cells in neonatally thymectomized and neonatally thymectomized-irradiated mice was also dependent on the dose of antigen employed. Sinclair and Elliott (1968) and Lemmel, Cooper and I (unpublished observations) carried out experiments which showed these relationships. The simplest interpretation of all the observations seemed to be that both thymus-independent and thymus-dependent cell lines contained antigen sensitive cells. It seemed to us only a matter of the manner in which antigens were delivered or presented to the antigen sensitive cells of the thymus-independent lymphoid system (Good and Rodey, 1970). If the antigen could be delivered effectively to the antigen sensitive cells of the thymus-independent lymphoid system antibody production would follow. If effective delivery of antigen to the antigen sensitive cells of the thymus-independent system required or was facilitated in some way by the expressions of cell-mediated immunity, then antibody response to that antigen would be thymus-system dependent.

Our persistence as a gad fly was strongly supported by observations that humans born lacking both thymus and parathyroid, patients with the DiGeorge syndrome regularly had ample amounts of all known immunoglobulins (Huber *et al.,* 1967; Lux *et al.,* 1970; Gatti and Good, 1970). Yet these patients sometimes formed antibody to a variety of antigens very well, even though their lymphocytes did not respond to PHA, and they could not reject skin allografts or develop cell-mediated delayed allergic responses either to natural infections or to skin sensitizing injections.

This previously unpopular view now seems the order of the day. Elaborate schemes are being developed which derive their energy from the postulate, attributable to both Mitchison (1969) and Cohn (1969) that the signal for an effective antibody response requires participation of two determinants on an antigen. Thus a carrier determinant is considered essential to engage the thymus-dependent cell, and an additional antigenic determinant to engage the antigen sensitive cell of the thymus-independent system. This popular view also derives substance from the requirements for multiple cell types in the *in vitro* systems (Fishman and Adler, 1967; Mishell and Dutton, 1967).

We are not, however, certain that the influence of the thymus-dependent cell line is always so direct as most are now picturing. Cell-mediated immunity may develop quite rapidly, and it may be the epiphenomena associated with the thymus-dependent delayed allergic reactions that play the telling role. Perhaps, for example, the angry macrophage or other non-specific consequences of delayed allergy are important in effecting presentation of certain antigens to antigen-sensitive cells of the thymus-independent antibody forming population.

We would like to raise the question of how essential biologically is interaction between thymus-dependent and bone marrow-derived cells in antibody production in bodily defense. One can look at this question in several ways. One could argue that the thymus-dependence of immunity to many viruses, fungi and to the facultative intracellular pyogenic bacterial pathogens is in reality a function of this cell-cell interaction. That might be the case and numerous experimental studies will be necessary to test the idea. However, even the complete elimination of the thymus-dependent system of cells does not compromise levels of any of the immunoglobulins thus far studied in mice, chickens, rats and man. Natural antibodies and antibodies to many antigens are present in the blood of such animals. If profound depression of antibody formation to a significant proportion of antigens were the consequence of genetic or experimental elimination of the thymus-dependent system decreased levels of immunoglobulins would surely be observed. This should be true with this kind of deprivation of antigenic stimulation, just as it is with the decreased effective antigenic stimulation in a germ free environment. Under the latter circumstances low levels of immunoglobulin reflect the low antigenic input. We do not deny that experimentally one can show co-operative functions between thymus-dependent and thymus-independent systems of lymphoid cells *in vitro* or under certain sets of conditions *in vivo*. We simply doubt that co-operation between thymus-dependent and thymus-independent systems plays an important or crucial role in most antibody responses as they are induced in nature in the intact animal. It seems to us highly unlikely, evolution being what it is, that antigen sensitive cells of the thymus-independent system, our bulwark against encapsulated high grade, extracellular pyogenic pathogens, would not be regularly approachable by pathways other than that which is the fad of the day. Remember that our scientific fad is based on highly contrived *in vitro* and *in vitro-in vivo* experimental systems. Further, this popularity of cell-cell interactions as a requirement for antibody synthesis is mainly based on responses to sheep red blood cells.

RECONSTITUTION OF IMMUNOLOGIC FUNCTIONS. SOME ACHIEVEMENTS AND RESTRICTIONS

It was most exciting when Cleveland *et al.* (1968) reported that DiGeorge syndrome could be corrected by a thymic transplant from an unrelated fetus. Even though search by biopsy failed to disclose the transplanted thymus the apparent confirmation by August *et al.* (1968) gave encouragement to the view that the thymus corrected the immune defect in DiGeorge syndrome. In the August case the question of persistence of the transplant was also raised, and the inordinately rapid restoration of lymphoid cell responses to PHA were of concern (Dempster, 1969; Good *et al.*, 1969b; Lischner and DiGeorge, 1969). This is a problem we have been addressing at the experimental level as well as at the clinical level. What restrictions, if any, are imposed by histocompatibility differences between recipient host and thymic transplant? These considerations seem all the more germane in the face of Jerne's provocative adaptation of the clonal selection theory to meet the demands imposed by the high proportion of lymphoid cells that react specifically in graft-*versus*-host reactions, implants of lymphocytes on chorioallantoic membranes and in mixed leukocyte cultures. Two percent of cells recognizing a single histocompatibility difference is of course just too much unless there has been prior immunization. Our experimental analysis of these relationships is still incomplete, but several points are clear. Fully differentiated peripheral lymphoid cells, being unable to return to the thymus do not, in the absence of the adrenals, produce significant destructive lesions in the thymus, even when they launch vicious graft-*versus*-host reactions in other lymphoid and non-lymphoid tissues throughout the body. Neonatally thymectomized mice even after being given near-lethal total body irradiation are fully reconstituted immunologically by syngeneic thymic transplants. The reconstitution need not involve lymphoid cells present in the thymus. Similarly long lasting reconstitution that is dependent on the differentiative influence of the thymus is observed when the donor of the thymus and the recipient differ at non-H_2 (weak) histocompatibility loci. In some instances similar reconstitution may be achieved when there are strong histocompatibility differences between the donor of the thymus and the host. Under the latter circumstances, especially if multiple thymus grafts are used, but also sometimes with single grafts, the thymus-derived population will set up destructive and even lethal graft-*versus*-host reactions in the recipient of the thymus transplant (Stutman *et al.,* 1969c). These reactions can be attributed to the development of the thymus cells of the graft and their immunologic function in the tissues of the recipient. Thymus grafts where donor and host have different strong (H_2) antigens often do not always reconstitute the host immunologically. In other circumstances such grafts do, indeed, reconstitute the recipient host immunologically, and the reconstitution may be attributable to the development of host cells in the thymus graft. These cells, circulating as they do to the thymus differentiate, there and then recognize the thymus graft as being foreign and launch a vigorous graft-*versus*-host reaction within the thymus itself. When this has occurred the immunologic reconstitution achieved by the foreign thymic transplant is relatively short lived. Beginning approximately 150 days after grafting the cell-mediated immunities wane and immunologic crippling again ensues (Stutman and Good, unpublished observations).

This set of observations requires careful consideration from a theoretical as well as a practical viewpoint. It places serious practical restrictions on therapeutic use of thymus transplants in reconstructing immune deficiencies, consequent to deficient thymus development, when matching at the major histocompatibility locus is not possible. It also seems to raise critical questions vis-a-vis Jerne's postulate concerning differentiation of immunologic functions based on obviating responses to host histocompatibility antigens. These and other questions dictate that extensive further research be carried out from this perspective.

Currently it seems popular in certain quarters to relate a thymus-dependent cell to the formation of certain rosettes with sheep red blood cells. Here again our models of immunodeficiency we believe are critical and must be attended to. Our observations indicate that if you really make agammaglobulinemia or extreme hypogammaglobulinemia in chickens by bursectomy plus irradiation you eliminate rosette forming cells (Cooper and Good, unpublished). All other thymus-dependent immune functions remain intact in such animals. These findings, together with the impressive association of θ antigen with certain rosette forming cells in mice, and observations of Sell and Gell (1965a, b, 1967) concerning anti-immunoglobulin antibody induction of lymphocyte transformation, raise the possibility that lymphocytes of the thymus-dependent system carry antibodies—perhaps cytophilic antibodies—produced not by themselves but by cells of the thymus-independent system. Alm and Peterson (1969) have clearly shown that chickens parallel rodents in having many lymphocytes in peripherical blood and peripheral lymphoid organs that can be transformed and induced to proliferate by antibodies directed toward immunoglobulin (Sell and Gell, 1965a, b, 1967). That the antibodies or immunoglobulin constituents on the peripheral blood lymphocytes are dependent on the differentiative influence of the bursa of Fabricius in chickens was shown by the elimination of such transformable cells from the circulation and tissues by treating the birds with irradiation and bursectomy to make them agammaglobulinemic. It is important to determine whether patients with the extreme form of infantile sex-linked agammaglobulinemia, who have essentially normal cell-mediated immunity, possess rosette forming lymphocytes related to the receptors responsible for their cell mediated immunities. We predict that they do not. We would like to believe that detecting IgX or the membrane-bound antibody responsible for cell-mediated immunity will be as easy as tracing material responsible for rosette formation, but it is very unlikely. It is much more likely that the antibody being detected by rosette formation is a product of the thymus-independent system of lymphoid cells, and is not the long sought antibody responsible for the specificity of cell-mediated immune responses.

We hold this same reservation about interference with cell-mediated immunities by antibodies directed against immunoglobulin constituents (Greaves et al., 1969).

Clearly current experiments linking θ antigen to certain rosette forming cells, and seeming to indicate that these cells function in what we all accept as cell-mediated immune responses, are attractive. We will be more ready to accept such experiments when they are carried out clinically or with chicken models in which the spectre of indirect influence of antibody production by the thymus-independent cells has been sharply reduced or eliminated.

THE BEGINNINGS OF CELLULAR ENGINEERING

Finally, we should like to draw attention to the first toddling steps of what could well develop into an era of cellular engineering. In 1967 it seemed to us likely that severe dual-system immunodeficiency of both the autosomal and X-linked recessive forms might be attributable to an abnormality of the lymphoid stem cell. Such a cell present in the bone marrow during post-natal life should be common to both the thymus-dependent and thymus-independent lymphoid systems. Abnormality or failure of development of this cell, it seemed, could underlie all those genetically determined diseases that involved severe dual-system immunologic deficiency. To test this hypothesis it seemed reasonable to attempt immunologic reconstitution of such patients with marrow or fetal liver from immunologically normal persons. In one instance fetal liver and in others bone marrow from an adult donor seemed to afford dramatic immunologic reconstitution of such children (Hong et al., 1968a). It was very distressing, however, that children treated in this way not only were reconstituted immunologically but were destroyed by fulminating graft-*versus*-host reactions (Hong *et al.*, 1968b). Our own extensive experience showed that certain histocompatibility barriers in mice could be manipulated fairly easily when donor and recipient were matched according to the major (H_2) (Martinez *et al.*, 1960; Simonsen, 1962) histocompatibility locus (Good *et al.*, 1964; Martinez *et al.*, 1960). Furthermore it has often been observed that graft-*versus*-host reactions produced by lymphoid cells are minimal when donor and host share the H_2 histocompatibility determinants. We therefore decided to try marrow transplants between children with dual system immunodeficiency disease using, as a source of stem cells, marrow from a sibling matched with the recipient at the HL-A histocompatibility locus. In man as in mice, rats, chickens, dogs and monkeys there is a major histocompatibility locus at which multiple allelles may operate. This means that matching for major determinants among siblings is feasible. Indeed, perfect match at the major histocompatibility determinants might be expected once in four times.

In fact matching the sibling donor with the recipient at the HL-A locus has resulted in successful marrow transplants which on several occasions have permitted:

1. Complete correction of the dual system immunodeficiency disease (Good *et al.*, 1969a; Gatti *et al.*, 1968; Amman *et al.*, 1970).

2. Complete correction with long lasting switching of blood type from A to O, to correct a complicating immunologically based aregenerative pancytopenia in a lymphopenic agammaglobulinemic child (Gatti *et al.*, 1968; Meuwissen *et al.*, 1969; Good, 1969). The latter manipulation dramatically created chimerism that has lasted more than two years, and involves tolerance of the host for donor cells and antigens.

3. A number of successful marrow transplantations have followed (de Koning *et al.*, 1969; Rosen, personal communication), several of which have corrected the inborn immunodeficiency disease that occasioned the transplant.

4. More recently encouraging results by Buckley and co-workers (personal communication) have suggested that even the HL-A barrier may be overcome by marrow transplantation in children with dual system immunodeficiency.

To be sure patients suffering from these dual system immunodeficiencies are rare and most difficult to study and treat. We would, however, contend that our ability to correct these deficiencies completely at this time, reflects in a dramatic way the progress of our understanding of the lymphoid system, the interaction of its cells with one another and their relationships to one another in developmental perspective.

We think these toddling steps can be expected, through continued work and development, to strengthen to a walk and then ultimately develop into a real run—an era of cellular engineering.

ACKNOWLEDGEMENTS

This work was aided by grants from the National Foundation-March of Dimes, American Heart Association, U.S. Public Health Service AI-08677, AI-00798, AI-00292 and HE-06314.

REFERENCES

Van Alten, P. J., Cain, W. A., Good, R. A. and Cooper, M. D. (1968). *Nature, Lond. 217*, 358-360.

Alm, G. V. and Peterson, R. D. A. (1969). *J. exp. Med. 129*, 1247-1259.

Ammann, A. J., Meuwissen, H. J., Good, R. A. and Hong, R. (1970). *Clin. exp. Immun.* (In press.)

Archer, O. K. and Pierce, J. C. (1961). *Fedn Proc. 20*, 26.

Archer, O. K., Sutherland, D. E. R. and Good, R. A. (1963). *Nature, Lond. 200*, 337-339.

Archer, O. K., Papermaster, B. W. and Good, R. A. (1964). *In* "The Thymus in Immunobiology" (R. A. Good and A. E. Gabrielsen, eds), pp. 414-431. Hoeber-Harper, New York.

Arnason, B. G., Janković, B. D. and Waksman, B. H. (1964). *In* "The Thymus in Immunobiology" (R. A. Good and A. E. Gabrielsen, eds), pp. 492-501. Hoeber-Harper, New York.

August, C. S., Rosen, F. S., Filler, R. M., Janeway, C. A., Markowski, B. and Kay, H. E. M. (1968). *Lancet ii*, 1210-1211.

Cain, W. A., Cooper, M. D., Van Alten, P. J. and Good, R. A. (1969). *J. Immun. 102*, 671-678.

Claman, H. N., Chaperon, E. A. and Triplett, R. F. (1966). *Proc. Soc. exp. Biol. Med. 122*, 1167-1171.

Cleveland, W. W., Fogel, B. J., Brown, W. T. and Kay, H. E. M. (1968). *Lancet ii*, 1211-1214.

Cohn, M. (1969). *In* "Immunological Tolerance" (M. Landy and W. Braun, eds), pp. 283-338. Academic Press, New York.

Cooper, M. D., Peterson, R. D. A. and Good, R. A. (1965). *Nature, Lond. 205*, 143-146.

Cooper, M. D., Perey, D. Y., McKneally, M. F., Gabrielsen, A. E., Sutherland, D. E. R. and Good, R. A. (1966a). *Lancet i*, 1388-1391.

Cooper, M. D., Schwartz, M. L. and Good, R. A. (1966b). *Science, N.Y. 151*, 471-473.

Cooper, M. D., Peterson, R. D. A., South, M. A. and Good, R. A. (1966c). *J. exp. Med. 123*, 75-102.

Cooper, M. D., Perey, D. Y., Peterson, R. D. A., Gabrielsen, A. E. and Good, R. A. (1968a). *In* "Immunological Deficiency Diseases in Man" (D. Bergsma and R. A. Good, eds), Vol. 4, pp. 7-12. National Foundation-March of Dimes.

Cooper, M. D., Perey, D. Y., Gabrielsen, A. E., Sutherland, D. E. R., McKneally, M. F. and Good, R. A. (1968b). *Int. Archs Allergy appl. Immun. 33*, 65-88.

Cooper, M. D., Cain, W. A., Van Alten, P. J. and Good, R. A. (1969). *Int. Archs Allergy appl. Immun. 35*, 242-252.

Cooper, M. D., Kincade, P. W., Lawton, A. R. and Bochman, D. E. (1970). *Fedn Proc. 29,* 492.
Davies, A. J. S., Leuchars, E., Wallis, V., Marchant, R. and Elliott, E. V. (1967). *Transplantation 5,* 222-231.
Dempster, W. J. (1969). *Lancet i,* 468.
Fahey, J. L. and Robinson, A. G. (1963). *J. exp. Med. 118,* 845.
Fishman, M. and Adler, F. L. (1967). *Cold Spring Harb. Symp. quant. Biol. 32,* 343-348.
Frommel, D. and Good, R. A. (1970). *Recent Adv. Ped.* (In press.)
Gabrielsen, A. E., Cooper, M. D., Peterson, R. D. A. and Good, R. A. (1969). *In* "Textbook of Immunopathology" (P. A. Miescher and H. J. Müller-Eberhard, eds), pp. 385-405. Grune and Stratton, New York.
Gatti, R. A. and Good, R. A. (1970). *New Engl. J. Med. 282,* 276-277.
Gatti, R. A., Meuwissen, H. J., Allen, H. D., Hong, R. and Good, R. A. (1968). *Lancet ii,* 1366-1369.
Glick, B., Chang, J. S. and Jaap, R. C. (1956). *Poult. Sci. 35,* 224-225.
Good, R. A. (1969). *Hospital Practice 4,* 41-47.
Good, R. A. and Finstad, J. (1968). *Trans. Am. clin. climat. Ass. 79,* 69-107.
Good, R. A. and Rodey, G. E. (1970). *Cellular Immun.* (In press.)
Good, R. A., Dalmasso, A. P., Martinez, C., Archer, O. K., Pierce, J. C. and Papermaster, B. W. (1962). *J. exp. Med. 116,* 773-796.
Good, R. A., Martinez, C. and Gabrielsen, A. E. (1964). *Adv. Pediatrics 13,* 93-127.
Good, R. A., Peterson, R. D. A., Perey, D. Y., Finstad, J. and Cooper, M. D. (1968). *In* "Immunologic Deficiency Diseases in Man" (D. Bergsma and R. A. Good, eds), pp. 17-34. National Foundation-March of Dimes.
Good, R. A., Gatti, R. A., Hong, R. and Meuwissen, H. J. (1969a). *Lancet i,* 1162.
Good, R. A., Gatti, R. A., Meuwissen, H. J. and Stutman, O. (1969b). *Lancet i,* 946-947.
Greaves, M. F., Torrigiani, G. and Roitt, I. M. (1969). *Nature, Lond. 222,* 885-886.
Hong, R., Kay, H. E. M., Cooper, M. D., Meuwissen, H. J., Allan, M. J. G. and Good, R. A. (1968a). *Lancet i,* 503-506.
Hong, R., Gatti, R. A. and Good, R. A. (1968b). *Lancet ii,* 388-389.
Hong, R., Ammann, A. J., Cain, W. A. and Good, R. A. (1969). *In* "Symposium on Local Antibody System". (In press.)
Huber, J., Cholnoky, P. and Zoethout, H. E. (1967). *Archs Dis. Childhood 42,* 190.
de Koning, J., Dooren, L. J., van Bekkum, D. W., van Rood, J. J., Dicke, K. A. and Rádl, J. (1969). *Lancet i,* 1223-1227.
Lischner, H. W. and DiGeorge, A. M. (1969). *Lancet ii,* 1044-1049.
Lux, S. E., Johnston, R. B., Jr., August, C. S., Say, B., Penchaszadeh, V. B., Rosen, F. S. and McKusick, V. A. (1970). *New Engl. J. Med. 282,* 234-236.
Martinez, C., Shapiro, F. and Good, R. A. (1960). *Proc. Soc. exp. Biol. Med. 104,* 256-259.
Meuwissen, H. J., Gatti, R. A., Terasaki, P. I., Hong, R. and Good, R. A. (1969). *New Engl. J. Med. 281,* 691-696.
Miller, J. F. A. P. (1961). *Lancet ii,* 748-749.

Miller, J. F. A. P. (1964). *In* "The Thymus in Immunobiology" (R. A. Good and A. E. Gabrielson, eds), pp. 436-460. Hoeber-Harper, New York.

Miller, J. F. A. P. and Mitchell, G. F. (1968). *J. exp. Med. 128,* 801-820.

Mishell, R. I. and Dutton, R. W. (1967). *J. exp. Med. 126,* 423-442.

Mitchell, G. F. and Miller, J. F. A. P. (1968). *J. exp. Med. 128,* 821-837.

Mitchison, N. A. (1969). *In* "Mediators of Cellular Immunity" (H. S. Lawrence and M. Landy, eds), pp. 71-141. Academic Press, New York.

Mosser, G., Good, R. A. and Cooper, M. D. (1970). *Int. Archs Allergy appl. Immun. 39,* 62-81.

Nossal, G. J. V., Cunningham, A., Mitchell, G. F. and Miller, J. F. A. P. (1968). *J. exp. Med. 128,* 839-853.

Parrott, D. M. V. and East, J. (1964). *In* "The Thymus in Immunobiology" (R. A. Good and A. E. Gabrielsen, eds), pp. 523-540. Hoeber-Harper, New York.

Perey, D. Y., Cooper, M. D. and Good, R. A. (1968a). *Science, N.Y. 161,* 265-266.

Perey, D. Y., Cooper, M. D. and Good, R. A. (1968b). *Surgery 64,* 614-621.

Perey, D. Y., Frommel, D., Hong, R. and Good, R. A. (1970). *Lab. Invest. 22,* 212-227.

Sell, S. and Gell, P. G. H. (1965a). *J. exp. Med. 122,* 423-440.

Sell, S. and Gell, P. G. H. (1965b). *J. exp. Med. 122,* 923-928.

Sell, S. and Gell, P. G. H. (1967). *J. exp. Med. 125,* 289.

Simonsen, M. (1962). *Progr. Allergy 6,* 349-467.

Sinclair, N. R. St C. and Elliott, E. V. (1968). *Immunology 15,* 325-333.

Stutman, O. and Good, R. A. (1969). *Exp. Hemat. 19,* 12-15.

Stutman, O., Yunis, E. J. and Good, R. A. (1967). *Lancet i,* 1120-1123.

Stutman, O., Yunis, E. J. and Good, R. A. (1968). *Exp. Hemat. 16,* 18-21.

Stutman, O., Yunis, E. J. and Good, R. A. (1969a). *J. exp. Med. 130,* 809-819.

Stutman, O., Yunis, E. J. and Good, R. A. (1969b). *J. natn. Cancer Inst. 43,* 499-508.

Stutman, O., Yunis, E. J. and Good, R. A. (1969c). *Transplantation 7,* 420-423.

Sutherland, D. E. R., McKneally, M. F., Kellum, M. J. and Good, R. A. (1970). *Int. Archs Allergy appl. Immun. 38,* 6-36.

Turk, J. L. (1969). "Immunology in Clinical Medicine". Appleton-Century-Crofts, New York.

Warner, N. L. and Szenberg, A. (1964). *In* "The Thymus in Immunobiology" (R. A. Good and A. E. Gabrielsen, eds), pp. 395-411. Hoeber-Harper, New York.

Warner, N. L., Szenberg, A. and Burnet, F. M. (1962). *Aust. J. exp. Biol. med. Sci. 40,* 373-388.

Wolf, J. K., Gokcen, M. and Good, R. A. (1963). *J. Lab. clin. Med. 61,* 230.

Graft Restoration of Primary Immunodeficiency

J. F. SOOTHILL, H. E. M. KAY and J. R. BATCHELOR

Department of Immunology, Institute of Child Health, London
Royal Marsden Hospital, London
and
McIndoe Memorial Research Unit, Queen Victoria Hospital
East Grinstead, Sussex, England

Evidence has been deduced from the study of immunodeficiency which supports the view that specific immunity may be divided into specific cellular and specific humoral responses, and that two cell populations underlie these functions. Patients may have virtually pure antibody deficiency, or specific cellular immunity deficiency, or frequently both may occur together—combined immunodeficiency (see Soothill, 1968 and W.H.O., 1970, for classification). Each may vary qualitatively or quantitatively, but the former may reasonably be ascribed to deficiency of bone marrow-derived cells (B cells), perhaps processed by a mammalian equivalent of the avian bursa of Fabricius. Cellular immunodeficiency may be ascribed to a deficiency of thymus-processed cells (T cells) and perhaps of the thymus itself, and the combined state to a deficiency of stem cells. The results of attempts at graft treatment have confirmed these views.

Following recognition that graft rejection was an immunological function, it was early appreciated that immunodeficiency might be amenable to radical treatment by grafts. Various organs have been used, depending on the theory of the grafter, the nature of the defect being treated, and therefore the effect intended. The experience to 1967 was reviewed by Kay (1968). At first the intention was to supply immunologically competent cells, and lymph nodes were grafted, which produced some effects of immunoglobulin and antibody formation. Many of the patients so treated probably had some degree of specific cellular immunity function, since the grafts were probably often rejected. This was fortunate, since those that were not would probably have killed the patient by graft-*versus*-host disease (GVH). Indeed there is reason to believe that some of these patients did have some GVH. This disease presents a major practical problem in all work in this field, and we outline some measures to combat it. Combined immunodeficiency (Hitzig *et al.*, 1968) was regarded by some early workers as thymus deficiency. Grafts of thymus with or without stem cells from

41

fetal liver were made, with only transient or fatal effects. Mature bone marrow has since been used with success. With the description of human thymus agenesis, as a syndrome of cellular immunity deficiency, with near normal humoral immunity function, hypoparathyroidism, and abnormalities of the great vessels (DiGeorge, 1968), a likely field for thymus grafts occurred, and has been realized in two patients.

These attempts were intended to achieve complete permanent reconstitution of the defective system, but it is possible that therapeutic effects of a more transient nature by cell transfer, rather than by grafting may also be valuable. This may have been achieved both in patients with progressive necrotic vaccinia (Fulginiti *et al.*, 1968) and in patients with chronic mucocutaneous candidiasis (Buckley *et al.*, 1968).

In addition to these examples of grafts for primary immunodeficiency diseases, bone marrow grafts have been used in patients with marrow suppression, accidental or iatrogenic (Mathé *et al.*, 1969) and in other fields, in which immune competence of the recipient or the donor presents a problem. This field is not reviewed here.

TISSUES GRAFTED

Both mature tissue—lymph nodes, blood, bone marrow and thymus—and thymuses and liver cell suspensions from fetuses of 8-16 weeks estimated gestation have been used. The handling of the fetal tissue has not been fully described, though the appearance of the lymphoid organ of such fetuses has been reported (Kay *et al.*, 1970). The thymuses, removed within 2 hr of delivery of the fetus and stored in TC199 medium at 0-4°C, for not more than 36 hr, were usually cut into fragments about 1 mm across, and inserted beneath the sheath of the rectus abdominis. A suspension of the fetal liver was prepared by gentle mincing in TC199 medium with scissors, and then repeated aspiration first through a No. 1 needle and then through a No. 17 needle into a syringe. This suspension is frozen by standard technique in ampoules containing 10% of dimethylsulphoxide and 5% human serum (Group AB) in TC199, and stored at $-130°C$. Each liver yields between 10^8 and 10^9 cells (only some of which are stem cells), and these have been given intravenously or intraperitoneally, either as a single dose or in divided doses. Collapse and hypotension has occurred in some patients after the injection, but this has been transient and not fatal with our material.

DETECTION OF EFFECTS

Effects, following grafts, can be demonstrated by immunological function tests related either to B or to T cells, which were abnormal before, by evidence of chimaerism, and by non-specific cellular effects. Interpretation of results is always difficult as we are increasingly recognizing that the defects are not absolute ones, and that all the various functions may vary very widely from patient to patient, and even within a single patient. It is virtually impossible to exclude spontaneous improvement in any particular case, and, though establishment of prolonged chimaerism is objective, female cells in a male patient may be maternal rather than graft. The following are the tests usually used:

a. B cell function: immunoglobulin concentration, presence of naturally occurring antibodies (isoagglutinin, ASO etc.), antibody response to administered antigens.

b. T cell function: lymphocyte count, lymphocyte thymidine uptake on PHA or antigen stimulus, skin delayed hypersensitivity to candida, streptodornase etc., or to DNCB. Skin graft rejection (possibly dangerous).

c. Chimaerism: red cell (by typing), lymphocytes (by sex chromosome).

d. Cellular changes: appearance of lymph nodes, or thymus shadow. Restoration of lymphoid cells in lymph node or gut biopsies, bone marrow changes, eosinophilia.

e. GVH: fever, rash, diarrhoea, positive Coombs' test, marrow aplasia or infiltration with lymphoid cells or histiocytes.

Effects of Grafts in Different Immunodeficiency Diseases

Sex linked agammaglobulinaemia. Though some transient effects were observed, no sustained effects, beneficial or harmful, were reported. This probably results from near normal T cell function in these children, who reject grafts nearly as fast as healthy subjects. They presumably are deficient in B cells only. Grafting has no obvious role in treating these patients.

Severe combined immunodeficiency. Grafts take readily in these patients, and it is likely that unmatched mature immunologically competent cells, whether given intentionally or unintentionally in blood transfusions, regularly kill the recipient with GVH disease (Miller, 1968). Severe forms of these diseases arising from such grafts have been consistently fatal, and these procedures are contraindicated completely.

Since the use of mature unmatched tissue was abandoned either fetal tissue, in the hope that the premature cells would become tolerant to the host, or bone marrow, either from tissue-matched sibling donors, or partly-matched donors, usually related, for whom some effort to combat GVH has been made, has been used. The stage is past when every such attempt is reported, but Table I gives some indication of the present position in this field. We include six unpublished new cases from the Hospital for Sick Children, London, and are grateful to others, especially Dr R. A. Good and Dr F. S. Rosen, for information which is included. It is certainly incomplete, but not misleading in principle.

Thymus grafts were given when this disease was thought to be T cell deficiency, and replacement was clearly incomplete (Hitzig *et al.*, 1965). Fetal tissue was usually used, to avoid GVH, but it certainly occurred in some cases. Particularly interesting was the development of two "monoclonal" immunoglobulins, as well as fatal GVH, in one such presumably two B cells existed in the thymus at the time (Harboe *et al.*, 1966).

TABLE I

Grafting in severe combined immunodeficiency

Tissue	No.	Effect				
		B	T	GVH	Restoration 3 months	Comment
Fetal thymus	1	+	+	+	−	monoclonal immunoglobulins
Fetal liver	1	−	?+	−	−	
Fetal thymus and liver	11	+	+	+	−	thymus matures
Matched sibling bone marrow + blood	1	++	++	+	+	marrow aplasia
Matched sibling bone marrow stem cells and fetal thymus	1	++	++	+	+	discrete immunoglobulins → heterogeneous
Matched sibling bone marrow	3	++	++	?+	1+	
Matched sibling bone marrow stem cells	5	1+	0	0	0	
Parental bone marrow + Ab	1	+	+	±	−	(thymus later)
Parental bone marrow, stem cells + Ab	1	+	−	−	+	

T = T cell function (see text), B = B cell function, GVH = graft-*versus*-host disease.

When the suggestion that this disease was a stem cell defect was made, intravenous fetal liver (a known source of stem cells at a stage of development before marrow exists), accompanied almost always by a thymus from the same fetus, was given. Figure 1 indicates some effects on both B and T cell function, in a child treated in this way; the biggest change is the eosinophilia. GVH disease occurred even though the fetus from which the graft was derived was only of 11

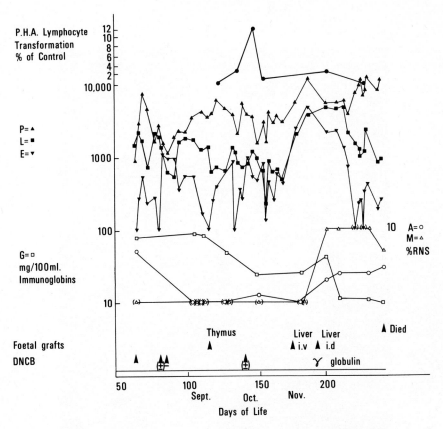

Fig. 1. Effect on blood cells and immunoglobulin concentrations of fetal thymus and liver transplantations in a boy of seven weeks with severe combined immunodeficiency.

weeks estimated age. The remarkable histiocytosis of the bone marrow (Fig. 2) which has been noted previously occasionally in these grafted patients, may have been a result of the GVH disease. Most of the effects could have been due to thymus or liver, but the observation of PHA response before the liver transplant is consistent with observed PHA responses following thymus transplant alone (Flad, Fliedner and Kay, unpublished observations). In this case, both an unmatched and, later, a partly tissue type-matched thymus was given; fetal tissue typing is now possible (Lawler and Klouda, personal communication) and the results of further typed transplants should be of interest. Experimental thymus grafts in inbred mice suggest that tissue type similarities between donor and recipient influence the function of a grafted thymus in thymectomized animals (Leuchars et al., 1970) as well as presumably minimizing the risk of GVH. It is clear, however, that the incompleteness of the effect of a thymus graft in these

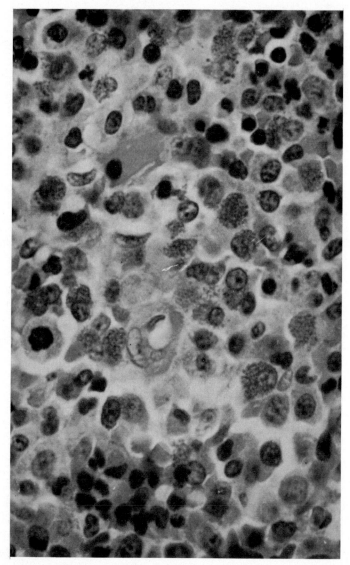

Fig. 2. Autopsy bone marrow from a child with severe combined immunodeficiency who received a thymus and bone marrow graft. There is marked proliferation of histiocytes. H. and E. ×480.

children is not due to a failure of the grafts to take. They have been shown to survive in almost all cases—for up to nine months—and they may mature in the host (Cottier et al., 1968). Hassall's corpuscles may be seen at autopsy, which would not be expected in the graft as given.

In contrast to these transient or clinically unhelpful effects, mature bone marrow has been successful in the special situation in which GVH disease can be virtually avoided—the use of matched sibling bone marrow (Gatti et al., 1968). Even here, some GVH and a period of marrow aplasia occurred, which was ameliorated by a second graft. One interesting effect was the development of a thymus shadow on X-ray. This procedure has achieved sustained reconstitution of B and T cell functions in five patients, with obvious clinical benefit in three (Table I) (Good, personal communication). An attempt to minimize GVH by separation of stem cells from mature lymphocytes in the marrow used from such a donor on an albumin gradient column was effective in one patient (de Koning et al., 1969). Here there was the interesting phenomenon, which was not observed in the ones receiving whole marrow, that the immunoglobulins were initially discrete and of restricted heterogeneity, and heterogeneity increased progressively. Presumably only a small range of B cells was initially established, whereas the inclusion of mature cells in the whole marrow produced full heterogeneity more rapidly. In five other attempts using marrow processed in this way (Rosen, personal communication; Good, personal communication) only one has produced any effect, and this was transient with B cells only.

Tissue-matched sibling marrow is rarely available, and the following techniques have been used in attempts to minimize GVH with incompletely matched parent donors. (There is usually at least one major tissue type antigen in the child that the parent lacks, as well as likelihood of other differences for which we do not have antisera. Mixed lymphocyte reactions may provide evidence of this, and of sensitization of the parent.) Though some progress has been made, no permanent success has yet been achieved.

    a. Stem cell preparations as above.
    b. Pre-treatment by cytotoxic drugs or ALS.
    c. Suppression by antibody.
    d. Isolation.

Suppression of GVH by antibody (immunological enhancement) has been achieved in mouse experiments by Batchelor and Howard (1965). An attempt was made in a girl with severe combined immunodeficiency, and no surviving siblings (one had died previously of the disease) to graft bone marrow from the father, who lacked antigens HLA 1 and 8 which the child had. The child had low levels of all immunoglobulins, and was grossly lymphopenic, though there was some PHA response. Immediately before an infusion of 10^9 marrow cells, 5 ml of a high titre human antiserum to these HLA antigens was injected intramuscularly, with no noticeable toxic effect. Following the graft, there was evidence of chimaerism (both male (40%) and female (60%) lymphocytes responding to PHA were detected eight days after the graft) and improved B and T cell function, without clinical evidence of GVH. Lymph nodes became palpable, but no thymus shadow was seen on X-ray. The child gained weight and improved clinically, and apart from six days of fever starting 13 days after the graft, there was no evidence of GVH. She was sent home, and maintained her better state, with some evidence of immunity reconstitution for 12 weeks, but these then waned and her condition deteriorated. In view of the possibility that the waning resulted from failure to wake up her own thymus, it was decided to give her a fetal thymus graft, four months after the first graft. It was intended to follow this with a further paternal marrow graft, suitably "enhanced", but she died before this was done. Her terminal illness had some of the features of GVH, but this was not established, and it is unlikely that it resulted from the parental marrow graft. The graft was a failure, but HLA incompatible cells were transferred with only a transient trivial disturbance during the period of GVH; this method of suppression requires further trial. The dose of serum was arbitrary, since there is no method of measuring the relevant activity. In mice a single injection of 4 ml/kg of a suitable antiserum was enough, but the equivalent volume of the human serum was not available. The dose used, 5 ml, was about half that equivalent to the mouse dose. Another attempt along these lines, using the non-donor parent as a source of antibody, has produced similar reduction of GVH, with more sustained take of the graft; repeated doses of serum were given (Buckley, personal communication).

There is evidence that GVH is more dangerous in normal animals than in germfree animals (Keast, 1968). This represents a theoretical case for laminar flow isolation care in such children. This was used in only one (de Koning *et al.,* 1969) of the four patients who have had successful sibling marrow transplants, and it does have risks in itself, but it may have a wider role in grafts from less ideal donors.

THYMUS APLASIA

In contrast to the lack of sustained effect of fetal thymus grafts in combined immunodeficiency, marked clinical improvement with evidence of restoration of cellular immune response followed grafting of fetal thymuses in two cases of thymus aplasia (Table II). A large and rapid rise in eosinophils and lymphocytes occurred. In contrast to some animal work (Davies *et al.,* 1968) it was the host cells that responded to PHA after the graft, and not the donors. There is evidence that one, and perhaps both of the grafts have been rejected, but the improved function has been maintained in spite of this, though the effect may be waning (August *et al.,* 1970). Unlike the experience of these grafts in severe combined immunodeficiency, there was no GVH. The rapid effect (Fig. 3)

TABLE II

Effects of thymus grafts in thymic aplasia (DiGeorge)

	Cleveland *et al.*, 1968	August *et al.*, 1968 and 1970
Clinical	++	++
Lymphocytes	+	(+)
Eosinophils	++	−
PHA	++	++
Reacting cells	Host	Host
Monilia skin test		+
Graft rejection	(+)	+
DNFB, DNCB	++	++
Paracortical lymphocytes	±	±

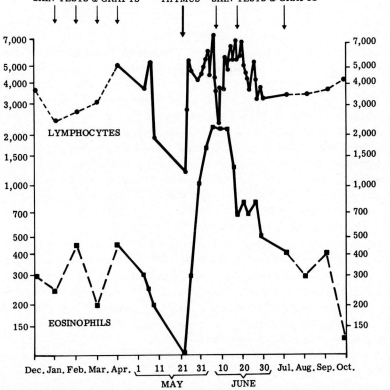

Fig. 3. Lymphocyte and eosinophil counts following transplantation of fetal thymus to a child with thymic aplasia (Cleveland *et al.*, 1968). Sensitization to DNCB and graft rejection were negative before the thymus graft and positive afterwards.

strongly points to its being due to the graft, but since partially improving forms of this condition have been observed (Rosen, personal communication), spontaneous improvement cannot completely be excluded.

PARTIAL COMBINED IMMUNODEFICIENCY

Abnormal response to vaccinia or candida infection may be associated with apparently partial cellular immunodeficiency, which may be to some extent antigen-specific. Possibly such patients, when infected with candida, may readily reach a state of high dose tolerance. Improvement following transient grafts of bone marrow has been reported, and it is possible that they may work by breaking tolerance through transfer of cells, even though a permanent graft was not achieved (Buckley *et al.*, 1968). Further work on this is needed, but such grafts must always be regarded as dangerous, and the patient's condition must be bad for the risk to be justifiable.

COMMENTS

Grafting has proved successful in two situations where grafting of the cell type actually deficient, whether thymus or stem cell, has been possible. Repeated attempts with inappropriate grafts have been unsuccessful. The big risk, GVH, is probably a quantitative phenomenon, which may possibly be avoidable in the face of partial tissue incompatibility.

Apart from the intended reconstitution, certain interesting and unexpected cytological effects have been noted. Widespread plasmacytosis may occur, with the production of immunoglobulin of restricted heterogeneity. Histiocytes sometimes appear in large numbers in marrow and lymph nodes; the significance of this is not known, but it may be a feature of GVH. Eosinophilia of blood, marrow and lymph nodes occurs rapidly after transplantation, in both severe combined immunodeficiency and in thymus aplasia.

Interpretation of effects is difficult because of the known heterogeneity of these deficiencies within and between patients. But basic differences occur even in patients with similar function, which may make a big difference to the outcome of graft treatment. For instance, some patients with severe combined immunodeficiency have lymph nodes with no lymphoid cells in them, whereas others have essentially no fixed lymphoid organs at all (Berry, 1970). Probably it will be easier to treat the former than the latter.

A large proportion of infants with severe combined immunodeficiency are probably missed, and this disease is responsible for 2% of the deaths at the Hospital for Sick Children, Great Ormond Street (Berry, 1968). Less severe forms are also probably more common than we now recognize. Grafting therefore has potential practical as well as theoretical importance, but the danger of GVH must always be considered carefully before grafts are given. With our increasing recognition of transient deficiency states, the theoretical safety factor—that if the graft is dangerous, it is needed, and if it isn't needed it isn't dangerous—which has sustained us in this work, is probably not completely true.

REFERENCES

August, C. S., Rosen, F. S., Filler, R. M., Janeway, C. A., Markowski, B. and Kay, H. E. M. (1968). *Lancet ii,* 1210-1211.

August, C. S., Levey, R. H., Berkel, A. I., Rosen, F. S. and Kay, H. E. M. (1970). *Lancet i,* 1080-1083.

Batchelor, J. R. and Howard, J. G. (1965). *Transplantation 3,* 161.

Berry, C. L. (1968). *Proc. R. Soc. Med. 61,* 867.

Berry, C. L. (1970). *J. clin. Path. 23,* 193-202.

Buckley, R. H., Lucas, Z. J., Hattler, B. G., Zmijewski, C. M. and Amos, D. B. (1968). *Clin. exp. Immun. 3,* 153-169.

Cleveland, W. W., Fogel, B. J., Brown, W. T. and Kay, H. E. M. (1968). *Lancet ii,* 1211-1214.

Cottier, K., Bürki, K., Hess, M. W. and Hässig, A. (1968). *In* "Immunologic Deficiency Diseases in Man" (R. A. Good, ed.), pp. 152-164. National Foundation-March of Dimes, New York.

Davies, A. J. S., Festenstein, H., Leuchars, E., Wallis, V. J. and Doenhoff, M. J. (1968). *Lancet i,* 183-184.

DiGeorge, A. M. (1968). *In* "Immunologic Deficiency Diseases in Man" (R. A. Good, ed.), pp. 116-121. National Foundation-March of Dimes, New York.

Fuliginiti, V. A., Kempe, C. H., Hathaway, W. E., Pearlman, D. S., Sieber, O. F., Jr., Eller, J. J., Joyner, J. J. and Robinson, A. (1968). *In* "Immunologic Deficiency Diseases in Man" (R. A. Good, ed.), pp. 129-144. National Foundation-March of Dimes, New York.

Gatti, R. A., Meuwissen, H. J., Allen, H. D., Hong, R. and Good, R. A. (1968). *Lancet ii,* 1366-1369.

Harboe, M., Pande, H., Brandtzaeg, P., Tveter, K. J. and Hjort, P. F. (1966). *Scand. J. Haemat. 3,* 351-374.

Hitzig, W. H., Kay, H. E. M. and Cottier, H. (1965). *Lancet ii,* 151-154.

Hitzig, W. H., Barandun, S. and Cottier, H. (1968). *Ergebn. inn. Med. Kinderheilk. 27,* 79.

Kay, H. E. M. (1968). *In* "Immunologic Deficiency Diseases in Man" (R. A. Good, ed.), pp. 168-172. National Foundation-March of Dimes, New York.

Kay, H. E. M., Doe, J. and Hockley, A. (1970). *Immunology 18,* 393-396.

Keast, D. (1968). *Immunology 15,* 237-245.

de Koning, J., Dooren, L. J., van Bekkum, D. W., van Rood, J. J., Dicke, K. A. and Rádl, J. (1969). *Lancet i,* 1223-1227.

Leuchars, E., Aird, J., Davies, A. J. S. and Wallis, V. (1970). *In* "Protides of Biological Fluids", p. 73. Pergamon Press, Oxford and New York.

Mathé, G., Nouza, K., Hrsak, I. and Kolar, V. (1969). *Antibiotica Chemother. 15,* 182.

Miller, M. E. (1968). *In* "Immunologic Deficiency Diseases in Man" (R. A. Good, ed.), pp. 257-263. National Foundation-March of Dimes, New York.

Soothill, J. F. (1968). *In* "Clinical Aspects of Immunology" (P. G. H. Gell and R. R. A. Coombs, eds), Second edition, pp. 540-572. Blackwell Scientific Publications, Oxford and Edinburgh.

World Health Organization (1970). Report of Meeting of Investigators of Primary Immune Deficiencies. *Bull. Wld Hlth Org.* In press.

Differentiation of Thymocytes and Thymus-dependent Lymphocytes

BYRON H. WAKSMAN and DANIEL G. COLLEY

Department of Microbiology, Yale University
New Haven, Conn. 06510, U.S.A.

The major immunologic role of the thymus in small mammals is to serve as the source of a population of long-lived small lymphocytes for the peripheral pool (reviewed in Miller and Osoba, 1967; Everett and Tyler, 1967; Osoba, 1968). These cells, while they remain within the thymus, will be referred to as thymocytes, their progeny in the periphery as T lymphocytes, and short-lived cells which enter the peripheral pool from the bone marrow, perhaps passing by way of some mammalian analogue of the bursa of Fabricius, as B lymphocytes (Roitt *et al.*, 1969). Our laboratory's interest has been directed to the inductive influences that determine the differentiation of thymocytes, and to their acquisition of cellular properties that may be responsible for "immunocompetence". We have at the same time tried to identify the specific role of thymocytes and T lymphocytes in certain immune responses to antigen, using inbred rat strains which differ by strong histocompatibility antigens: Lewis (Le), dark agouti (DA), and brown Norwegian (BN) (Elkins and Palm, 1966).

In the work reviewed here, Le animals were thymectomized at approximately five weeks and irradiated with 900 R (central axis dose) at eight weeks. Syngeneic or hemiallogeneic bone marrow (BM) cells were given intravenously within a few hours following irradiation, and half-lobes of 5-7 week old thymus grafted subcutaneously along the ventrolateral surface of the abdomen within 18 hr (Malakian *et al.*, 1970). An immunofluorescence procedure using living lymphoid cells, suitable isoantisera, and labeled rabbit anti-rat globulin (Lubaroff and Waksman, 1968b) was used to identify histocompatibility antigens and lymph node (LN), spleen, or other cells originating in the hemiallogeneic marrow or thymus grafts. The same procedure was used to study thymus-specific antigens identifiable with rabbit anti-rat thymus serum and fluorescent goat anti-rabbit globulin (Order and Waksman, 1969). For quantitative study of these antigens, we used cytotoxicity and absorption techniques modified from the original method of Gorer and O'Gorman (Colley and Waksman, 1970).

DIFFERENTIATION OF THYMOCYTES

Lymphoid cells of the thymus are derived from BM stem cells and, after leaving, become recirculating peripheral T lymphocytes, whose immuno-competence may be judged by their ability to restore function in thymectomized animals or produce graft-*versus*-host (GVH) reactions (Ford and Micklem, 1963; Moore and Owen, 1967; Weissman, 1967). Questions regarding the cause and nature of changes that they undergo within this organ appear central to understanding the development of immunocompetence and the generation of diversity.

The influences that determine thymocyte differentiation are not understood. Lymphoid stem cells mature in the presence of epithelial reticular cells which function under the inductive influence of the mesenchyme (Auerbach, 1961; Owen and Ritter, 1969). This process is comparable to differentiation in other organs such as the lens. This may also require three cell types, and is a special case of the general embryologic problem of induction. In human subjects, congenital absence of the epithelial component leads to the well-known diGeorge syndrome, abnormality of the mesenchymal component to ataxia

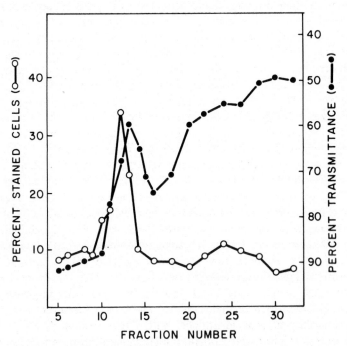

Fig. 1. Natural antibody in normal rabbit serum against antigen present in rat thymocytes. Graph compares cytotoxicity (without added complement) of fractions obtained by filtering serum through Sephadex G-200 (○) with percent transmittance at 280 mμ (●). Activity is found in fractions containing γM-globulins. From Colley and Waksman (1970).

telangiectasia, and absence of BM stem cells to reticular dysgenesis (Owen and Ritter, 1969).

Induction must depend on a phenomenon associated with cell contact, whether electrical or chemical, or on the local release of a diffusible mediator. The existence of a local hormone has been postulated by many authors on the basis of morphologic observations or apparent restoration of neonatally thymectomized or thymectomized, irradiated animals by thymus in diffusion chambers or thymus extracts (see review in Osoba, 1968). The best known candidate for a hormonal role, "thymosin" extracted from calf thymus, was ineffective in our hands in producing restoration of thymus-dependent areas of the spleen and nodes in thymectomized, irradiated, BM reconstituted rats and, more important, failed to induce antigenic changes in BM-derived cells in the spleen of such animals comparable to those observed during their differentiation in the thymus (Krüger et al., 1970). The highly active protein isolated from mouse submaxillary gland (Naughton et al., 1969) provides a suggestive new candidate for a hormonal function. An alternative mechanism is "modulation", the rapid change in surface components of thymocytes under the influence of antibody against one of these (Old et al., 1968). Natural antibodies directed to

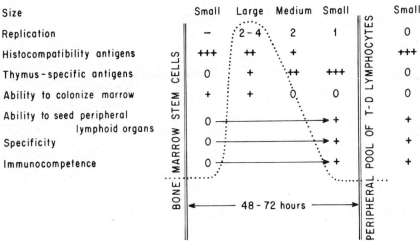

Fig. 2. Scheme showing properties of thymocytes that change during differentiation and have been investigated thus far. Dotted line symbolizes cell size.

thymocyte antigens and apparently not strictly species-specific are widely present in adult laboratory animals (Kidd and Friedewald, 1942; Herberman, 1969; Colley and Waksman, 1970; Boyse et al., 1970) (Fig. 1). They may effect changes in lymphoid cells as they enter or leave the thymus. The role of a possible "second messenger", such as cyclic AMP within the cell itself, in the initiation of differentiation remains to be investigated.

Thymocytes (Fig. 2) remain in the mouse or rat thymus 48-72 hr as judged by ^{3}H-thymidine labeling (Weissman, 1967; Craddock *et al.*, 1964). They enter as small stem cells, soon after entry become large, rapidly dividing cells, then diminish progressively in size, undergoing at the same time as many as 5-7 cell divisions (Weismann, 1967; Sainte-Marie and Leblond, 1965). They migrate from the cortex to the medulla, acquiring some of the properties of peripheral T lymphocytes as they do so (Metcalf, 1966). With the change in size, thymocytes acquire new surface components identifiable as characteristic thymus antigens (Boyse *et al.*, 1968). In the rat, with the use of specific immunofluorescence staining (Order and Waksman, 1969), we found partial replacement of histocompatibility antigen by thymus antigen in some of the largest (most immature?) cells and all medium cells (Order and Waksman, 1969; Grabar *et al.*, 1968; Colley *et al.*, 1970a). In small cells, the dominant population within the thymus, this rat thymus antigen (RTA) appears to replace histocompatibility antigen almost completely. By quantitative absorption techniques (Colley *et al.*, 1970a) normal rat thymocytes were found to possess approximately 30 times more RTA than LN cells (Fig. 3). The use of refined morphologic methods (Aoki *et al.*, 1969) should soon permit precise topographic mapping of such surface components on cells at different stages of maturation.

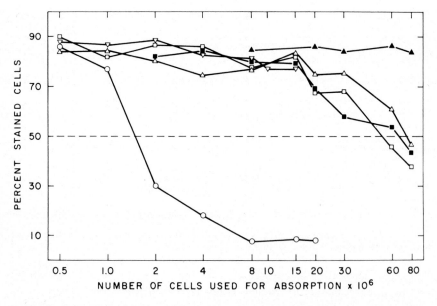

Fig. 3. Comparison of RTA content in adult rat thymocytes and other lymphoid cells by their ability to absorb cytotoxic activity from dilute rabbit anti-rat thymus serum. Absorption with thymocytes ○, LN cells □, spleen cells △, peripheral blood lymphocytes ▽, bone marrow ■, and erythrocytes ▲ Absorbed serum tested on rat thymocytes. From Colley *et al.* (1970a).

Correlated with these striking changes in cell size and surface character is a rapid loss in the ability of thymocytes to return to bone marrow, if the cells are infused into a suitable irradiated recipient (Order and Waksman, 1969). Those that do return, presumably the most immature cells, remain lymphoid in character over more than a week. This finding implies that stem cells, by the time they enter the thymus, are committed to a lymphoid line of differentiation. Conversely, most of the cells, during their sojourn in the thymus, lack the surface properties of peripheral T lymphocytes that permit homing into LN and spleen. In the mouse 1.3-1.6% of normal thymocytes migrate into LN if transferred to normal recipients, as compared with 7.9-12.5% of normal LN cells (Goldschneider and McGregor, 1968; Lance and Taub, 1969). In the rat the proportion is even smaller, 0.7-1.1% as compared with 15.0-16.3% (Everett *et al.*, 1964; Colley *et al.*, 1970b). These numbers are similar to the number found to be immunocompetent. Restoration of thymectomized mice requires approximately 4-10 times more thymocytes than peripheral lymphocytes (Yunis *et al.*, 1965) and restoration of neonatally thymectomized rats more than 20 times as many (Isaković *et al.*, 1965). The proportion of thymocytes competent to produce a GVH reaction, in mice and rats respectively, has been estimated as seven times and 20-60 times less than the proportion of similarly competent spleen or LN cells (Cantor *et al.*, 1970; Billingham *et al.*, 1962).

Immunocompetences implies the possession not only of the ability to recirculate but also of specificity, i.e. antibody-like receptors at the cell surface. Actual uptake of antigen by thymocytes *in vitro* is minimal, compared to that seen with peripheral lymphocytes (Sulitzeanu, 1968; Byrt and Ada, 1969), and immunoglobulins are not found on thymocytes by immunofluorescence, or immunoautoradiography (Raff *et al.*, 1970). Exposure of thymocytes in culture to antibody against the known classes of immunoglobulin fails to induce a significant degree of blastogenesis (Daguillard and Richter, 1969; Kaplan and Thorbecke, 1969). Nevertheless the fact that specific tolerance may be induced in these cells by antigen (see below) and that a certain subpopulation can respond to antigenic stimulation by histocompatibility antigen in the mixed lymphocyte reaction (MLR) (Schwarz, 1967; Daguillard and Richter, 1969) strongly suggests that they have specific receptors. We have shown with cells separated by centrifugation in discontinuous BSA gradients (Colley *et al.*, 1970b), that the properties mentioned as characteristic of peripheral cells, i.e. ability to seed lymph nodes, GVH competence, and the ability to give an MLR, are found together in cells within a single fraction (Table I). In some rats these cells possess a degree of reactivity for PHA while in others they apparently do not. In the latter case, they may be regarded as lacking the full complement of properties that characterize T lymphocytes. Most but not all of the cells in this fraction are positive by immunofluorescence for RTA. We have not determined to what extent a shift from RTA to histocompatibility antigen is correlated with

TABLE I

Peripheral properties in rat thymocyte subpopulations

Thymus fraction	Average yield (percent)	Average percentage medium and large cells	Spontaneous DNA synthesis	Localization of ^{51}Cr labeled cells in lymph nodes (percent of injected dose)	GVH activity	Mitotic response (average E/C)	
						MLR	PHA
A	1.0	24	++	0.7	0	2.4	1.0
B	3.0	15	++	0.9	0	3.7	2.2
C	7.6	2	+	2.7	+	8.5	3.1
D	13.0	2	0	0.5	0	2.0	1.8
P	30.0	1	0	0.3	0	1.0	0.8

Bands of increasing density and pellet separated on discontinuous BSA density gradients. From Colley *et al.* (1970b).

the other peripheral properties under discussion. These cells are small (Schwarz, 1967), but are less dense than competent T lymphocytes in a LN cell suspension. Immunocompetence is determined by the possession of specificity and the ability to recirculate. The latter property, in turn, depends upon protein constituents of the surface membrane, as shown in brilliant studies by Woodruff and Gesner (1968). An intriguing possibility is that the protein in question is one of the antigens or receptors demonstrated by immunofluorescence or by blastogenesis with PHA. The overwhelming majority of small thymocytes lack all the peripheral properties tested thus .far: nothing can be said at present about their functional attributes.

TRANSITION OF THYMOCYTES TO T LYMPHOCYTES

The time at which thymocytes become T lymphocytes has not been determined and the mechanism of this transition, called "peripheralization" (Lance et al., 1970), is as yet entirely unknown. The number of peripheral cells possessing thymus-specific antigens is negligible in the case of Tla in the mouse (Boyse et al., 1968), less than 7% in the case of RTA in the rat (Colley et al., 1970a), and less than 1% of lymph node or thoracic duct cells in the calf (Williams et al., 1970). Therefore the transition must be abrupt. Using immunofluorescence (Williams et al., 1970) we could show that 3-8% of cells in thymus vein blood and thymus efferent lymphatics of calves possess thymus-specific surface antigen (Table II). Evidence obtained in adult guinea pigs suggests that approximately one-third of cells in the thymus vein have entered

TABLE II

Thymus-specific and histocompatibility antigens in calf lymphoid cells

	Percent fluorescent cells	
Cell source	Thymus-specific antigen (+, ++ or +++)	Histocompatibility antigen (++ or +++)
Thymus	74-90	23-32
Thymic vein blood	3-8	78-94
Thymic lymph	4	88-96
Thymic lymph node	3-5	93-97
Arterial blood	<2	91-100
Thoracic duct lymph	<1	91-99
Prescapular lymph node	<1	100

Data obtained in three calves.
From Williams et al. (1970).

the blood from the thymus parenchyma (Ernström and Larsson, 1969). We therefore estimate that of such cells (in calves) 10-25% have not completed the process of peripheralization, while the others must have done so within the short period of passing from tissue to blood or lymph. Lance *et al.* (1970) report that in the mouse, thymocytes localizing in lymph nodes after being injected intravenously lose all their Tla within a few hours ("forced peripheralization"). Modulation under experimental conditions can occur in a period of less than 1 hr (Old *et al.*, 1968) and suggests itself as a possible mechanism for this change.

The actual number of cells leaving the thymus has been estimated (again in guinea pigs) as 1.2×10^7/day via lymph (Kotani *et al.*, 1966) and 8.8×10^7/day via veins (Larsson, 1967), a total of 10^8 cells. Since the total pool of T lymphocytes is between 10^9 and 10^{10}, this figure implies replacement of the pool in weeks or months and agrees with other data showing that T lymphocytes are long-lived cells. There is evidence that a large proportion of cells formed in the thymus may never leave (Metcalf, 1966; Metcalf and Brumby, 1966). One can speculate that tolerance to self-antigens induced in the thymus (see below) may account for cell death or some other deviation from the physiologic pathway.

The quantitative data given above apply only to young adult animals. During late fetal and early neonatal life, there is a greatly increased tempo of lymphopoiesis. The thymus contains a substantial number of lymphocytic cells lacking RTA (Table III); presumably these are stem cells that have not initiated their thymic differentiation (Colley *et al.*, 1970a). At this time a very large proportion of lymphocytes in the peripheral pool contain RTA, probably because as recent arrivals they have not completed "peripheralization". These findings agree with the preponderance of larger cells in the fetal thymus and with the demonstration that many large immature thymocytes migrate to the

TABLE III

Change in RTA distribution in Le rats with age

Age	Percent of cells bearing RTA		
	Thymus	Spleen	Lymph node
Fetal (18 days)	74	62	ND
Neonatal	90	28	ND
4 weeks	100	4.5	3.5
8-10 weeks	100	2.5	1.0
6 months	100	2.5	0.5

ND: Not determined.
From Colley *et al.* (1970a).

periphery at this time (Weissman, 1967; Sherman and Auerbach, 1966). A comparable picture is seen in the regenerating thymus following irradiation (Order and Waksman, 1969) and in regenerating thymus grafts (Malakian *et al.*, 1970).

T Lymphocytes in Peripheral Pool

Peripheral small lymphocytes include two populations easily distinguished on the basis of size (Metcalf and Brumby, 1966; Sherman and Auerbach, 1966), electrophoretic mobility (Rühenstroth-Bauer and Lücke-Huhle, 1968), and life span (Everett *et al.*, 1964) which probably correspond to T and B lymphocytes. Tentative identification appears to be provided by the presence on the former of a retained thymus-specific antigen θ (Raff, 1969) and on the latter of conventional immunoglobulins (Sell and Gell, 1965; Raff *et al.*, 1970). Absolute identification is possible only by the use of chromosomal markers (Davies, 1969) or histocompatibility antigens (Malakian *et al.*, 1970) in suitably prepared chimaeras. T lymphocytes, estimated by these techniques, make up approximately 90% of thoracic duct lymphocytes, 70-80% of blood lymphocytes, and 60-80% and 30-40% of lymph node and spleen cells respectively in various species. Monocytes, which cannot be distinguished by light microscopy from small lymphocytes, contribute an additional 3-7% to the peripheral pool (Whitelaw, 1966). The estimated numbers of T lymphocytes presumably include precursor cells and memory cells or sensitized cells of delayed (cell-mediated) hypersensitivity. There is no information to show whether the latter differ from precursor cells merely in number or also in such cellular properties as the number and combining properties of antibody-like receptors. The B lymphocytes also include precursor cells, and probably a second type of memory cells. These terms have a rough correspondence with X and Y in the classical XYZ scheme, the precursors also being designated ARC (antigen-reactive cells) in various laboratories. It is clear that new, more precise terms such as T-ARC and B-ARC are now needed.

The relative numbers of T lymphocytes, their rate of production in the thymus, and their rate of recirculation are determined in part by metabolic rate and appear to be under pituitary control. A discussion of this control and of various agents that affect their numbers and/or distribution (age, vitamin B6 deficiency, irradiation, radiomimetic agents, corticosteroids, neonatal thymectomy, thoracic duct drainage, antilymphocyte serum, pertussis, etc.) is beyond the scope of this paper.

T lymphocytes apparently possess specific antibody-like receptors. There have been reports of their ability to take up antigen specifically, in animals with "pure" delayed hypersensitivity (see, e.g. Steffen and Rosak, 1963). More convincing perhaps is their ability to give MLR *in vitro* (Schwarz, 1967; Daguillard and Richter, 1969), to produce GVH *in vivo* (Woodruff and Gesner, 1968; Lance *et al.*, 1970), or to be rendered specifically tolerant (Follett *et al.*, 1966;

Scott and Waksman, 1969). The additivity of MLR responses to different antigens (Colley and de Witt, 1969; Wilson and Nowell, 1970) appears decisive (Table IV). There is as yet no direct proof, however, that any of these observations involve T lymphocytes exclusive of other cell types. The presence of light chains on T lymphocytes can be inferred from data obtained with "sensitized" cells (see below). In both MLR and GVH involving strong

TABLE IV

Additivity of MLR responses of Lewis LN cells against BN and DA antigens

Disintegrations per minute x 10^3			Le + BN* + DA*
Le + BN*	Le + DA*	Le + BN* + DA*	Le + BN* + Le + DA*
14.9	49.0	66.3	1.04
66.5	152.0	200.0	0.91
11.4	47.1	61.0	1.04
13.2	30.0	62.0	1.43
27.2	75.2	86.7	0.85
45.0	44.7	84.3	0.94
32.5	91.2	96.7	0.78
29.1	118.7	126.0	0.86
13.8	14.2	18.5	0.66
20.5	22.7	64.2	1.49
			1.00 ± 0.15 (2SE)

4×10^6 Le cells cultured with 2×10^6 cells of BN and/or DA (treated with 1100 r prior to culture)*.

All counts adjusted by subtracting half the count obtained in tubes containing Le cells alone.

From Colley and de Witt (1969).

histocompatibility antigens in the rat, mouse and chicken, as many as 2% of cells appear capable of reacting against a single "antigen" (Wilson and Nowell, 1970; Szenberg et al., 1962). This number may be considerably inflated by the multiple antigenic specificities controlled by a single gene in this system and possibly by unrecognized priming. It implies nevertheless that the number of distinct specificities may be of the order of 10^2-10^3 in cells of this class. The comparable number of B lymphocytes, estimated from uptake of various protein antigens by spleen or LN cells, is 10^3-10^4 (Sulitzeanu, 1968; Byrt and Ada, 1969). The frequency of successful "hits" by any given antigen, if T and B cell interaction is required would be approximately one in 10^5-10^7, a figure that agrees with those obtained by several authors (see *Transplant. Rev.*, 1969).

CELL-MEDIATED HYPERSENSITIVITY

The mononuclear cells of the perivascular infiltrate in delayed skin reactions elicited with tuberculoprotein or other protein antigens were found to be largely or entirely hematogenous (Kosunen et al., 1963). A substantial proportion of the earliest cells infiltrating skin test sites are T lymphocytes, as shown by immunofluorescence in thymectomized, irradiated, syngeneic marrow restored Le rats with Le/DA thymus grafts (Williams and Waksman, 1969). In similar rats given hemiallogeneic BM and adoptively sensitized with Le LN cells (Lubaroff and Waksman, 1968b), most of the infiltrating cells by 24 hr are monocytes coming more or less directly from the BM (Table V). This result explains the

TABLE V

Cell populations in 24 hr tuberculin reactions

| Bone marrow donor | Lymph node donor | PPD reaction | Fluorescent cells (%) | | |
			Spleen	Mesenteric node	Bone marrow
Le/DA	Le	75.2	22.2	7.4	84.7
Le	Le/DA	25.6	12.5	91.4	12.5

Reactions elicited in thymectomized irradiated Le rats infused with normal bone marrow and tuberculin-sensitized LN cells and tested 24 hr later.
From Lubaroff and Waksman (1968b).

finding of McCluskey et al. (1963) that in tuberculin-sensitive guinea pigs, specific cells make up a small proportion of the total infiltrate at the peak of the local cutaneous response. It accounts for the radiosensitivity of tuberculin-type sensitivity since, in adoptive transfer in rats, irradiation of the recipient abolishes sensitivity (Coe et al., 1966) and provision of adequate numbers of BM cells restores it (Lubaroff and Waksman, 1968a).

Specifically sensitized T lymphocytes appear to react with antigen by virtue of possessing conventional light chains on their surface; none of the known types of heavy chain has been shown to be present (Greaves et al., 1969; Mason and Warner, 1970). It is difficult, however, to believe that the exquisite specificity of peripheral cell-mediated responses, directed to both determinant and carrier (Gell and Benacerraf, 1961), is thus adequately accounted for. One looks to the possibility that a new type of heavy chain will be found, that an additional co-operating cell type (also specific) will be identified in the reactions, or that co-operative effects between two or more T lymphocytes reacting with different parts of the eliciting antigen are required to initiate a response. There is a strong suggestion, from recent *in vitro* work, that "processing" or "presentation" of antigen by a macrophage is essential to its initial reaction with a T lymphocyte (Oppenheim et al., 1968).

Specifically sensitized T lymphocytes act by releasing a variety of mediators such as migration-inhibitory factor (MIF), chemotactic factor, and transfer factor. These may play at least three distinct roles in the developing lesion. First, additional T lymphocytes are stimulated to react by the direct mitogenic action of certain mediators. Those acted on by transfer factor may acquire the ability to react specifically with antigen and then undergo triggering. The cells that initially release mediators subsequently (and independently) undergo blast transformation and divide. All these mechanisms provide for enlargement of the pool of reacting cells and amplification of the local response. Secondly, monocytes (macrophages) are rendered sticky by MIF (David et al., 1964) and show activation by a number of criteria (Waksman and Matoltsy, 1958; Mooney and Waksman, 1970). These changes may account for their diapedesis at the local site, their ability to damage parenchymal elements in the zone of infiltration (Waksman, 1960), and their enhanced bactericidal activity, so-called "cell-mediated immunity" (Mackaness, 1970). Finally, released mediators are cytotoxic for certain cells (Ruddle and Waksman, 1968) and may produce direct damage of the vessel wall and possibly of extravascular tissue elements.

The range of possible cell-mediated responses has by no means been fully explored (Table VI). Sensitized T lymphocytes produce rejection of skin homografts without requiring the assistance of macrophages (Lubaroff, 1970), presumably by exerting a direct cytotoxic effect on vascular and other components of the graft. Similarly rejection of transplanted tumors of Gorer's type II, by immune lymph node cells, in the hamster, was not augmented by macrophages (Nomoto et al., 1970). Two new types of delayed skin reactions, both identified as "Jones-Mote" reactions, have recently been described. In one "cutaneous basophilic hypersensitivity" the principal infiltrating cell is a hematogenous basophil (Richerson et al., 1969; Dvorak et al., 1970) and in the other a plasma cell (Martins and Raffel, 1964). Both conventional protein antigens and conjugates capable of inducing antibody formation give rise to cutaneous basophil hypersensitivity. This is nevertheless a cell-mediated response, since it can be elicited only with immunogenic molecules. It is not known whether it is thymus-dependent. B lymphocytes rather than monocytes may react co-operatively with the antigen and T lymphocytes and produce some type of antibody. Complexes of such antibody with the test antigen might attract basophils to the site. The second type of reaction can be elicited with polysaccharide antigens completely lacking peptide (Gerety et al., 1970; Wick et al., 1970). Intense "delayed" reactions are also readily elicited with large proteins such as keyhole limpet hemocyanin (KLH) in thymectomized, irradiated, syngeneic BM reconstituted rats (Table VII). These, if they are not due to residual T lymphocytes, must be thymus-independent; perhaps they

TABLE VI

Classes of "cell-mediated" reaction

Antigens	Thymus dependence	Participating cells	Prominent histologic elements	Common names
Insoluble Ags?	+	T alone	Lymphocytes	Rejection of homograft, tumor
Proteins, conjugates	+	T + M	Histiocytes	Delayed skin reaction
Proteins, other?	+	T + M	Hemorrhage	Hyperacute delayed reaction
Persistent, insoluble Ags	?	T + M	Epithelioid cells	Granuloma, GVH
Proteins	+	T + B	Plasma cells	Jones-Mote skin reaction
SIII, large proteins	−	B alone	Plasma cells	Jones-Mote skin reaction
Proteins, conjugates	?	T ? + B ?	Basophils	Cutaneous basophil hypersensitivity
Proteins, at retest site	?	T ? + B ?	Eosinophils	Retest skin reaction

T: Thymus-dependent lymphocytes.
B: Bone marrow-derived lymphocytes.
M: Bone marrow-derived monocytes.

TABLE VII

Thymus-independence of Arthus and delayed skin reactivity to KLH in Lewis rats

Group	No. of rats	Average diameter of skin reaction (mm)			
		10 days		20 days	
		3 hr	24 hr	3 hr	24 hr
Normal	5	17.6	30.0	15.6	26.2
TxXBM, 4 weeks	5	16.2	28.4	16.4	27.6
TxXBM, 6 months	4	14.8	24.0	16.0	22.8
Unimmunized	5	11.2	9.8		

Sensitization: 200 μg KLH in complete Freund's adjuvant in footpad.
Skin testing: 50 μg KLH intradermally.
TxXBM = thymectomized, irradiated, BM reconstituted.
Unpublished data of J. Krüger.

involve circulating B lymphocytes directly. They may be peripheral reactions comparable to thymus-independent antibody-responses elicited by similar antigens in the spleen or lymph nodes, as suggested by Raffel. Still another class of lesion, the "retest reaction", is a predominantly eosinophilic infiltrate at the site of a previous delayed skin reaction re-exposed to antigen (Arnason and Waksman, 1963). Here conventional antigens such as tuberculoprotein or diphtheria toxoid are effective. Again some type of local antibody-formation may prove to be responsible. A similar skin reaction is often elicited in the late phase of cutaneous basophil sensitization.

ANTIBODY FORMATION

In rats, as in other species, the antibody response to certain antigens is thymus-dependent (Arnason et al., 1964). In mice T lymphocytes co-operate with B lymphocytes, undergoing blast transformation and proliferation in response to the primary or secondary antigenic stimulus, while the B lymphocytes proliferate somewhat later and are transformed into plasma cells producing antibody (see Transplant. Rev., 1969). The same relationship holds in the rat (Table VIII). There is suggestive evidence that the two types of cell may react with carrier and determinant portions of the antigen molecules (Taylor, 1969). It is important that they be physically near each other, since co-operative effects are not observed when carrier and determinant are on different molecules.

One may envisage two different functions of the T lymphocyte in this process. There may be production and release of mediators comparable to those

TABLE VIII

Partial thymus-dependence of PFC response in Lewis rats

Experimental group	PFC per spleen at 5 days		
	>150	50-150	<50
Normal	3	0	0
TxXBM	2	1	8
TxXBM + T graft	8	4	4
Ab against T	2	0	2
Ab against BM	0	1	2
Ab against both	0	0	7

Grafts of Le x DA thymus or BM.
Intravenous SRBC 11 days after grafting.
Spleen cells treated immediately prior to plating with antibody against donor of T graft or BM, or against Lewis histocompatibility antigen.
TxXBM = thymectomized, irradiated, BM reconstituted.
From Malakian and Waksman, unpublished.

identified in cell-mediated reactions involving these cells. Perhaps a mitogenic action of such mediators on B lymphocytes is required to initiate their transformation to plasma cells. Blast transformation of the T lymphocytes may not be essential here, since release of mediators in *in vitro* systems occurs long before blastogenesis and is not inhibited by agents that suppress the latter (Granger, 1969). However transformation and replication of T lymphocytes is probably required to provide the increased pool of cells of a given specificity that we call "memory".

TOLERANCE

Tolerance is induced by contact of protein antigens with thymocytes, as shown by direct injection of small amounts of BγG into the thymus of normal or irradiated rats (Staples *et al.*, 1966; Horiuchi and Waksman, 1968). Presumably antigen interacts with cells that possess antibody-like surface combining sites. This finding in fact provides solid proof that most or all thymocytes must develop their definitive specificity before leaving this organ. Since the thymus does not contain B lymphocytes, it appears improbable that any co-operating cell is needed for tolerance induction.

Protein antigens can also react with peripheral lymphocytes to induce tolerance, even in the absence of the thymus (Follett *et al.*, 1966; Scott and Waksman, 1969). In this case, however, some form of cell interaction appears essential. Injection of antigen directly into the spleen and lymph nodes *in vitro* produces tolerance within 2 hr, as shown by transferring the lymphocytes to thymectomized, irradiated, BM reconstituted recipients and challenging. Incubation of the same antigen with the same cells in suspension does not do so (Scott and Waksman, 1969). This finding is in accord with the observation of Gershon and Kondo (1970) that the presence of T lymphocytes is essential for the induction of tolerance to the thymus-dependent antigen—SRBC in mice. With a mainly thymus-independent antigen, such as endotoxin, tolerance can be produced by incubating antigen with peripheral lymphoid cells (Britton, 1969). Presumably cell co-operation is not required here.

The possibility that there may be differences in the specificity of tolerance to protein antigens induced centrally (no cell interaction) and peripherally (cell interaction required) has not been looked into. One might anticipate that tolerance induced within the thymus would encompass a range of antigens having the same carrier function while peripheral tolerance would be highly specific.

ACKNOWLEDGEMENTS

This work was supported by USPHS research grants AI-06112 and AI-06455 and training grant AI-00291. D.G.C. is a Postdoctoral Fellow of the USPHS, grant AI-31712.

SUMMARY

Recent studies of the immune response have focused attention of the family of cells that mature within the thymus and recirculate as long-lived T lymphocytes. Particular interest must now be directed to the distinct proliferative and differentiational events they undergo, first within the thymus and later in response to antigen.

REFERENCES

Aoki, T., Hämmerling, U., de Harven, E., Boyse, E. A. and Old, L. J. (1969). *J. exp. Med. 130*, 979-1001.

Arnason, B. G. and Waksman, B. H. (1963). *Lab. Invest. 12*, 737-747.

Arnason, B. G., DeVaux, St-Cyr C. and Relyveld, E. H. (1964). *Int. Archs Allergy appl. Immun. 25*, 206-224.

Auerbach, R. (1961). *Devl Biol. 3*, 336-354.

Billingham, R. E., Defendi, V., Silvers, W. K. and Steinmuller, D. (1962). *J. natn. Cancer Inst. 28*, 365-435.

Boyse, E. A., Old, L. J., Stockert, E. and Shigeno, N. (1968). *Cancer Res. 28*, 1280-1287.

Boyse, E. A., Bressler, E., Iritani, C. A. and Lardis, M. (1970). *Transplantation 9*, 339-341.

Britton, S. (1969). *J. exp. Med. 129*, 469-482.

Byrt, P. and Ada, G. L. (1969). *Immunology 17*, 503-516.

Cantor, H., Mandel, M. A. and Asofsky, R. (1970). *J. Immun. 104*, 409-413.

Coe, J. E., Feldman, J. D. and Lee, S. (1966). *J. exp. Med. 123*, 267-281.

Colley, D. G. and Waksman, B. H. (1970). *Transplantation 9*, 395-404.

Colley, D. G. and de Witt, C. W. (1969). *J. Immun. 102*, 107-116.

Colley, D. G., Malakian, A. and Waksman, B. H. (1970a). *J. Immun. 104*, 585-592.

Colley, D. G., Shih Wu, A. and Waksman, B. H. (1970b). *J. exp. Med. 132*, 1107-1121.

Craddock, C. G., Nakai, G. S., Fukata, H. and Vanslager, L. M. (1964). *J. exp. Med. 120*, 389-412.

Daguillard, F. and Richter, M. (1969). *J. exp. Med. 130*, 1187-1208.

David, J. R., Al-Askari, S., Lawrence, H. S. and Thomas, L. (1964). *J. Immun. 93*, 264-282.

Davies, A. J. S. (1969). *Transplant. Rev. 1*, 43-91.

Dvorak, H. F., Dvorak, A. M., Simpson, B. A., Richerson, H., Leskowitz, S. and Karnevsky, M. J. (1970). *J. exp. Med. 132*, 558-582.

Elkins, W. L. and Palm, J. (1966). *Ann. N.Y. Acad. Sci. 129*, 573-580.

Ernström, U. and Larsson, B. (1969). *Nature, Lond. 222*, 279-280.

Everett, N. B. and Tyler, R. W. (1967). *Int. Rev. Cytol. 22*, 205-237.

Everett, N. B., Caffrey, R. W. and Rieke, W. O. (1964). *Ann. N.Y. Acad. Sci. 113*, 887-897.

Follett, D. A., Battisto, J. R. and Bloom, B. R. (1966). *Immunology 11*, 73-76.

Ford, C. E. and Micklem, H. S. (1963). *Lancet i*, 359-362.

Gell, P. G. H. and Benacerraf B. (1961). *Adv. Immun. 1*, 319-343.

Gerety, R. J., Ferraresi, R. W. and Raffel, S. (1970). *J. exp. Med. 131*, 189-206.

Gershon, R. K. and Kondo, K. (1970). *Immunology 18*, 723-737.
Goldschneider, I. and McGregor, D. D. (1968). *Lab. Invest. 18*, 397-406.
Grabar, P., Tadjebakche, H. and Buffe, D. (1968). *Annls Inst. Pasteur, Paris 114*, 159-172.
Granger, G. A. (1969). *In* "Mediators of Cellular Immunity" (H. S. Lawrence and M. Landy, eds), p. 330. Academic Press, New York.
Greaves, M. F., Torrigiani, G. and Roitt, I. M. (1969). *Nature, Lond. 222*, 885-886.
Herberman, R. B. (1969). *Transplantation 8*, 813-820.
Horiuchi, A. and Waksman, B. H. (1968). *J. Immun. 101*, 1322-1332.
Isaković, K., Waksman, B. H. and Wennersten, C. (1965). *J. Immun. 95*, 602-613.
Kaplan, R. E. and Thorbecke, G. J. (1969). *Fedn Proc. 28*, 680.
Kidd, J. G. and Friedewald, W. F. (1942). *J. exp. Med. 76*, 543-578.
Kosunen, T. U., Waksman, B. H., Flax, M. H. and Tihen, W. S. (1963). *Immunology 6*, 276-290.
Kotani, M., Seiki, K., Yamashita, A. and Horii, I. (1966). *Blood 27*, 511-520.
Krüger, J., Goldstein, A. L. and Waksman, B. H. (1970). *Cellular Immunology 1*, 51-61.
Lance, E. M. and Taub, R. N. (1969). *Nature, Lond. 221*, 841-843.
Lance, E. M., Cooper, S., Buchhagen, D. and Boyse, E. A. (1970). *Fedn Proc. 29*, 436.
Larsson, B. (1967). *Acta path. microbiol. scand. 70*, 385-389.
Lubaroff, D. M. (1970). *Fedn Proc. 29*, 651.
Lubaroff, D. M. and Waksman, B. H. (1968a). *J. exp. Med. 128*, 1425-1436.
Lubaroff, D. M. and Waksman, B. H. (1968b). *J. exp. Med. 128*, 1437-1449.
Mackaness, G. B. (1970). *In* "Infectious Agents and Host Reactions" (S. Mudd, ed.), pp. 61-75. Saunders, Philadelphia.
Malakian, A., Williams, R. M. and Waksman, B. H. (1970). *Lab. Invest. 22*, 260-265.
Martins, A. B. and Raffel, S. (1964). *J. Immun. 93*, 948-952.
Mason, S. and Warner, N. L. (1970). *J. Immun. 104*, 762-765.
McCluskey, R. T., Benacerraf, B. and McCluskey, J. W. (1963). *J. Immun. 90*, 466-477.
Metcalf, D. (1966). *In* "The Thymus". Springer Verlag, Berlin.
Metcalf, D. and Brumby, M. (1966). *J. cell. comp. Physiol. 67*, Suppl. 1, 149-168.
Miller, J. F. A. P. and Osoba, D. (1967). *Physiol. Rev. 47*, 437-520.
Mooney, J. J. and Waksman, B. H. (1970). *J. Immun.* In press.
Moore, M. A. S. and Owen, J. J. T. (1967). *J. exp. Med. 126*, 715-726.
Naughton, M. A., Koch, J., Hoffman, H., Bender, V., Hagopian, H. and Hamilton, E. (1969). *Expl Cell Res. 57*, 95-103.
Nomoto, K., Gershon, R. K. and Waksman, B. H. (1970). *J. natn. Cancer Inst. 44*, 739-749.
Old, L. J., Stockert, E., Boyse, E. A. and Kim, J. H. (1968). *J. exp. Med. 127*, 523-539.
Oppenheim, J. J., Leventhal, B. G. and Hersh, E. M. (1968). *J. Immun. 101*, 262-270.
Order, S. E. and Waksman, B. H. (1969). *Transplantation 8*, 783-800.
Osoba, D. (1968). *In* "Regulation of the Antibody Response" (B. Cinader, ed.), pp. 232-275. Charles C Thomas, Springfield, Ill.

Owen, J. J. T. and Ritter, M. A. (1969). *J. exp. Med. 129*, 431-442.

Raff, M. C. (1969). *Nature, Lond. 224*, 378-379.

Raff, M. C., Sternberg, M. and Taylor, R. B. (1970). *Nature, Lond. 225*, 553-554.

Richerson, H. B., Dvorak, H. F. and Leskowitz, S. (1969). *J. Immun. 103*, 1431-1434.

Roitt, I. M., Greaves, M. F., Torrigiani, G., Brostoff, J. and Playfair, J. H. L. (1969). *Lancet ii*, 367-371.

Ruddle, N. H. and Waksman, B. H. (1968). *J. exp. Med. 128*, 1255-1279.

Rühenstroth-Bauer, G. and Lücke-Huhle, C. (1968). *J. Cell Biol. 37*, 196-199.

Sainte-Marie, G. and Leblond, C. P. (1965). *Blood 26*, 765-783.

Schwarz, M. R. (1967). *Am. J. Anat. 121*, 559-570.

Scott, D. W. and Waksman, B. H. (1969). *J. Immun. 102*, 347-354.

Sell, S. and Gell, P. G. H. (1965). *J. exp. Med. 122*, 923-928.

Sherman, J. and Auerbach, R. (1966). *Blood 27*, 371-379.

Staples, P. J., Gery, I. and Waksman, B. H. (1966). *J. exp. Med. 124*, 127-139.

Steffen, C. and Rosak, M. (1963). *J. Immun. 90*, 337-346.

Sulitzeanu, D. (1968). *Bact. Rev. 32*, 404-424.

Szenberg, A., Warner, N. L., Burnet, F. M. and Lind, P. E. (1962). *Bri. J. exp. Path. 43*, 129-136.

Taylor, R. B. (1969). *Transplant. Rev. 1*, 114-149.

Transplant Rev. (1969). *1.*

Waksman, B. H. (1960). *In* "Cellular Aspects of Immunity" (G. E. W. Wolstenholme and M. O'Connor, eds), pp. 280-322. Ciba Fdn Symp.

Waksman, B. H. and Matoltsy, M. (1958). *J. Immun. 81*, 220-234.

Weissman, I. L. (1967). *J. exp. Med. 126*, 291-304.

Whitelaw, D. M. (1966). *Blood 28*, 455-464.

Wick, G., Kite, J. H., Jr., Cole, R. K. and Witebsky, E. (1970). *J. Immun. 104*, 45-53.

Williams, R. M. and Waksman, B. H. (1969). *J. Immun. 103*, 1435-1437.

Williams, R. M., Chanana, A. D., Cronkite, E. P. and Waksman, B. H. (1970). *J. Immun.* Submitted.

Wilson, D. B. and Nowell, P. C. (1970). *J. exp. Med. 131*, 391-407.

Woodruff, J. and Gesner, B. M. (1968). *Science, N. Y. 161*, 176-178.

Yunis, E. J., Hilgard, H. R., Martinez, C. and Good, R. A. (1965). *J. exp. Med. 121*, 607-632.

The Use of Alkaline Phosphatase for Distinguishing Thymocytes in the Guinea Pig

KAUKO KOUVALAINEN

Departments of Anatomy and Pediatrics, University of Turku
Turku, Finland

Though there may be minor morphological differences between thymocytes and peripheral lymphocytes, they cannot be used for classification. Isotope and chromosome labels are therefore widely used in kinetic studies of thymocytes and other lymphocytes. The re-utilization of DNA and its breakdown products is a disadvantage in the former method, and both of them are quite laborious. Thymic-specific antigens are one endogenous label for thymocytes.

Törö (1957) first observed that the thymocytes of the guinea pig have a high alkaline phosphatase (AP) activity. A comprehensive study of the species and organ dependence of the AP activity of lymphatic tissues (Kouvalainen, 1970) confirmed Törö's finding, and suggested the use of AP as an endogenous label for thymocytes in the guinea pig. The present paper describes the usefulness of the AP method in distinguishing thymocytes from other lymphoid cells.

DEMONSTRATION OF AP ACTIVITY

AP is easily demonstrated in cell smears and tissue sections by histochemical methods. The calcium-cobalt method was used in the present study (Pearce, 1960). AP was also extracted from tissues and the enzyme activity estimated quantitatively in the extract (Bessey *et al.*, 1946; Sussman *et al.*, 1968). AP isoenzymes were analysed by acrylamide disc gel electrophoresis (Smith *et al.*, 1968).

RESULTS AND CONCLUSIONS

The cortex of the guinea pig's thymus had a very high AP activity (Fig. 1). In the medulla some of the Hassall's corpuscles were strongly stained. All the cortical cells showed AP activity, but the activity of the small thymocytes was most prominent (Fig. 2). Most lymphocytes in blood and peripheral lymphoid

71

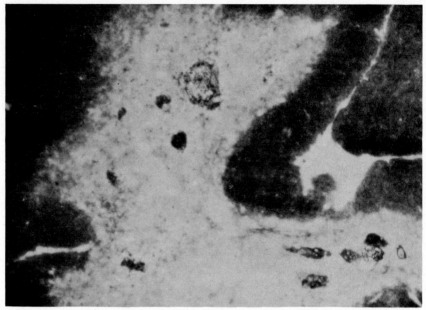

Fig. 1. Histological section of thymus of guinea pig. AP activity is shown by the dark colour in the cortex and some Hassall's corpuscles. x40.

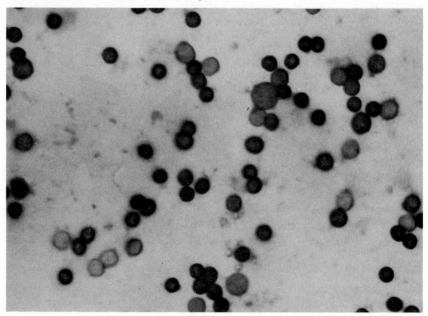

Fig. 2. Smear of guinea pig thymic cells. All the cells have AP positive cytoplasm though the staining of small thymocytes is more intense. x400.

organs were AP negative (Fig. 3). Only single cells or small accumulations of cells were AP positive. The circular layer around germinal centres quite often, but not regularly, contained AP positive cells. Chemical estimation of AP in lymphoid organs:

Thymus 33 (range 19-86) μM/60 min/mg protein.
Spleen 0.6 (range 0.26-1.1) μM/60 min/mg protein.
Lymph nodes 1.0 (range 0.4-2.3) μM/60 min/mg protein.

In acrylamide disc gel electrophoresis thymic AP moved as fast or a little slower than the main fraction of AP in the serum.

The results of this study indicate that the high AP activity is a useful label for thymocytes in the guinea pig, especially in experimental studies of the kinetics of thymocytes. For most studies the histochemical demonstration of AP in cell smears and a differential count of AP positive and negative cells seems to be a reliable method. Whether the AP positive lymphocytes in the blood and peripheral lymphoid organs directly derive from the thymus is obscure. Their rarity, however, clearly indicated that very few thymocytes leave the thymus, or if they leave it in larger amounts, they are rapidly destroyed or transformed into AP negative cells.

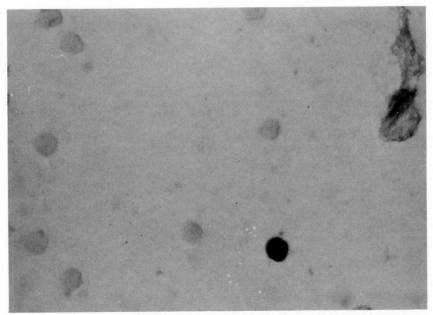

Fig. 3. Smear of guinea pig lymph node lymphocytes. Only one AP positive cell is seen. x400.

ACKNOWLEDGEMENT

This work was aided by a grant from the Finnish National Committee for Medical Research.

REFERENCES

Bessey, O. A., Lowry, O. H. and Brock, M. J. (1946). *J. biol. Chem. 164*, 321-329.

Kouvalainen, K. (1970). *Histochem. J.* In press.

Pearce, C. (1960). "Histochemistry: Theoretical and Applied". J. and A. Churchill Ltd., London.

Smith, I., Lightstone, P. J. and Perry, J. D. (1968). *Clinica chim Acta 19*, 499-505.

Sussman, H. H., Small, P. A., Jr. and Cotlove, E. (1968). *J. biol. Chem. 243*, 160-166.

Törö, J. (1957). *Bull. Ass. Anat. 92*, 1312-1325.

Membrane Receptors for Antigen–Antibody–Complement Complexes on Lymphocytes

V. NUSSENZWEIG, C. BIANCO and P. DUKOR

*Department of Pathology, New York University School of Medicine
New York, New York 10016, U.S.A.*

Several independent lines of evidence suggest that different types of cells may participate in the initial steps of the immune response. It would obviously be desirable to characterize these cells in order to clarify their function. Attempts to use physical methods to separate them have met with only limited success. For this reason, some time ago we began to study the possibility of using specific membrane properties for this purpose. It seemed reasonable to suppose that cells that have different functions during the immune process should have specific receptors for substances, such as immunoglobulins or complement components (C) which may be produced, released or modified after antigenic stimulation. The purpose of this presentation is to summarize some recent evidence (Lay and Nussenzweig, 1968; Bianco *et al.*, 1970) which demonstrates that a distinct lymphocyte population can be differentiated by the presence on the plasma membrane of receptors for modified C3 (these cells have been provisionally called CRL).

THE *IN VITRO* ASSAY FOR CRL

CRL can be distinguished from other lymphoid cells (non-CRL) because they bind antigen-antibody-complement (Ag-Ab-C) complexes. When these complexes consist of sheep erythrocytes sensitized with antibodies and complement (EAC), their interaction with CRL can be visualized and measured because it leads to the formation of clusters ("rosettes") seen in Fig. 1. The assay for the detection of CRL consists of mixing washed lymphoid cells and EAC, incubating at 37°C for 30 min with agitation, and counting "rosettes" and free lymphocytes under the microscope. Most of the work that will be discussed was done using fresh mouse serum as a source of C. This is very poorly lytic under the conditions of the test.

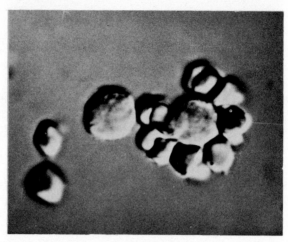

Fig. 1. "Rosette" formed by small lymphocyte from mouse lymph node, after interaction with EAC. (Differential interference, x2800.)

COMPLEMENT COMPONENTS WHICH PARTICIPATE IN THE INTERACTION

It appears probable that C3 is the component required for the interaction since it has been demonstrated (Bianco *et al.*, 1970) that EAC43 (but not EAC4) binds to lymphocytes. These experiments were performed with human peripheral blood lymphocytes and purified human C components. In addition, treatment of fresh mouse serum with cobra venom, which destroys C3, also inactivates its ability to sensitize EA for "rosette" formation on lymphocytes (Bianco *et al.*, 1970). It appears that the fixation of components subsequent to C3 may be irrelevant to the process of adhesion. Fresh sera from strains of mice deficient in C5 or from rabbits deficient in C6 appear to be as good as normal serum when used in the standard assay as a source of C (Lay and Nussenzweig, 1968).

PREPARATION OF CRL-DEPLETED POPULATIONS OF LYMPHOCYTES

As "rosettes" have a higher density than free lymphocytes, CRL-depleted populations can be obtained by density gradient centrifugation of mixtures of lymphocytes and EAC that have been preincubated for the formation of "rosettes". The supernatant of such preparations is much enriched in non-CRL. For example, in a series of experiments suspensions of mouse lymph node lymphocytes contained 22.1 and 77.9% of CRL and non-CRL respectively, whereas these numbers after depletion were 3.7 and 96.3%. In addition, the recovery of non-CRL after ultracentrifugation is close to 80%, and the recovery of CRL is only 10%.

The nature of the receptor site. Three hypotheses could be made about the nature of the interaction between EAC and CRL: (a) binding of CRL and EAC could be the consequence of a non-specific modification of the erythrocyte membrane secondary to the fixation of C. This possibility is rendered very unlikely by the finding that aggregated IgG, or AgAb complexes after fixation of C, bind to the same receptor site and inhibit the formation of EAC "rosettes" on CRL (Bianco and Nussenzweig, 1970); (b) a binding site could develop on the antibody molecule after C fixation, and this modified antibody would bind to CRL. This cannot be excluded, but again it does not appear very likely because EAC preparations, equally effective for "rosette" formation on mouse CRL, can be made with mouse C and anti-SRBC antibodies of different classes (IgM and IgG) from different animal species (rabbit and mouse) (Bianco and Nussenzweig, 1970). On the other hand, samples of EAC made with the same anti-SRBC antibody preparation but with C from different animal species can be quite different in their binding properties. For example, EAC43 prepared with human C does not adhere to mouse lymphocytes, but binding occurs if the C is mouse (Bianco *et al.*, 1970); (c) the interaction between EAC and CRL is made through a split product or modified C3 bound to the membrane of EAC. This is of course the simplest hypothesis. The idea that *native* C3 is not involved is a consequence of the observation that the binding of EAC to CRL occurs even in the presence of large concentrations of fresh normal serum (Bianco *et al.*, 1970). Needless to say, from the physiological point of view, this would be a receptor ideally suited for the detection of AgAb interactions. On the one hand it would never bind Ag or Ab alone, and on the other, through the fixation of C, it would bind the product of the union of most Ag and Ab, independently of their specificities. The function of such a receptor on lymphocytes is not clear (see discussion below). However, we think that it is important to understand the nature of the site on the Ag-Ab-C complex that binds to the cells, because of the clues that it would offer in the search for inhibitors of this interaction.

PROPERTIES OF CRL

The properties of CRL are summarized in Table I. Although CRL were defined operationally, there are a number of reasons to suppose that they constitute a distinct population of lymphoid cells (Bianco *et al.*, 1970). (a) CRL bind preferentially to nylon wool in the presence of mouse serum at 37°C. (b) CRL and non-CRL can be distinguished by their relative distribution following ultracentrifugation in a discontinuous BSA density gradient. A significantly higher proportion of CRL are found in the lightest layers of the gradient. (c) It appears that CRL may coincide with the population of lymphocytes that has immunoglobulin (Ig) determinants on their membranes (Raff *et al.*, 1970). This has been ascertained by experiments in which the specific depletion of CRL from mouse lymphoid cells induced the simultaneous depletion of most of the

TABLE I

Characteristics of CRL:

Have a receptor for modified C3

Are present in different mammalian species including man

Can be physically separated from non-CRL

Adhere preferentially to nylon wool

Have lower density than non-CRL

May coincide with the lymphocyte population that have gamma-globulin determinants on the membrane

Can be short and long lived

Are absent from "thymus-dependent areas" in spleen and lymph nodes of mice

---

lymphocytes bearing Ig (Bianco *et al.,* 1970). (The lymphocytes with Ig are detected by means of a cytotoxic test employing rabbit anti-mouse Ig and guinea pig C.) This coincidence raises the important question of whether the receptor for C3 is not the membrane-bound Ig itself. Recent observations (Bianco and Nussenzweig, 1970) suggest that this is not the case because mouse lymphocytes can be treated *in vitro* with rabbit anti-mouse Ig (or its $F(ab')_2$ fragment) without affecting their capacity to bind EAC. The proof that in these experiments the Ig sites on the membrane of the lymphocytes had actually bound the $F(ab')_2$, was the observation that these lymphocytes were not subsequently killed by mixtures of intact anti-Ig antibody and guinea pig complement. In addition, it would not be likely that the Ig molecules on the lymphocyte membrane are the receptors for C3 because they would have to bind C3 directly without previously interacting with C1, C4 and C2. As stated earlier, "rosette" formation on CRL takes place with EAC43, in the absence of bound C1, and it does not take place with EAC1 or EAC14 cells. (d) CRL can be long and short lived (Bianco *et al.,* 1970). In mice injected with ^{3}H-thymidine, using schedules that would label short or long lived cells, no differences in the distribution of label were found between CRL and non-CRL.

Localization of CRL in Lymphoid Organs

The frequency of CRL among suspensions of lymphocytes obtained from different lymphoid organs of the mouse is seen in Table II. They are not found among thymocytes and may constitute for 20 to 40% of the lymphocytes from the spleen. A smaller proportion is found in the lymph nodes or among the lymphocytes from the thoracic duct.

TABLE II

Percentage of CRL in different organs of the mouse[a]

Thymus	0
Lymph node	10-25%
Spleen	20-40%
Bone marrow	5- 8%
Thoracic duct	10-20%

[a] Data summarized from Bianco *et al.* (1970).

In addition, the pattern of localization of CRL in lymphoid organs is not random (Dukor *et al.,* 1970a). This has been determined by studying the binding of E, EA or EAC to frozen sections of spleen, lymph nodes and thymus of normal mice. While E or EA were found to adhere only to areas containing macrophages (for example, marginal sinus of spleen and lymph nodes), EAC localized in the so-called thymus-independent areas of the lymph nodes or spleen. In the spleen they adhere to the follicular areas of the white pulp and the surrounding marginal zone, but spared both the hematopoietic regions of the red pulp and the periarteriolar lymphocyte sheaths (Fig. 2). In the lymph nodes EAC was bound to the cortical but not the paracortical regions. No EAC were ever found to adhere to sections of thymus.

CRL and antigen localization. Follicular localization of antigen (White, 1963; Mitchell and Abbot, 1965; Nossal *et al.,* 1968; Ada *et al.,* 1967; Nossal *et al.,* 1965; Cohen *et al.,* 1966) appears to be a mechanism designed to bring foreign

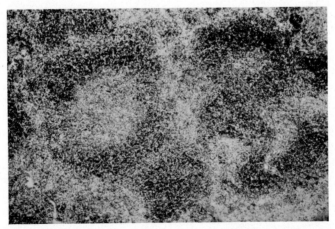

Fig. 2. Adherence of EAC to cryostat sections of mouse spleen. The overlayered EAC appear as darker areas. Note the absence of adherent erythrocytes from periarteriolar lymphocyte sheaths. (Phase, x80.)

substances to the proper areas in lymphoid organs in their pathway for the maintenance and/or induction of antibody synthesis (or tolerance). The following reasons lead us to believe that CRL may be directly implicated in this process:

1. Follicular localization of antigen is definitely an antibody mediated phenomenon (Nossal *et al.*, 1965; Miller *et al.*, 1968; Lang and Ada, 1967; McDevitt *et al.*, 1966; Humphrey and Frank, 1967). It is enhanced in the presence of passively administered or actively produced antibody (Nossal *et al.*, 1965; Lang and Ada, 1967), and less efficient in germ free (Miller *et al.*, 1968) or completely tolerant animals (McDevitt *et al.*, 1966; Humphrey and Frank, 1967). Localization is increased when antigen is injected complexed with antibodies (Lang and Ada, 1967). The mechanism of localization is unknown. However, participation of C seems possible, not only because it has been demonstrated that the immunoglobulin fragment that mediates C fixation (Fc) is necessary for the retention of Ag-Ab complexes, but also in view of the fact that in most cases C fixation to Ag-Ab complexes would be unavoidable *in vivo*. The hypothesis of C participation is in agreement with the puzzling observation that homologous or autologous γ globulin, but not other serum proteins, can localize in the follicles (Ada *et al.*, 1964a, b). In these experiments labeled purified γ globulins were employed; it is reasonable to believe that they might have contained some aggregated material which could fix C.

2. Examination of sections of follicles by electron-microscopic autoradiography showed that the antigen was in part associated with membranes of the dendritic reticular cells and with their extensions (Mitchell and Abbot, 1965; Nossal *et al.*, 1968; Ada *et al.*, 1967). However, in these same areas, lymphocytes were so densely packed that in many cases it was not possible to decide whether the labeled antigen was on the lymphocyte membrane or on dendritic processes of the reticular cells. In other instances the antigen was seen between lymphoid cells where *no* separating cell processes could be resolved (Ada *et al.*, 1967). In spite of these findings, most investigators only stress the role of the dendritic reticular cells in the phenomenon. Our findings show of course that the areas in which antigen localizes coincide entirely with the areas of distribution of CRL, which *do* have the ability to bind Ag-Ab-C complexes. In addition, direct binding of EAC to lymphoid cells and dendritic reticular cells in these areas has been recently demonstrated by electron microscopy (Dukor *et al.*, 1970a). The idea that small lymphocytes are partly responsible for Ag localization is also in agreement with the observation that depletion of lymphocytes by thoracic duct drainage reduces the uptake of antigenic material by primary follicles, and that antibody injections only partly restore localization (Williams, 1966). Thus it appears reasonable to suggest that both CRL and dendritic reticular cells participate in Ag retention in the follicular areas of lymphoid organs. It is quite

pertinent in this respect to point out that several investigators have reported that lymphoid cells bound antigen after it was injected into normal or immune animals (Cheng *et al.,* 1961; Coons *et al.,* 1951; Han and Johnson, 1966; Roberts, 1966). Sometimes a surprisingly large proportion of lymphocytes contained Ag. However, in most instances, these findings were difficult to interpret, in relation to the specificity of the binding and its relevance to the immune process.

Another interesting point in our findings deserves some comments: the possible co-existence in the same lymphoid compartment of two cell types (CRL and dendritic reticular cells) capable of reacting with Ag-Ab-C complexes. We propose that this may be the basis for the accumulation of lymphocytes in the follicles. CRL could be either specifically bound to Ag-Ab-C complexes previously deposited on the process of the dendritic cells or, alternatively, CRL that had previously bound Ag-Ab-C to their membranes while in the circulation could be trapped within the reticular framework of the follicles. Of course, these possibilities are not mutually exclusive. In either case, a modified product of C3 would provide the adhesive responsible for the contact between the membranes of both types of cells.

Origin of CRL

From the evidence obtained so far it appears that CRL may be part of a non-thymus-dependent lymphocyte population originating perhaps directly in the bone marrow. The reasons for this are as follows: CRL are distributed in lymphoid organs in areas that are not considered to contain thymus-dependent lymphocytes. Also, the frequency of CRL among suspension of lymphocytes obtained from different organs of the mouse is remarkably complementary to that of lymphocytes with the antigen θ on their membranes. These are supposed to be thymus-derived (see data of M. Raff, in this volume). In addition, we have recently found that 6-8 week old neonatally thymectomized mice show a significant increase in the proportion of CRL in spleen and lymph nodes (Dukor *et al.,* 1970b). Of course, the actual proof that CRL are in fact directly bone marrow-derived would be the demonstration of their presence in lymphoid organs of adult mice that have been thymectomized and lethally irradiated and have received bone marrow transplants. These experiments are now in progress in our laboratory.

In summary, it appears that some of the lymphocytes from man and many other mammalian species, have receptors for Ag-Ab-C complexes. These receptors provide a non-specific but very efficient mechanism to trap antigen in the presence of circulating antibody. The relevance of this process to the immune mechanism is still to be determined.

ACKNOWLEDGEMENTS

This investigation was supported by a National Institutes of Health grant No. AI-08499. V.N. is a Career Investigator of the Health Research Council of the City of New York, Contract I-558. C.B. is supported by a training fellowship from the World Health Organization and from the Fundacao de Amparo a Pesquisa, Sao Paulo, Brazil.

REFERENCES

Ada, G. L., Nossal, G. J. V. and Austin, C. M. (1964a). *Aust. J. exp. Biol. med. Sci. 42,* 331-346.

Ada, G. L., Nossal, G. J. V. and Pye, J. (1964b). *Aust. J. exp. Biol. med. Sci. 42,* 295-310.

Ada, G. L., Parish, C. R., Nossal, G. J. V. and Abbot, A. (1967). *Cold Spring Harb. Symp. quant. Biol. 32,* 381-393.

Bianco, C. and Nussenzweig, V. (1970). In preparation.

Bianco, C., Patrick, R. and Nussenzweig, V. (1970). *J. exp. Med. 132,* 702-720.

Cheng, H. F., Dicks, M., Shellhamer, R. H., Brown, E. S., Roberts, A. N. and Haurowitz, F. (1961). *Proc. Soc. exp. Biol. Med. 106,* 93-97.

Cohen, S., Vassali, P., Benacerraf, B. and McCluskey, R. T. (1966). *Lab. Invest. 15,* 1143-1155.

Coons, A. H., Leduc, E. H. and Kaplan, M. H. (1951). *J. exp. Med. 93,* 173-188.

Dukor, P., Bianco, C. and Nussenzweig, V. (1970a). *Proc. Natn. Acad. Sci. U.S.A.* In press.

Han, S. S. and Johnson, A. G. (1966). *Science, N.Y. 153,* 176-178.

Humphrey, J. H. and Frank, M. M. (1967). *Immunology 13,* 87-100.

Lang, P. G. and Ada, G. L. (1967). *Immunology 13,* 523-534.

Lay, W. H. and Nussenzweig, V. (1968). *J. exp. Med. 128,* 991-1009.

McDevitt, H. O., Askonas, B. A., Humphrey, J. H., Schechter, I. and Sela, M. (1966). *Immunology 11,* 337-351.

Miller, J. J., III, Johnsen, D. O. and Ada, G. L. (1968). *Nature, Lond. 217,* 1059-1061.

Mitchell, J. and Abbot, A. (1965). *Nature, Lond. 208,* 500-502.

Nossal, G. J. V., Ada, G. L., Austin, C. M. and Pye, J. (1965). *Immunology 9,* 349-356.

Nossal, G. J. V., Abbot, A., Mitchell, J. and Lummus, Z. (1968). *J. exp. Med. 127,* 277-290.

Raff, M. C., Sternberg, M. and Taylor, R. B. (1970). *Nature, Lond. 225,* 553-554.

Roberts, A. N. (1966). *Am. J. Path. 49,* 889-910.

White, R. G. (1963). *In* "The Immunologically Competent Cell" (G. E. W. Wolstenholme and J. Knight, eds), Ciba Fdn Symp., Vol. 16, pp. 6-16. Churchill, London.

Williams, G. M. (1966). *Immunology 11,* 475-488.

The Use of Surface Antigenic Markers to Define Different Populations of Lymphocytes in the Mouse

MARTIN C. RAFF

National Institute for Medical Research
Mill Hill, London, England

The very fact of this symposium on "Cell Interactions in Immune Responses" underlines the need for methods to identify and separate the various cells involved in immunity. Surface antigens are becoming increasingly important as markers for lymphocyte identification and it is likely that in the future they will prove to be equally valuable for cell separation.

I have been primarily interested in the theta (θ) alloantigen of mice and would like to consider several aspects of work done with this lymphocyte surface antigen. I will briefly review the evidence that θ can serve as a marker for thymus-derived lymphocytes in mice, and discuss how it has been used to study the development and distribution of these cells. I will then discuss the relationship between θ-bearing and surface-immunoglobulin-bearing lymphocytes and finally review some of the functional studies that have been done with anti-θ serum.

THETA AS A MARKER FOR THYMUS-DERIVED LYMPHOCYTES

Reif and Allen first described the θ alloantigen in 1963 and have subsequently shown that it is determined by a single locus with two alleles: θ AKR found in AKR, RF and some related sublines, and θ C3H found in all other inbred strains of mice tested (Reif and Allen, 1964). Although they found the antigen mainly in thymus and brain, some θ appeared to be present on lymph node and spleen lymphocytes (Reif and Allen, 1964). It was not clear whether the antigen was present in small amounts on all peripheral lymphocytes, or whether there was a sub-population that carried θ and another that did not.

Using ^{51}chromium and dye exclusion cytotoxic testing, it was found that whereas all thymocytes could be killed by anti-θ serum, only a proportion of lymph node and spleen lymphocytes could be killed (Raff, 1969). The fact that the cytotoxic curves reached a plateau at high concentrations of the antiserum suggested that there was in fact a discrete subpopulation of θ-bearing lymphocytes in lymph node and spleen (Raff, 1969). This has been confirmed by immunofluorescence (Raff, 1970b) and has been found to be true for all of the peripheral lymphoid tissues of mice (Raff and Wortis, 1970; Raff and Owen, 1971).

To determine if the population of θ-bearing cells was thymus-dependent, the percentage of θ-positive cells in the peripheral lymphoid tissues of normal mice was compared to that in mice known, on other grounds, to be depleted of thymus-derived lymphocytes (Table I). Mice treated with antilymphocytic serum (ALS), mice thymectomized as newborns or thymectomized as adults, lethally irradiated and reconstituted with fetal liver, and congenitally athymic "nude" mice, all had very few θ-bearing cells in the periphery. It seemed clear then that the θ-bearing population was thymus-dependent and that θ could be used as a marker for thymus-derived lymphocytes.

TABLE I

Percentage of θ-bearing cells in peripheral lymphoid tissues of immunologically incompetent mice[a]

Mice	Lymph nodes	Spleen	Blood
Normal (CBA)	65	35	70
ALS treated (CBA)	10	5	5
Neonatal thymectomy (CBA)	10	5	—
Adult thymectomy + 900 R + fetal liver (CBA)	20	5	20
Nude (congenitally athymic)	<10	—	<5

[a] Figures are approximate averages compiled from experiments using ^{51}Cr and dye exclusion cytotoxic testing as well as immunofluorescence.

DISTRIBUTION OF THYMUS-DERIVED CELLS IN LYMPHOID TISSUES OF ADULT MICE

Using anti-θ serum it was possible for the first time to directly determine the distribution of thymus-derived cells in the various lymphoid tissues of normal mice (Table II). The figures obtained using ^{51}Cr and dye exclusion cytotoxic testing, as well as indirect immunofluorescence, have all been in good agreement and are generally in accord with the estimated distribution of long-lived (Everett *et al.*, 1964; Denman *et al.*, 1968) and recirculating lymphocytes (Gowans and Knight, 1964; Lance and Taub, 1969; Zalz and Lance, 1970), which for the most part appear to be thymus-dependent (Miller *et al.*, 1967).

TABLE II

Distribution of θ-bearing cells in the peripheral lymphoid tissues of adult Balb/c mice

Tissue	θ-positive cells (%)
Thymus	100
Thoracic duct[a]	80
Blood lymphocytes	70
Lymph node	65
Spleen	35
Peritoneal lymphocytes	35
Peyer's patches	25

Figures are approximate averages compiled from experiments using ^{51}Cr and dye exclusion cytotoxic testing as well as immunofluorescence.

[a] In (CBA × C57B1/6)F$_1$ mice.

DEVELOPMENT OF THYMUS-DERIVED LYMPHOCYTES

The thymus is the first tissue to become lymphoid in the mouse, and typical small lymphocytes first appear in the thymus around day 15 or 16 of fetal life. It has been clearly shown that the thymocytes develop from stem cells that migrate into the thymus from the yolk sac and fetal liver (Moore and Owen, 1967; Owen and Ritter, 1969). The stem cells first migrate to the thymus on day 11, and appear to be large blast-like cells, which do not carry thymocyte alloantigens such as θ or TL on their surface (Owen and Raff, 1970). By day 15 or 16, some of the stem cells have differentiated to become typical small lymphocytes, and it is at this time that θ and TL can first be detected on thymus cells (Owen and Raff, 1970). In the adult mouse, the majority of stem cells migrate to the thymus from the bone marrow (Ford, 1966), but it is likely that the differentiation to thymocytes is similar to that in the fetus.

It is now generally accepted that thymus cells migrate from the thymus to the peripheral lymphoid tissues where they make up the majority, if not all, of the thymus-dependent population of lymphocytes (Miller, 1963; Weissman, 1967; Davies, 1969). Thus the term thymus-derived seems appropriate for these cells. The first thymus cells appear to migrate to the peripheral lymphoid tissues sometime before birth, since there are some θ-bearing cells in the blood and mesenteric lymph node of newborn mice (Raff and Owen, 1971). However, the peripheral lymphoid tissues are very poorly developed at birth in the mouse, and it is only around the third or fourth day of life that there is a striking increase in the number of lymphocytes in the periphery. At the same time, the percentage of θ-bearing cells rapidly increases, reaching adult proportions by day 5 or 6 (Raff and Owen, 1971). The parallel increase in the absolute numbers of lymphocytes and the percentage of θ-positive cells beginning on the third day of life, indicates that the early development of the peripheral lymphoid tissues is largely the result of the inflow and/or proliferation of thymus-derived cells.

Although thymus cells migrate to the periphery to become thymus-derived lymphocytes, it is clear that thymocytes and peripheral thymus-derived lymphocytes are very different. They differ in their immunological competence, and they differ in their surface antigen composition. For example, thymocytes have more θ (Reif and Allen, 1964) and less H2 antigen (Winn, 1960), than do peripheral lymphocytes and, in TL-positive strains of mice, only thymocytes normally express the TL antigen (Aoki et al., 1969). Thus, there is clearly another differentiation step between thymocyte and thymus-derived lymphocyte.

One can reproduce this migration of thymocytes experimentally by injecting ^{51}Cr-labelled thymus cells intravenously into lethally irradiated syngeneic hosts, and harvesting them sometime later from the peripheral lymphoid tissues, such as spleen. Without further labelling, the cells can be subjected to cytotoxic testing, and therefore only those ^{51}Cr-labelled cells in the original thymus inoculum that have migrated to the spleen are being tested. It is gratifying to find that the cells harvested from the spleen, as early as 3 hr after injection, behave exactly like peripheral lymphocytes rather than thymocytes (Raff, 1971). They are resistant to anti-TL, and are less susceptible to anti-θ and more susceptible to anti-H2 than are the great majority of thymus cells. Lance et al. (1970) have obtained similar results looking at the susceptibility of migrated thymocytes to anti-TL and anti-H2.

This type of experiment did not settle whether the change from thymocyte to thymus-derived cell occurred within the thymus, in the periphery, or both. However, the finding that the change had already occurred 3 hr after injection suggested that a population of peripheral-like lymphocytes already existed in the thymus, and these were the cells that migrated to the spleen. The finding that only 80% of thymocytes from TL-positive strains of mice could be killed by anti-TL serum (Raff and Owen, unpublished observations) suggested a way of partly testing this hypothesis. It is known that 1-2% of mouse thymocytes will migrate to a group of peripheral lymph nodes (Lance and Taub, 1969). If the cells that are migrating are already partly or wholly differentiated and thus resistant to anti-TL serum, then killing off the TL-positive cells should increase the percentage of the remaining cells that will migrate. In fact, that is what happens. In one such experiment, 75% of the cells were killed by the anti-TL serum (the dead cells releasing their ^{51}Cr). Therefore the surviving cells should have been four times enriched for cells that would migrate to the lymph nodes, if all the migrating cells were resistant to anti-TL serum. The actual enrichment was 3.3 (Raff, 1971).

It has recently been shown that those thymus cells that are capable of producing a graft-*versus*-host reaction are also resistant to anti-TL serum (Leckband, 1970). It seems likely that these TL-negative, migrating, immunologically competent thymus cells arise within the thymus by differentiation from less mature thymocytes, but at the moment one cannot exclude the possibility that these cells have returned to the thymus as part of the recirculating pool of lymphocytes.

Dr John Owen and I have recently grafted 14-day fetal CBA thymus under the kidney capsule of AKR mice that had been thymectomized, lethally irradiated and reconstituted with AKR bone marrow, creating a modified "Davies-type" mouse (Davies, 1969). After four or five weeks we found lymphocytes in the blood, lymph node and spleen that carried θ C3H of the peripheral type (i.e. 6-8 times less θ than thymocytes) (Owen and Raff, 1970). Since the 14-day thymus graft was pre-lymphoid and thus contained only stem cells, this type of experiment clearly establishes the pathway of stem cell → thymocyte → thymus-derived lymphocyte.

IMMUNOFLUORESCENCE STUDIES: THE RELATIONSHIP BETWEEN θ-BEARING AND SURFACE-IMMUNOGLOBULIN-BEARING LYMPHOCYTES

Drs Taylor, Sternberg and I have recently reported our observations on the use of fluorescein-labelled and radio-labelled anti-immunoglobulin (Ig) to demonstrate Ig determinants on the surface of mouse lymphocytes (Raff *et al.*, 1970). The striking thing about these experiments was that both methods demonstrated the same proportion of Ig-bearing cells in a given lymphocyte population, despite the fact that in our hands immunofluorescence was about 10,000 times less sensitive than immunoautoradiography. This indicated that mouse lymphocytes did not form a spectrum of cells containing varying amounts of surface Ig. There was a distinct population with readily demonstrable surface Ig and another population with very much less, or no, surface Ig. There are three lines of evidence that suggest that we were only demonstrating Ig on the surface of marrow-derived lymphocytes.

Firstly, the distribution of the surface-Ig-bearing cells was inversely related to the distribution of θ-bearing lymphocytes (Table III). This immediately suggested that we were not staining the majority of thymus-derived lymphocytes. Secondly, in adult thymectomized, irradiated and fetal liver-reconstituted mice, and in mice treated with ALS, the percentage of surface-Ig-bearing cells was greatly increased (Raff, 1970b). Thirdly, the surface-Ig-bearing cells could be distinguished from the θ-bearing cells by the nature of the

TABLE III

Distribution of surface-Ig-bearing cells compared with distribution of θ-bearing cells

Tissue	Surface-Ig-bearing cells[a] (%)	θ-bearing cells[b] (%)
Thymus	0	100
Thoracic duct	5	80
Blood lymphocytes	10	70
Lymph node	15	65
Spleen	35	35

[a] As determined by immunofluorescence (for detailed data see Raff, 1970b).

[b] Figures are approximate averages compiled from experiments using ^{51}Cr and dye exclusion cytotoxic testing as well as immunofluorescence.

fluorescence. When peripheral lymphocytes were treated with anti-Ig-fluorescein, the great majority of the stained cells showed a clearly defined polar fluorescent "cap" usually overlying a pseudopod-like structure (Raff, 1970b; Raff et al., 1970). If the cells were first incubated in anti-θ serum and then in anti-Ig-fluorescein, the percentage of cells showing fluorescent caps remained unchanged, but another population of cells showing "ring" staining appeared. The percentage of ring cells was the same as the percentage of θ-bearing cells determined by cytotoxic testing (Raff, 1970b). As it happens, the cap-like staining appears to be an induced change in the cell and can be prevented by high protein concentration in the medium, in which case the staining is non-polar and spotted (Pernis and Raff, unpublished observations). Nonetheless, the cap appearance was only seen with the surface-Ig-bearing cells and proved to be very helpful.

The failure to detect Ig on the surface of thymus-derived cells is disturbing in view of the likelihood that these cells have antigen-specific receptors on their surface, and the increasing evidence that the function of these cells can be inhibited by anti-Ig sera (Greaves et al., 1969; Mason and Warner, 1970). If Ig determinants are present on thymus-derived cells, which seems likely, they must be relatively few and/or relatively inaccessible compared to those on marrow-derived lymphocytes.

FUNCTIONAL STUDIES WITH ANTI-θ

Since anti-θ serum acts selectively on thymus-derived cells, it can serve as a useful tool for studying the function of these cells. To my knowledge, no one has succeeded in getting the antiserum to work convincingly *in vivo*, but this has not been a serious drawback. The lack of effect *in vivo* is unexplained, but may be related to the fact that anti-θ is not cytotoxic in the presence of mouse complement (Raff, unpublished observations). On the other hand it may be simply a dose phenomenon. Anti-θ used *in vitro* with guinea pig complement has no effect on immune (E. Möller and Greaves, 1970) or background (Raff and Dresser, unpublished observations) plaque-forming cells. However, it inhibits the *in vitro* primary response to sheep erythrocytes (SRBC) (Schimpl and Wecker, 1970), a proportion of SRBC rosette-forming cells (E. Möller and Greaves, 1970), the transfer of skin contact sensitivity (E. Möller and Greaves, unpublished observations), and graft-*versus*-host response (E. Möller, unpublished observations).

I have used anti-θ *in vitro* to study the function of thymus-derived lymphocytes in the secondary humoral response to hapten-protein conjugates. In an adoptive transfer system, anti-θ and complement treated primed spleen cells showed a markedly reduced response (about 80% reduction) when boosted in the adoptive host with 4-hydroxy-3-iodo-5-nitrophenyl acetic acid-bovine serum albumin (NIP-BSA), BSA, 2,4-dinitrophenylated chicken gamma-globulin (DNP-CGG), NIP-CGG, DNP-BSA or NIP-haemocyanin (Raff, 1970a). The anti-hapten and anti-protein responses were equally affected.

To determine what type of role the thymus-derived cells are playing in the humoral response, the effect of anti-θ on a cell co-operation system was studied. In this system, hapten-carrier-primed cells fail to produce a significant anti-hapten response when boosted with the hapten conjugated to a different carrier (carrier effect), but do so if cells primed with the second carrier (helper cells) are given at the same time (Mitchison, 1968). Thus NIP-CGG-primed cells are helped by BSA-primed cells to produce an anti-NIP response when boosted with NIP-BSA. It was found that the anti-NIP response was markedly suppressed when the BSA-primed helper cells were treated with anti-θ and complement, but the anti-NIP response was unaffected when the NIP-CGG cells (which are known to produce the anti-NIP antibody) were so treated (Raff, 1970a). Thus the helper cell in this system is a thymus-derived lymphocyte and the anti-hapten producing cell is not.

It is clear then that thymus-derived lymphocytes do play an important role in the secondary humoral response, at least for some antigens at some dose levels, and their role appears to be one of antigen-specific antigen handling rather than secreting antibody.

ACKNOWLEDGEMENTS

I am grateful to Drs N. A. Mitchison and J. J. T. Owen for helpful discussion, and to Miss P. Chivers for competent technical assistance. I am supported by a Post-doctoral Fellowship from the National Multiple Sclerosis Society of the United States.

SUMMARY

There is now very good evidence that the theta isoantigen of mice is found only on those peripheral lymphocytes that are thymus-dependent or thymus-derived. Using theta as a marker for thymus-derived cells, one can determine the distribution of these cells in the various lymphoid tissues of mice, and to some extent study their function. Using immunofluorescence it is possible to distinguish theta-bearing cells from an apparently different population of lymphocytes that carry demonstrable naturally-occurring immunoglobulin determinants on their surface.

REFERENCES

Aoki, T., Hämmerling, U., de Harven, E., Boyse, E. A. and Old, L. J. (1969). *J. exp. Med. 130*, 979-1002.

Davies, A. J. S. (1969). *Transplant. Rev. 1*, 43-91.

Denman, A. M., Denman, E. J. and Embling, P. H. (1968). *Lancet i*, 321-325.

Everett, N. B., Caffrey, R. W. and Rieke, W. O. (1964). *Ann. N.Y. Acad. Sci. 113*, 887-897.

Ford, C. E. (1966). *In* "Thymus" (A. E. W. Wolstenholme, ed.), Ciba Fdn Symp., p. 131. J. and A. Churchill Ltd., London.

Gowans, J. L. and Knight, E. J. (1964). *Proc. R. Soc. B 159*, 257.

Greaves, M. F., Torrigiani, G. and Roitt, I. M. (1969). *Nature, Lond. 222*, 885-886.

Lance, E. M. and Taub, R. N. (1969). *Nature, Lond. 221*, 841-843.

Lance, E. M., Cooper, S., Buchhagen, D. and Boyse, E. A. (1970). *Fedn Proc. 29*, 436.

Leckband, E. (1970). *Fedn Proc. 29*, 621.

Mason, S. and Warner, N. L. (1970). *J. Immun. 104*, 762-765.

Miller, J. F. A. P. (1963). *Lancet i*, 43-45.

Miller, J. F. A. P., Mitchell, G. F. and Weiss, N. S. (1967). *Nature, Lond. 214*, 992-997.

Mitchison, N. A. (1968). *In* "Differentiation and Immunology" (K. B. Warren, ed.), p. 29. Academic Press, New York.

Möller, E. and Greaves, M. F. (1970). This symposium.

Moore, M. A. S. and Owen, J. J. T. (1967). *J. exp. Med. 126*, 715-726.

Owen, J. J. T. and Ritter, M. A. (1969). *J. exp. Med. 129*, 431-437.

Owen, J. J. T. and Raff, M. C. (1970). *J. exp. Med. 132*, 1216-1232.

Raff, M. C. (1969). *Nature, Lond. 224*, 378-379.

Raff, M. C. and Wortis, H. H. (1970). *Immunology 18*, 931-942.

Raff, M. C., Sternberg, M. and Taylor, R. B. (1970). *Nature, Lond. 225*, 553-554.

Raff, M. C. (1970a). *Nature, Lond. 226*, 1257-1258.

Raff, M. C. (1970b). *Immunology 19*, 637-650.

Raff, M. C. and Owen, J. J. T. (1971). *Europ. J. Immun.* In press.

Raff, M. C. (1971). *Nature, Lond.* In press.

Reif, A. E. and Allen, J. M. V. (1964). *J. exp. Med. 120*, 413-433.

Schimpl, A. and Wecker, E. (1970). *Nature, Lond. 226*, 1258-1259.

Weissman, I. L. (1967). *J. exp. Med. 126*, 291-304.

Winn, H. J. (1960). *Natn Cancer Inst. Monograph 2*, 113-138.

Zalz, M. M. and Lance, E. M. (1970). *Cell Immun. 1*, 3.

Surface IgM on Lymphoid Cells

EVA KLEIN and TROND ESKELAND

Institute for Tumor Biology, Karolinska Institutet, Sweden,
Anatomical Institute and Institute for Experimental Medical Research
University of Oslo, Oslo, Norway

Both normal and neoplastic lymphoid cells seem to synthesize and accumulate immunoglobulin structures on the cell membrane. The evidence is provided by experiments in which anti-immunoglobulin antibodies were observed to bind to the surface of lymphoid cells or to influence the performance of lymphoid cells (for review see Singhal and Wigzell, 1970). Although these membrane structures have some of the antigenic determinants of serum immunoglobulins, it is not clear how far they resemble serum immuno-globulins and what is the nature of their location on the cell membrane.

The biological significance of mu and kappa immunoglobulin structures on the surface of cells from Burkitt lymphomas and chronic lymphocytic leukaemias (Johansson and Klein, 1970; Klein *et al.,* 1967, 1968, 1970; Nadkarni *et al.,* 1969) is not understood, but their study might throw some light on similar structures on normal lymphoid cells. The mu and kappa structures can be demonstrated by anti-mu and anti-kappa antibodies in fluorescence, cytotoxicity, or agglutination tests. This report is focused on: (1) determination of the amount of the surface-bound immunoglobulin structures, and (2) characterization of the molecular size.

1. With viable cells exposed to fluorescein-labelled anti-mu and anti-kappa antibodies it was evident that the quantity of surface-bound mu and kappa structures, as reflected by the intensity of fluorescence, varied between cells from different patients. This was confirmed by the difference in sensitivity to cytolysis (Johansson and Klein, 1970; Klein *et al.,* 1968, 1970). Daudi and T.P. cells were among the most reactive, and these cells were therefore mostly used for the experimental studies. The Daudi cells were derived from a Burkitt tumour (Klein *et al.,* 1967, 1968; Nadkarni *et al.,* 1969), and kept in culture for more than one year. The T.P. cells were chronic lymphocytic leukaemia cells obtained fresh from the blood of the patient (Johansson and Klein, 1970). The leukaemia cells were usually purified by Dextran sedimentation and NH_4Cl-lysis of the red cells.

The absolute amount of immunoglobulin structures on cells was determined by comparing the ability of cells and known amounts of purified IgM to bind anti-mu and anti-kappa antibodies. The residual activity was evaluated by visual observation of immunofluorescence on T.P. cells, by agglutination of IgM-coated formalinized sheep red blood cells (passive haemagglutination) or by cytotoxic effect on Daudi cells. Similar results were obtained with the three methods. 10^9 Daudi or T.P. cells seemed to carry an amount of mu and kappa structures equivalent to 25 μg of IgM. The other cells, in accordance with their lower reactivity in immunofluorescence and cytotoxicity tests, were found to contain lower amounts of mu and kappa structures (Klein *et al.*, 1970) (Fig. 1).

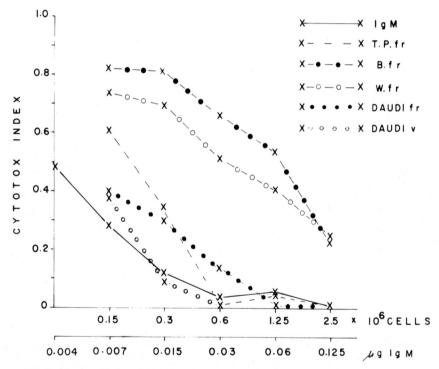

Fig. 1. Quantitative estimation of mu structures carried by various cells.
Varying amounts of IgM, 0.004-0.125 μg, or varying amounts of cells, 0.15-2.5 x 10^6, in 50 μl veronal buffer with gelatin were incubated with 50 μl anti-IgM serum (1 : 150) for 30 min at 37°C. After centrifugation, 25 μl of the supernatant was added to 5 x 10^4 Daudi cells in 5 μl, incubated for 30 min at 37°C, washed once and 25 μl unabsorbed 1 : 2 diluted guinea pig complement was added. After incubation for 60 min at 37°C the percentage of trypan blue stained cells was evaluated. Cytotoxic index: percentage of unstained cells in the control minus percentage of unstained cells in the sample, divided by the former figure. Cells used for absorption: frozen (fr) Daudi cells and cells from patients T.P., B. and W., and viable (v) Daudi cells.

Repeated samples of leukaemia cells from the same patient showed the same intensity of immunfluorescence. On the other hand, the Daudi tissue culture cells seemed to vary slightly. This was confirmed by daily quantitations of growing cultures (Fig. 2). After an initial rise the amount of mu structures declined as the culture aged. The decline was reflected by resistance to cytolysis using anti-mu antibodies (Fig. 2). The fate of the mu structures in the ageing cells is at present not known.

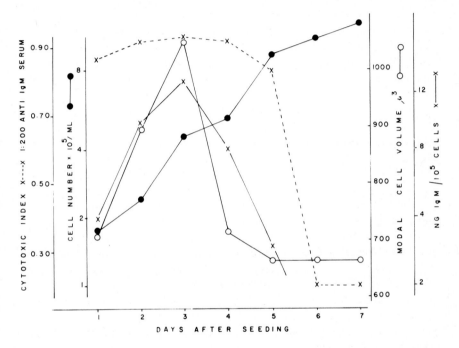

Fig. 2. Changes in the amount of mu structures of Daudi cells. The culture was seeded with 10^6 cells in 10 ml. Each day the cell number, modal cell volume, cytotoxic sensitivity towards anti-IgM antibodies and amount of IgM was determined. IgM was determined by collecting 1.5 and 3×10^5 cells in 50 μl veronal buffer with gelatin. Samples were kept frozen until the end of the experimental period. Quantitations were then performed simultaneously as described in Fig. 1.

2. In order to characterize the mu and kappa structures in more detail, we tried to liberate these structures from T.P. cells. About 2.5×10^9 cells in 0.05 M tris buffer, pH 8.0, containing 0.005 M $MgCl_2$ were homogenized in a Dounce or Potter-Elvehjem homogenizer. After centrifugation at 1000 g for 10 min, which was sufficient to sediment nuclei and unbroken cells, the supernatant (supernatant I) was found to contain mu and kappa structures by using the assay

methods described above. The amount of liberated mu structures seemed equivalent to the amount on viable cells.

To sediment immunoglobulin structures bound to any cellular particles, supernatant I was subjected to ultracentrifugation for 2 hr at 100,000 g. After centrifugation the supernatant (supernatant II) still contained mu structures, but the amount was reduced by about 50%. The sediment was therefore rehomogenized and subjected to another ultracentrifugation for 2 hr at 100,000 g. Mu structures were also found in the supernatant.

We considered that we were dealing with surface-derived immunoglobulin structures, because no intracellular immunoglobulin had been detected in experiments designed to reveal such structures. These experiments included comparison of the absorbing efficiency of viable and killed cells (by freezing and thawing three times) and immunofluorescent staining of fixed cells (Klein et al., 1970).

The fact that immunoglobulin structures were liberated from the surface membranes by homogenization agrees with the results of experiments on Daudi cells, in which the amount of mu structures was quantitated on dead and living washed cells. When cells killed by freezing and thawing were washed the amount of mu structures was reduced by about 50%, while viable cells could be washed six times without any apparent loss. Consequently it seemed that only non-covalent bonds kept the immunoglobulin structures in the membrane, and that an intact membrane was of importance to these bonds.

The size of the immunoglobulin structures in supernatant II was determined by sucrose density gradient ultracentrifugation and gel filtration on Sephadex G-200. Before this the supernatant was concentrated about 10 times. ^{125}I-labelled 7S rabbit gamma globulin was added to the supernatant as reference substance. For ultracentrifugation, 19S IgM was also used as reference substance by running normal human serum together with ^{125}I-labelled 7S rabbit gamma globulin in a parallel tube. The fractions obtained after ultracentrifugation (Fig. 3) or gel filtration (Fig. 4) were analysed for mu and kappa structures by inhibition of passive haemagglutination. A more detailed description of these experiments is in preparation. In both types of experiments all the mu and some of the kappa activity in supernatant II coincided with the peak activity of 7S rabbit gamma globulin. Some kappa structures were apparently of smaller size, because the sedimentation was slower and the elution later than the 7S rabbit gamma globulin. We concluded that the mu and some of the kappa chains were linked to form 7S IgM molecules, while some kappa chains were free. 7S IgM has previously been found in small quantities in normal serum and intracellularly in a mouse plasma cell tumour secreting 19S IgM (Parkhouse and Askonas, 1969). 7S IgM is also found when 19S IgM is reduced, and the subunits consist of two mu and two light, kappa or lambda, chains (Miller and Metzger, 1965) and have a molecular weight at about 180,000. If the molecular structure of the IgM from T.P. cells is similar, there were more kappa than mu chains in the cell extract

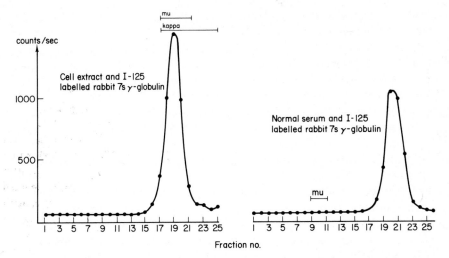

Fig. 3. Sucrose density gradient ultracentrifugation of supernatant II or normal serum and ^{125}I-labelled rabbit 7S gamma globulin. Bottom fraction is No. 1. · = amount of 7S rabbit gamma globulin (counts/sec). Kappa and mu structures were localized by the fraction's ability to inhibit passive haemagglutination.

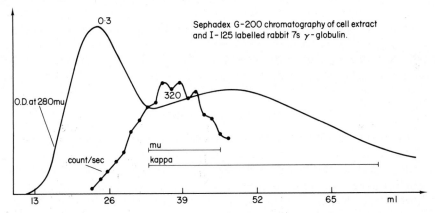

Fig. 4. Sephadex G-200 filtration of supernatant II and ^{125}I-labelled 7S rabbit gamma globulin. Abscissa shows elution volume. Amount of protein is expressed as optical density at 280 mμ. 7S rabbit gamma globulin, mu and kappa structures were determined as in Fig. 3.

used for density gradient ultracentrifugation and gel filtration.

A number of about 80,000 IgM molecules per T.P. cell was found by calculation (molecular weight 180,000, 25 μg IgM on 10^9 cells). Estimation of the distances between the molecules is difficult because immunofluorescence showed that the molecules were unevenly distributed on the cell surface (Johansson and Klein, 1970).

The immunoglobulin concentration on T.P. cells is probably close to the maximal concentration on cells, because T.P. cells were among the most brilliantly stained cells. Previously we reported (Klein *et al.*, 1970) that some cells in normal organs reacted with fluorescein-conjugated anti-mu and anti-kappa antibodies, in a way similar to the most reactive neoplastic cells, i.e. showing strong circular or sectorial staining. One to 10% of the cells in thymus, liver and bone marrow suspensions obtained from human fetuses reacted. Also, in two samples of adult bone marrow, 2-10% of the cells showed small dots and aggregates of varying intensity. No stained cells were seen with anti-IgA and anti-lambda antibodies. Anti-IgG antibodies were not tested, because we had reason to doubt the specificity of our reagent. In this context, it is notable that among 50 Burkitt tumours and 30 other lymphoid malignancies, about a quarter of the Burkitt tumours and some chronic lymphocytic leukaemias reacted strongly with anti-mu and anti-kappa antibodies. There was no reaction with anti-IgA and anti-lambda antibodies. Some of the chronic lymphocytic leukaemias gave a weak reaction with anti-IgG antibodies, but as mentioned, the specificity of this reaction is questionable. On the other hand, anti-IgG antibodies reacted strongly with some Burkitt tumours (Klein, 1970; Klein *et al.*, 1969). However, this reaction was considered to be due to IgG absorbed by the cells *in vivo,* because its degree varied during the clinical course, and was not maintained in the cell lines established in tissue cultures. These observations may indicate that IgM with kappa chains occurs in higher concentrations than other immunoglobulins on the surface of lymphoid cells. The function of the normal cells with the high densities of IgM is unknown, but the large number of immunoglobulin molecules should enable them to bind antigens efficiently.

ACKNOWLEDGEMENTS

We are grateful to Mrs Karin Kvarnung for excellent technical assistance.

This work has been supported by grants from the Swedish Cancer Society, contract No. 69-2005 within the Special Virus-Cancer Programme of the National Cancer Institute, NIH, PHS, the Medical Research Council, the Jane Coffin Childs Memorial Fund for Medical Research and the Damon Runyon Memorial Fund for Cancer Research DRG-1064. T.E. works at the Anatomical Institute and Institute for Experimental Medical Research, University of Oslo, Norway.

DISCUSSION

GOOD. I was very much interested in the fact that when you identified this molecule it seemed to be a 7S IgM. There are clinical and experimental models where there occurs a tremendous increase in 7S IgM—as for example in

Trypanosomiasis. Have you per chance looked at cells from patients from East Africa who have very high levels of 7S IgM. Do they have increased numbers of cells that stain in this way. You can also produce nice experimental models of Trypanosomiasis as Dr Frommel has done in our laboratory in rats. He has animals with very high 7S IgM. It would be interesting to look at such models to see whether the 7S IgM has a propensity to form this kind of cellular staining.

KLEIN. I have not looked at any of the systems, but I have to point out that these chronic lymphocytic leukaemia patients who have monoclonal immunoglobulins do not have cells which stain.

REFERENCES

Johansson, B. and Klein, E. (1970). *Clin. exp. Immun. 6,* 421-428.

Klein, G. (1970). *Israel J. med. Sci.* In press.

Klein, E., Nadkarni, J. S., Klein, G., Nadkarni, J. J., Wigzell, H. and Clifford, P. (1967). *Lancet ii,* 1068-1070.

Klein, E., Klein, G., Nadkarni, J. S., Nadkarni, J. J., Wigzell, H. and Clifford, P. (1968). *Cancer Res. 28,* 1300-1310.

Klein, G., Clifford, P., Henle, G., Henle, W., Geering, G. and Old, L. J. (1969). *Int. J. Cancer 4,* 416-421.

Klein, E., Eskeland, T., Inoue, M., Strom, R. and Johansson, B. (1970). *Expl Cell Res. 62,* 133-148.

Miller, F. and Metzger, H. (1965). *J. biol. Chem. 240,* 3325-3333.

Nadkarni, J. S., Nadkarni, J. J., Clifford, P., Manolov, G., Fenyö, E. M. and Kelin, E. (1969). *Cancer 23,* 64-79.

Parkhouse, R. M. E. and Askonas, B. A. (1969). *Biochem. J. 115,* 163-169.

Singhal, S. K. and Wigzell, H. (1970). *Progr. Allergy 14.* In press.

ANTIGEN-RECEPTORS ON THYMUS-DEPENDENT (T) AND
THYMUS-INDEPENDENT (B) LYMPHOCYTES

On the Thymic Origin of Antigen Sensitive Cells

ERNA MÖLLER and M. F. GREAVES

Division of Immunobiology, Karolinska Institutet, Wallenberglaboratory,
10405 Stockholm, Sweden
and
National Institute for Medical Research
Mill Hill, London, England

INTRODUCTION

The immune response of mice to sheep red blood cells (SRBC) has been studied extensively during recent years. Single antibody secreting cells can be enumerated by the local haemolysis in gel assay of Jerne and Nordin (1963). With this technique, cells secreting immunoglobulins of various classes can be enumerated (Wortis *et al.*, 1969). Antibody secreting cells or plaque forming cells (PFC) divide rapidly and most likely represent differentiated cells with a limited life span (Perkins *et al.*, 1969; Möller, 1968). The antibody secreting cells originate from precursor cells which by themselves do not necessarily secrete antibodies, but which acquire the ability after antigenic stimulation. These precursor cells are usually termed antigen reactive cells or antigen sensitive cells (ASC) (Kennedy *et al.*, 1966; *Transplantation Reviews*, 1969).

Antibody secreting cells can be detected *in vitro*, whereas ASC are usually detected by the ability of lymphoid cells mixed with antigen to give rise to antibody production after transfer into irradiated recipients.

Certain *in vitro* tests for immunocompetent cells do not necessarily depend on secretion of antibodies, but rather on binding of antigen to the surface of lymphocytes. The immunocytoadherence test (Reiss *et al.*, 1950) was modified to allow detection of cells binding erythrocytes (Nota *et al.*, 1964; Zaalberg *et al.*, 1968). The antigen binding lymphoid cells reacting with sheep erythrocytes are usually referred to as rosette-forming cells (RFC).

At the peak immune response of mice to SRBC, the proportion of antigen-binding cells or RFC, may be as high as 1-2% (Zaalberg *et al.*, 1968), whereas the peak response of antibody secreting cells (PFC) for each single immunoglobulin class under normal circumstances does not exceed 0.1% (Wigzell *et al.*, 1966). Thus, the antigen binding cells considerably outnumber the antibody secreting cells at the peak response. This difference may be due to (a) different sensitivity of the two *in vitro* tests employed, (b) antigen binding cells being different from antibody secreting cells, or (c) antigen binding cells comprising both secreting and non-secreting immunocompetent cells.

We have investigated the possibility that some antigen binding cells may be related to antigen sensitive cells. Two different experimental approaches were used to study these problems, in which it was possible to dissociate antibody secreting cells from antigen sensitive cells, namely by (a) passive antibody inhibition of the primary immune response and (b) priming with low doses of antigen.

a. It is well established that passively administered antibody suppresses the appearance of antibody secreting cells whilst leaving antigen sensitive cells or memory cells relatively unaffected (Uhr and Baumann, 1961; Uhr and Möller, 1968). We have studied the effect of passive antibodies on the formation of antibody secreting cells and of antigen binding cells in mice immunized to sheep red blood cells. The presence of antigen sensitive cells was determined by transfer of spleen cells from antibody treated or non-antibody treated immunized mice into irradiated syngeneic recipients, which were also challenged with SRBC.

b. We have also attempted to correlate antigen binding cells with antigen sensitive cells by studying the quantitative relationships of RFC, PFC and ASC in the splenic response of mice to different doses of antigen (SRBC), since Šterzl and Jílek (1967) have observed that priming with low doses of antigen leads to the formation of memory cells, demonstrable by transfer into irradiated recipients in the absence of a primary immune response.

EXPERIMENTAL PROCEDURES AND RESULTS

Effect of passive antibodies on the development of antibody secreting cells, antigen binding cells and antigen sensitive cells (memory cells). CBA mice were injected with 4×10^8 SRBC intravenously. At various intervals after immunization the mice were divided into two groups. One group was treated with hyperimmune mouse anti-SRBC antiserum, the other group received normal mouse serum. Two to five days later the mice were killed and cell suspensions prepared from the spleens. The number of antibody secreting cells in the spleens was determined using Jerne and Nordin's plaque assay (Jerne and Nordin, 1963) with modifications (Dresser and Wortis, 1965; Šterzl and Říha, 1965) for the determination of indirect plaque forming cells. Antigen binding cells were tested by essentially the same technique as described by McConnell *et al.* (1969). At least 20-50,000 spleen cells were scanned in each sample for the determination of the proportion of antigen binding cells. The presence of antigen sensitive cells was tested by adoptive transfer of the spleens from passive antibody treated or non-antibody treated immune animals into lethally irradiated syngeneic recipients. The adoptive immune response in the irradiated recipients was tested on day 7 after transfer with regard to direct and indirect plaque forming cells (D.PFC and ID.PFC) and antigen binding cells (RFC). For a further description of experimental procedures see Greaves *et al.* (1970). Table I shows the pooled results of more than 10 similar experiments. It can be seen that passive antibody treatment leads to a very marked suppression of antibody

TABLE I

Relationship between antigen binding cells and antigen sensitive (memory) cells to SRBC in mice treated with anti-SRBC antibody

Passive antibody	% suppresion of			Memory cells (adoptive transfer, relative response)		
	D.PFC/10^6	ID.PFC/10^6	RFC/10^6	D.PFC/10^6	ID.PFC/10^6	RFC/10^6
+	92.7±1.5	84.7±3.2	32.8±10.5	2.1±0.3	4.6±3.1	1.3±0.1
−	−	−	−	1.0	1.0	1.0

secreting cells of both IgM and IgG type, whereas the suppression of antigen binding cells was less marked. The adoptive transfer experiments showed that the spleens of passive antibody treated animals contained considerable numbers of antigen sensitive cells, since spleens from antibody treated animals as a rule gave rise to a higher immune response in the irradiated recipients than the spleen from non-antibody treated mice. The reason for this discrepancy is discussed elsewhere (Greaves *et al.*, 1970).

These experiments demonstrated a correlation between antigen binding cells and antigen sensitive cells whereas there was no such correlation between antibody secreting cells and antigen sensitive cells, indicating that some of the antigen binding cells may in fact be antigen sensitive cells.

Effect of various doses of SRBC on the development of antibody secreting cells, antigen binding cells and antigen sensitive cells. An optimal dose of SRBC in mice (4×10^8) gives a peak primary immune response of IgM PFC at day 4 or 5. Lower doses give rise to a peak response at day 6 or 7 (Wigzell *et al.*, 1966). We have immunized mice with SRBC in doses varying from 4×10^8 to 4×10^4 cells intravenously. At day 7 the immune response with regard to PFC, RFC and ASC was investigated. Table II shows the pooled results of five experiments. It can be seen from this table that the primary immune response is strongest with the highest antigen dose used. 4×10^4 SRBC did not give rise to a significantly elevated number of direct or indirect plaque forming cells. However, the number of antigen binding cells induced by this low dose was significantly increased.

TABLE II

Relationship between antigen binding cells and antigen sensitive (memory) cells in mice primed with various doses of SRBC

Antigen dose	Immune response—day 7			Memory cells (adoptive response)	
	D.PFC/10^6	ID.PFC/10^6	RFC/10^6	D.PFC/10^6	ID.PFC/10^6
4 x 10^8	1.99±0.14	1.96±0.44	3.79±0.38	2.76±0.38	2.90±0.42
4 x 10^6	1.84±0.35	1.51±0.46	3.48±0.14	3.00±0.23	3.42±0.19
4 x 10^4	0.80±0.16	0.64±0.57	2.87±0.11	2.94±0.26	3.21±0.14
–	0.65±0.10	0.56±0.11	2.47±0.04	2.04±0.45	1.55±0.58

Furthermore, transfer experiments demonstrated that even the lowest dose of SRBC induced the formation of an increased number of antigen sensitive cells, demonstrated by the capacity of spleens from low dose primed mice to give rise to antibody secreting cells after transfer into syngeneic irradiated recipients. We also confirmed Šterzl and Jílek's finding that a lower antigen priming dose give a more efficient development of memory cells than an optimal dose for a primary immune response (Šterzl and Jílek, 1967). These experiments, like the one cited above, showed a correlation between antigen binding cells and antigen sensitive cells in the absence of antibody secreting cells and thus imply that some antigen binding cells may in fact be antigen sensitive cells.

Origin of antigen binding cells in mice immune to SRBC. In view of the above findings of a correlation between the number of antigen binding cells and antigen sensitive cells in two different experimental models, we were interested in determining the origin of antigen binding cells. Studies during recent years have clearly established that a primary humoral immune response in mice to certain antigens is dependent upon co-operation between two different types of lymphoid cell populations (for review see *Transplantation Review,* 1969). Thymus-dependent or processed cells act as helper cells to induce antibody formation and secretion in bone marrow-dependent lymphocytes. Antibody secretion appears to be an exclusive property of the bone marrow population (Mitchell and Miller, 1968; Davies *et al.,* 1967).

The function of the thymus-dependent lymphocytes is not known, but these cells are known to respond specifically to antigenic stimulation (Davies *et al.,* 1966). Evidence suggesting that thymus-dependent lymphoid cells also carry specific immunoglobulin determinants has accumulated (for review see Greaves, 1970). Since we found a correlation between antigen binding cells and antigen sensitive cells in our studies, we wanted to investigate whether the antigen binding cells were of thymic origin. For that purpose, we investigated whether antigens could be used as markers on thymus-dependent lymphocytes. For these studies we made use of the isoantigenic markers, H-2 (Greaves and Möller, 1970a) and of the theta (θ) antigen, originally described by Reif and Allen (1964). It has been recently established that the theta antigen can be regarded as a marker of thymus-dependent lymphocytes (for ref. see Raff, 1969; Schlesinger and Hurvitz, 1969; Raff and Wortis, 1970).

Since the theta (θ) antigen is considered to be a relevant antigenic marker for thymus-dependent lymphoid cells in the peripheral lymphoid organs, we investigated whether antigen binding or antibody secretion could be inhibited by pretreatment of spleen cells with anti-θ serum and complement. Normal serum served as a control. For further details of experimental procedure see Greaves and Möller (1970b).

TABLE III

Effect of AKR anti-CBA thymus (anti-θ-C3H) serum on RFC and PFC in the spleens of CBA mice immunized with SRBC

Days after SRBC immunization	Immune response					
	RFC/10^{6} [a]			D.PFC/10^{6} [b]		
	anti-θ	NS	% inhibition	anti-θ	NS	% inhibition
0	528	889	41	NT		
0	376	723	48	NT		
3	1111	1852	40	10.1± 1.8	9.6±1.3	− 5.2
5	5343	8628	38[c]	114.8±17.1	97.3±3.4	−17.9
6	4282	5613	24	NT		
25	2199	3125	29	NT		

[a] Between 20,000 and 100,000 spleen cells were scanned in each sample.
[b] Mean ± S.E.
[c] In all, 11 experiments have been carried out with spleen cells from animals injected 5-6 days earlier; the mean inhibition of RFC with anti-θ serum was 36±2 · NS-normal AKR serum.

Table III shows the results of a few experiments of this design. The formation of antigen binding cells could be inhibited considerably by anti-θ serum. The degree of inhibition varied with the dose of antigen used for immunization and time after immunization (Greaves and Möller, 1970b). No similar effect on antibody secreting cells was found. These tests indicated that some antigen binding cells, but no antibody secreting cells are thymus-dependent lymphocytes.

The proportion of θ positive RFC was higher in lethally irradiated animals, reconstituted with graded doses of syngeneic thymus and bone marrow cell mixtures and thereafter immunized with SRBC. Table IV shows the results of some experiments of this type. The higher proportion of θ positive RFC in such animals could either be due to the fact that proportionally more thymus-derived cells localize in the spleens of such animals, or, alternatively, to the possibility that the θ antigen is not a stable characteristic of thymus-derived lymphocytes (Schlesinger and Yron, 1970).

TABLE IV

Effect of AKR anti-CBA thymus (anti-θ-C3H) serum on RFC and PFC in the spleens of lethally irradiated CBA mice, treated with 15 x 10^6 syngeneic bone marrow cells and 50 x 10^6 thymus cells and immunized with SRBC (day 7 response)

Exp. no.	RFC/10^6			D.PFC/10^6 ± S.E.		
	Anti-θ^a	NS^c	% inhibition with anti-θ	Anti-θ	NS	% inhibition with anti-θ
1	239^b	657	64	NT		
2	286	800	64	40.3±9.6	34.1±7.5	−18
3	309	607	49	NT		
4	278	1296	78	NT		
5	509	972	48	NT		

^{a} Treatment with anti-θ serum (No. 1) *in vitro* before assay.

^{b} In this and subsequent anti-θ experiments the number of RFC/10^6 given is derived from the total RFC counted in a pool of spleen cells from five mice. Between 20,000 and 80,000 spleen cells were scanned in each individual sample.

^{c} NS-normal AKR serum.

Percentage theta positive antigen binding cells in immune responses to SRBC. Since we found that the proportion of thymus-dependent antigen binding cells can vary, we wanted to study the proportion of theta positive antigen binding cells in experimental situations where antigen binding cells were selectively activated. Tables III and V show that 40 to 60% of the background antigen binding cells in normal animals were sensitive to anti-θ serum, i.e. were of thymic origin. In animals primed with a high dose of SRBC proportionally less antigen binding cells were of thymic origin at the peak response, than in low dose primed animals. Furthermore, passive antibody inhibition of the primary immune response resulted in a higher proportion of theta positive antigen binding cells.

Contribution of θ positive cells to memory. In the light of the above findings we also studied whether thymus-dependent spleen cells were necessary for the induction of a primary immune response in mice to SRBC. Furthermore, the possibility that some antigen sensitive cells (memory cells) were thymus-

TABLE V

Percentage θpositive antigen binding cells in immune responses to SRBC

Treatment	% θ positive RFC
–	60
4×10^8 SRBC	36±2
8×10^4 SRBC	54±3
4×10^8 SRBC + anti-SRBC ab	68

dependent lymphocytes was investigated. Spleen cells from normal or immune mice were treated *in vitro* with anti-θ serum in the presence of complement. Normal serum served as a control. After antiserum treatment and washing, graded doses of cells were transferred into lethally irradiated syngeneic recipients. The immune response was measured at day 7 after transfer. Table VI shows the pooled results of several such experiments. It can be seen that anti-θ serum caused an inhibition of the primary immune response amounting to 60.3% for D.PFC and 67.5% for ID.PFC, indicating that thymus-dependent cells serve as "helper" cells in the induction of a humoral immune response. Other groups of mice were primed with low doses of SRBC before transfer. Spleens of such mice contained increased numbers of antigen sensitive cells, demonstrable in the transfer system. Anti-θ serum treatment of such spleens reduced the capacity of the cells to transfer secondary reactivity, indicating that some cells, necessary for the expression of immunological memory, are sensitive to anti-θ antibodies, and thus most probably are thymus dependent lymphocytes (Table VI).

The inhibitory effect of our anti-θ serum was probably not due to a non-specific toxic effect since antiserum treatment did not reduce the capacity of spleen cells to give a humoral immune response to a thymus independent antigen, such as endotoxin derived from *E. coli* O55:B5 (E. Möller, unpublished).

TABLE VI

Effect of anti-θ serum on adoptive transfer of immunity by spleen cells from animals primed with different antigen doses

Antigen dose	% inhibition by anti-θ of		RFC/10^6
	D.PFC/10^6	ID.PFC/10^6	
4×10^5	51.9±10.4	35.5±20.6	
4×10^4	66.5	85.1	
–	60.3±14.6	67.5±14.0	49.5±5.5

Recently, Raff (1970) has found that anti-θ serum treatment of spleen cells *in vitro* abolishes the helper effect after adoptive transfer into lethally irradiated recipients. Our findings substantiate the conclusion that thymus-derived lymphocytes act as helper cells in the induction of a primary immune response.

DISCUSSION

The purpose of the experiments described above has been to examine whether the antigen binding cells in mice immune to sheep red blood cells are in fact antigen sensitive. The quantitative relationships between antibody secreting cells, antigen sensitive or memory cells and antigen binding or rosette forming cells was investigated in two different experimental situations. Injection of mice with hyperimmune serum (passive antibody) is known to depress specifically the development of antibody secreting cells without impairing the development of immunological memory (Uhr and Baumann, 1961; Uhr and Möller, 1968). Suppression is thought to operate by rendering antigenic determinants relatively inaccessible for triggering of antigen sensitive cells. We found that injection of mice with hyperimmune anti-SRBC serum led to a considerable reduction in the level of splenic antibody secreting cells. In contrast, antigen binding cells were less affected in antibody treated animals, although they too could be suppressed to a certain extent. The degree of suppression of RFC varies with time of passive antibody injection (Greaves *et al.,* 1970). The transfer experiments indicated, as anticipated, that antigen sensitive cells were not reduced by passive antibody treatment. We have also found that the antigen binding cells in animals treated with hyperimmune serum were of thymic origin more often than were antigen binding cells in non-antibody treated animals. It seems likely that passive antibody injection may select out immature secreting cells by preventing their terminal differentiation and subsequent death (Sercarz and Coons, 1962; Šterzl and Silverstein, 1967). Our experiments with anti-θ serum suggested that one major effect of passive antibody may be to select for thymus-dependent antigen binding cells. We suggest that the antigen binding cells in mice injected with hyperimmune serum are (a) thymus-dependent cells associated with cellular immunity and/or the co-operation phenomenon, which may escape inhibition by responding early to antigen and/or being relatively long-lived cells, that might not be dependent upon the continuous presence of antigen for the expression of cell surface antibody activity, and (b) bone marrow-dependent precursors .of antibody secreting cells whose terminal differentiation has been halted by removal of antigen.

The absolute and relative numbers of these two cell types will depend on the timing of antibody injection relative to antigen.

Recent experiments have shown that low zone tolerance may involve selective inactivation of thymus-dependent lymphocytes (Rajewsky and Mitchison, personal communication). These results suggest that thymus-dependent cells may have a lower threshold than bone marrow cells (the precursors of antibody secreting cells) for the induction of both tolerance and immunity. The reason for this difference is the subject of further analysis.

ACKNOWLEDGEMENTS

The skilful technical assistance of Mrs Lill-Britt Andersson is gratefully acknowledged. The experimental work reported in this paper was supported by grants from the Swedish Medical Research Council, the Swedish Cancer Society and the Wallenberg Foundation.

SUMMARY

The present experiments have demonstrated a correlation between antigen sensitive (memory) cells and antigen binding cells in animals immune to SRBC. No similar correlation existed between antigen sensitive cells and antibody secreting cells however. Studies involving the use of an isoantigenic marker (theta), showed that a certain proportion of antigen binding cells in normal and immune animals were thymus-derived cells. The proportion of thymus-derived antigen binding cells varied with time after immunization and with dose of antigen. Experimental situations that resulted in a selective activation of antigen sensitive cells without a concomitant increase in the number of antibody secreting cells, also led to a greater proportion of thymus-derived antigen binding cells. Finally, it was found that treatment of normal or immune spleen cells with anti-θ serum and complement *in vitro*, led to a decreased humoral immune response after transfer to irradiated animals. This indicates that thymus-derived cells can serve as both helper cells for the induction of humoral immunity and as memory cells in the secondary response.

REFERENCES

Davies, A. J. S., Leuchars, E., Wallis, V. and Koller, P. C. (1966). *Transplantation 4*, 438-541.

Davies, A. J. S., Leuchars, E., Wallis, V., Marchant, R. and Elliott, E. V. (1967). *Transplantation 5*, 222-231.

Dresser, D. W. and Wortis, H. H. (1965). *Nature, Lond. 208*, 859-861.

Greaves, M. F. (1970). *Transplantation Rev. 5*, 45-75.

Greaves, M. F. and Möller, E. (1970a). In "Developmental Aspects of Antibody Formation and Structure", Symposia CSAV, 627, (J. Sterzl, ed.).

Greaves, M. F. and Möller, E. (1970b). "Cellular Immunology", vol. 4. In press.

Greaves, M. F., Möller, E. and Möller, G. (1970). "Cellular Immunology", vol. 4. In press.

Jerne, N. K. and Nordin, A. A. (1963). *Science, N.Y. 140*, 405.
Kennedy, J. C., Till, J. E., Siminovich, L. and McCulloch, E. A. (1966). *J. Immun. 96*, 973-980.
McConnell, I., Munro, A., Gurner, B. W. and Coombs, R. R. A. (1969). *Int. Archs Allergy appl. Immun. 35*, 209-227.
Mitchell, G. F. and Miller, J. F. A. P. (1968). *J. exp. Med. 128*, 821-837.
Möller, G. (1968). *J. exp. Med. 127*, 291-306.
Nota, N. R., Liacopoulos-Briot, M., Stiffel, C. and Biozzi, G. (1964). *C.r. Acad. Sci., Paris 259*, 1277.
Perkins, E. H., Sado, T. and Makinodan, T. (1969). *J. Immun. 103*, 668-678.
Raff, M. C. (1969). *Nature, Lond. 224*, 378-379.
Raff, M. C. (1970). *Nature, Lond. 226*, 1257-1258.
Raff, M. C. and Wortis, H. H. (1970). *Immunology.* (In press.)
Reif, A. E. and Allen, J. M. V. (1964). *J. exp. Med. 120*, 413-433.
Reiss, E., Mertens, E. and Ehrich, W. E. (1950). *Proc. Soc. exp. Biol. Med. 74*, 732-735.
Schlesinger, M. and Hurvitz, D. (1969). *Transplantation 7*, 132-141.
Schlesinger, M. and Yron, I. (1970). *J. Immun. 104*, 798-804.
Sercarz, E. E. and Coons, A. H. (1962). *In* "Mechanisms of Immunological Tolerance" (M. Hašek, A. Lengerová and M. Vojtíšková, eds), pp. 73-83. Czech. Acad. Sci. Press, Prague.
Šterzl, J. and Jílek, M. (1967). *Nature, Lond. 216*, 1233-1235.
Šterzl, J. and Říha, I. (1965). *Nature, Lond. 208*, 858-859.
Šterzl, J. and Silverstein, A. M. (1967). *Adv. Immun. 6*, 337-459.
Transplantation Rev. (1969). *1.*
Uhr, J. W. and Baumann, J. B. (1961). *J. exp. Med. 113*, 935-957.
Uhr, J. W. and Möller, G. (1968). *Adv. Immun. 8*, 81.
Wigzell, H., Möller, G. and Andersson, B. (1966). *Acta path. microbiol. scand. 66*, 530-540.
Wortis, H. H., Dresser, D. W. and Anderson, H. R. (1969). *Immunology 17*, 93-110.
Zaalberg, O. B., van der Meul, V. A. and van Twisk, M. J. (1968). *J. Immun. 100*, 451.

Rosette Formation, a Model for Antigen Recognition

JEAN-FRANÇOIS BACH, FELIX REYES,
MIREILLE DARDENNE, CATHERINE FOURNIER and
JEAN-YVES MULLER

Hôpital Necker, Paris 15e France

INTRODUCTION

Several "immunocytoadherence" techniques have been devised in order to detect antigen-binding cells. The most extensively studied have been bacterio-adherence, the rosette test and labelled-antigen binding. When these techniques were first described, they were generally thought to detect mostly antibody-forming cells. More recently it has been shown that at least some antigen-binding cells are antigen sensitive cells, that is cells involved in the first, antigen recognition, phase of the immune response (Ada and Byrt, 1969; Bach *et al.,* 1970a; Humphrey and Keller, 1969; Wigzell and Mäkelä, 1970).

The rosette test is the simplest of the immunocytoadherence techniques. This paper will describe the main characteristics and the functional significance of the spontaneous rosette-forming cells (RFC) found in normal animals. This situation seemed to us particularly interesting, since few antibody-forming cells should be found in normal animals. Conversely many of the RFC should be antigen sensitive cells (ASC) committed to the primary response, if ASC can bind antigen.

ANTIGEN RECOGNITION BY SPONTANEOUS RFC

Antigen recognition by lymphocyte receptors is the first step in the immune response in Burnet's clonal selection theory (1959). These receptors are supposed to be antibody-like molecules. The problem is to determine whether spontaneous RFC, or at least some of them, belong to this population of committed cells.

To prove this, we have demonstrated that normal spleen cells depleted of their RFC against an antigen are specifically unresponsive to this particular antigen (Bach *et al.,* 1970a). In fact this only shows that RFC are necessary for the immune response against the antigen that they can bind. However, the fact that rosette formation occurs within 5 min strongly suggests that RFC are more precisely related to the process of antigen recognition.

Normal spleen cells have been depleted of their RFC against chicken or sheep red blood cells (SRBC) by passage on a Ficoll-Triosil gradient. In this gradient, prepared as described by Harris and Ukaejiofo (1969), lymphocytes remain on the top and red cells sink through the gradient. When rosettes are formed before putting the spleen cells on the gradient, rosettes go through the gradient and sediment with the red cells at the bottom of the tube. The spleen cells are thus specifically depleted of their RFC against the type of RBC used for rosette formation. Depleted spleen cells are injected intravenously together with chicken and sheep RBC. Haemagglutinins and RFC are determined seven days later (Fig. 1). Passage on the gradient induced a specific unresponsiveness to the type of red cells used for rosette formation, both at the humoral and cellular level (Table I). Moreover, the depletion reached the background itself, that is

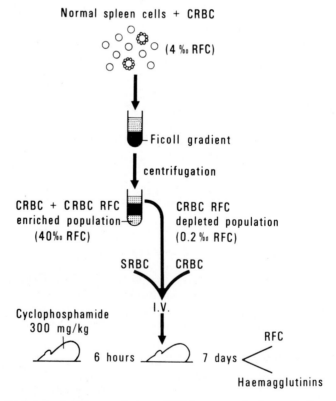

Fig. 1. Depletion of chicken erythrocyte RFC from normal spleen cells. Restoration of cyclophosphamide-treated mice.

TABLE I

Cellular and humoral responses of cyclophosphamide-treated mice to sheep (SRBC) and chicken erythrocytes (CRBC) after immunological restoration with normal or RFC-depleted spleen cells

Cells used for restoration	No. of RFC/spleen $(\times 10^{-3})$		Haemagglutinin titres $(\log_2)$	
	Sheep	Chicken	Sheep	Chicken
Normal spleen	316 ± 12	403 ± 12	6.5 ± 0.54	5.3 ± 0.31
SRBC RFC-depleted spleen	55 ± 42	295 ± 27	2.9 ± 0.55	5.2 ± 0.51
CRBC RFC-depleted spleen	411 ± 61	24 ± 22	5.6 ± 0.51	2.15 ± 0.22

Means $\pm$ SE

significantly lower than that observed in fully reconstituted mice $(P < 0.01)$. Unresponsiveness was not induced by incubating spleen cells and heterologous red cells without passage on the gradient or by passage on the gradient without rosette formation.

The results of this experiment are in keeping with those of Ada and Byrt (1969) who obtained a similar specific unresponsiveness in normal spleen cells using heavily radioactively-labelled flagellin, and those of Wigzell and Mäkelä (1970), who obtained similar results by passing normal spleen cells on antigen-coated columns. Radiation-induced tolerance, without cell elimination, might affect the experiments of Ada and Byrt. However, our experiments show specific cell elimination. Thus it has been proved that RFC, or at least some of them, are involved in antigen recognition at the initial phase of the immune response.

CHARACTERISTICS OF SPONTANEOUS RFC

a. *Morphology.* Spontaneous RFC have been studied by optical microscopy (after injection of colloidal carbon) and electron microscopy after rosette isolation by micromanipulation (Reyes and Bach, 1970). In normal mice, two-thirds of the spleen and bone marrow spontaneous RFC are small lymphocytes (Fig. 2), and one-third are macrophages (Figs 3, 4 and 6). No other cell type has been found. The small lymphocytes are typical small lymphocytes with very little endoplasmic reticulum (Fig. 5). This homogeneity is in contrast with the marked heterogeneity observed for RFC in immunized animals (Storb *et al.*, 1969). It confirms observations on flagellin-binding cells (Mandel *et al.*, 1969). Macrophages are eliminated by 90 min incubation at 37°C. Therefore they did not interfere in subsequent experiments, which all included an incubation step. Rosette-forming lymphocytes do not adhere to glass even during 3 hr at 37°C.

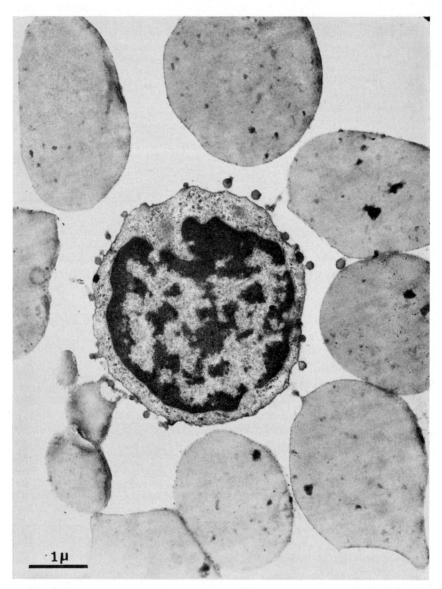

Fig. 2. Rosette-forming lymphocyte against SRBC in a normal mouse spleen. Typical small lymphocyte. Small erythrocyte fragments are seen against the lymphocyte membrane. x20,000.

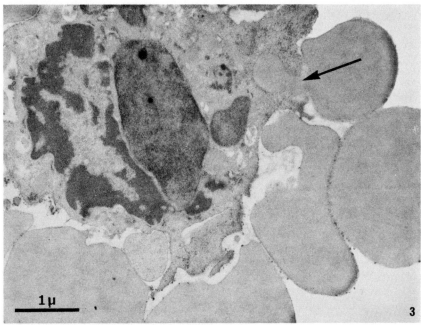

Fig. 3. Rosette-forming macrophage with juxtanuclear phagosomes. An erythrocyte is being phagocytosed. x21,000.

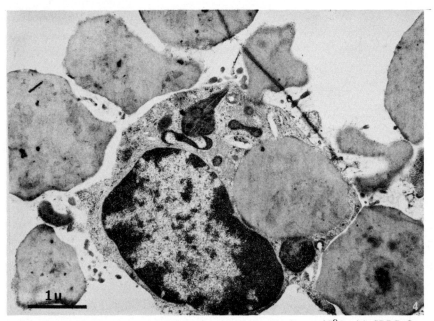

Fig. 4. Rosette-forming macrophage after 90 min incubation at 37°C with SRBC. One SRBC has been phagocytosed. x19,000.

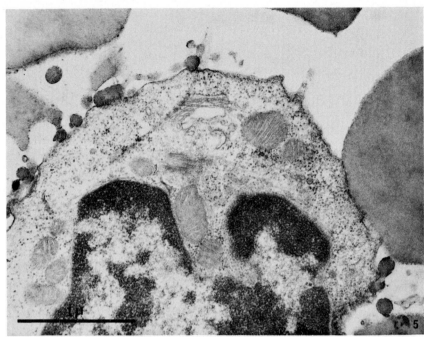

Fig. 5. Details of rosette-forming lymphocyte. Golgi zone and monoribosomes can be seen. Contact between the lymphocyte and SRBC is linear and discontinuous. x40,000.

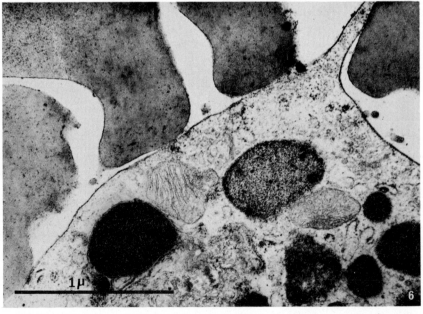

Fig. 6. Details of rosette-forming macrophage with erythrophagocytose phagolysosome. x40,000.

b. *Number.* The number of RFC is constant in one mouse strain. However, it differs very much with the strain and the lymphoid organ. The highest figure is found in the spleen. Antigen-binding cells are also found in the bone marrow and lymph nodes, but very few are found in the thymus. Table II gives the number of RFC against SRBC in the various lymphoid organs of C57B1/6 mice.

TABLE II

Number of rosette-forming cells per 1000 nucleated cells in non-immunized C57B1/6 mice (±SD)

Bone marrow	0.4 ± 0.05
Spleen	1.1 ± 0.1
Lymph node	0.4 ± 0.06
Thymus	<0.02
Peripheral blood	$0.95^a \pm 0.1$

[a] RFC per 1000 lymphocytes

c. *Receptors responsible for rosette formation.* These receptors are probably *immunoglobulins,* because rosette formation is inhibited by preincubation for 90 min at 37°C with anti-immunoglobulin antisera. However, the class of the immunoglobulin is not known. Greaves and Hogg (1971) showed that the receptors responsible for spontaneous rosette formation had immunoglobulin light chain immunogenicity, but did not have heavy chain determinants of any of the known classes of immunoglobulins. Similar findings have been reported for the recognition receptors for delayed hypersensitivity and graft-*versus*-host reactions (Mason and Warner, 1970; Greaves *et al.,* 1969). However, Warner *et al.* (1970) recently reported that flagellin-binding cells present in normal mice were inhibited by anti-heavy chain sera. Bone marrow and spleen RFC are both inhibited by anti-immunoglobulin antibodies, whereas only spleen flagellin-binding cells can be inhibited by anti-immunoglobulin antibodies (Byzt and Ada, 1969). One interpretation may be that with the labelled flagellin technique there is some non-specific binding by bone marrow cells which is not inhibited by anti-immunoglobulin antibodies.

1. *The specificity.* The specificity of the receptors has been demonstrated for spleen (Laskov, 1968; Bach, 1970) and bone marrow (Bach, 1970). When the cell suspension is incubated for 2 hr at 37°C to eliminate the macrophages, the specificity is perfect. This means, for example, that the cells forming rosettes against SRBC are different from those forming rosettes against chicken RBC. This specificity is a property only of the lymphocytes and is not found in the macrophages. We have proved this absence of specificity for the macrophages,

TABLE III

Specificity of spontaneous bone marrow RFC

	SRBC	CRBC	Mixed rosettes SRBC + CRBC
SRBC	422 ± 55	0	0
CRBC	0	1045 ± 45	0
SRBC then CRBC	342 ± 60	1121 ± 60	0
CRBC then SRBC	405 ± 60	1090 ± 55	0

Bone marrow cells from normal mice were incubated 120 min at 37°C in Hanks medium in order to eliminate macrophages, then mixed with sheep red blood cells (SRBC) or chicken red blood cells (CRBC) or both, successively. In the latter case rosettes were formed with the first species (by centrifugation) before the second species was added. (RFC/10^6 cells ± SD).

although no double rosettes are found when erythrocytes from two different species are added together to spleen cell preparations. This absence of double rosettes is due to a pre-emption effect, i.e. the preferential formation of a rosette by the macrophage with the first antigen to be presented.

2. Antimetabolites. Drugs such as azathioprine, 6-mercaptopurine, 6-thioguanine, 5-fluorouracil and amethopterin inhibit rosette formation at non-cytotoxic concentrations (Bach *et al.* 1969a). With azathioprine the inhibition is reversible within 20 min when low concentrations are used. This indicates that the receptors are secreted by the lymphocytes and not secondarily fixed on them, and that the turnover is short, perhaps as short as 20 min. The inhibition of rosette formation by some inhibitors of cellular respiration and by puromycin at non-cytotoxic concentrations (Bach, 1970; Bach *et al.*, 1969a) confirms this hypothesis. However, one cannot exclude the possibility that antimetabolites act on rosette formation *in vitro* by membrane modification.

It is unusual to speak of the *affinity* of a cell for an antigen. However, with flagellin the more antigen that is placed in contact with the cell the more antigen-binding cells are detected. This suggests a wide range of affinities among antigen-binding cells in normal animals. We have made similar findings for RFC and realize that the interpretation of such results is more difficult for red cells than for soluble antigens. However, the number of rosettes gradually increases when the ratio of erythrocytes to lymphocytes varies from 1 to 100 in normal animals, whereas the variation is only from 1 to 20 in immunized animals. This suggests that the population of RFC found in immunized animals is less heterogeneous with respect to affinity than the population found in normal animals. Moreover when rosettes formed with unimmunized lymphocytes are suspended for a long time, for instance 30 min, the number of rosettes remains stable after 10 min, at the same level, whatever the ratio of erythrocytes to leucocytes used for rosette formation.

d. *Relation to plaque-forming cells (PFC).* Liacopoulos *et al.* (1970) have shown that spontaneous RFC found in normal animals did not form plaques when complement was added. Moreover, the number of RFC is about 1000 times lower than the number of PFC in normal animals.

e. *Ontogeny.* RFC against SRBC only appear at a high level in the mouse between the 7th and the 14th day after birth, but in Swiss mice RFC against chicken RBC are present before birth and are at their adult level at birth.

f. *Thymus dependence.* We have shown (Bach *et al.*, 1970a) that RFC found in normal animals are inhibited by treatment with anti-θ serum (kindly provided by Dr M. Raff). θ antigen is known to be characteristic of thymus-dependent cells (Raff, 1969). Rosettes are inhibited by concentrations cytotoxic for thymus-dependent cells but not for spleen cells. Bone marrow RFC are less sensitive than spleen RFC to azathioprine and anti-lymphocyte serum (ALS) *in vitro* (Bach and Dardenne, 1969). Neonatally thymectomized mice have a normal level of spontaneous RFC (Bach and Dardenne, 1969). We have shown that spleen RFC of neonatally thymectomized mice are significantly less sensitive to these immunosuppressive agents than spleen RFC from normal animals (Bach *et al.*, 1970b) (Table IV). Similar findings have been made in deprived mice (adult thymectomized irradiated mice restored with bone marrow) provided by Dr A. J. S. Davies (1969). We have also confirmed these findings in Nude mice (provided by Dr N. Muller-Berat), which are born without thymuses. Spleen RFC of these mice behave like bone marrow cells, that have not been activated by the thymus. More recently we have shown, in collaboration with Dr A. J. S. Davies, that spleen cells from adult thymectomized mice are unexpectedly also resistant to azathioprine (Table IV), which suggests the possible role of some thymic humoral factor. It is interesting to note that Schlesinger and Yron (1970) found that θ antigen disappeared six weeks after adult thymectomy.

TABLE IV

Influence of thymus on rosette inhibition by azathioprine

	Azathioprine—minimal inhibitory conc. (γ/ml)	
	Spleen cells	Bone marrow cells
Normal mouse	0.5 ± 0.1	40 ± 4
Neonatal thymectomy	35 ± 10	35 ± 5
Adult thymectomy	25 ± 5	40 ± 7
Nude mice	40 ± 5	50 ± 10
Deprived mice	20 ± 10	30 ± 5

Means $\pm$ SD

Conclusions

a. *Receptors responsible for rosette formation are produced by spontaneous RFC.* Antigen-binding cells are for the most part small lymphocytes, specific for the antigen. They are not suppressed when the cells are washed several times before the test. Passive fixation of cytophilic antibodies is therefore unlikely. Further, and this is probably the major argument, antimetabolites such as azathioprine and amethopterin inhibit rosette formation at very low, non-cytotoxic concentrations in a reversible way.

b. *Spontaneous RFC are not all natural antibody producers.* This has been controversial for a long time because it has been clearly demonstrated that PFC in the normal animal are mostly natural antibody forming cells. More precisely they are producing antibodies against antigens encountered in the first days of life. The three arguments in favour of this function for the background PFC are: (a) the absence of PFC in young germ-free piglets fed with synthetic antigen-free diets (Šterzl, 1967), (b) the cross-reactions between antigens of *Escherichia coli* and SRBC demonstrated by the PFC against SRBC produced by immunization with *E. coli* (Cheng and Trentin, 1967), (c) the fact that removing background PFC does not suppress the competence of spleen cells (Cunningham 1969). None of these has been demonstrated for antigen-binding cells. On the contrary antigen-binding cells are found at birth in the mouse, in germ-free mice (Byrt and Ada, 1969), and in young germ-free piglets on an antigen-free diet (Šterzl, personal communication). Lastly, RFC against synthetic antigens such as DNP have been found in normal animals (Biozzi *et al.*, 1967).

c. *Spontaneous RFC probably include three types of cells.* These are antigen sensitive cells as demonstrated above, antibody-forming cells (some of them producing natural antibodies) and macrophages, as shown by electron microscopy and after uptake of carbon particles. The same cell types probably exist in immunized mice since Greaves and Möller (1970a, b) have shown that a high proportion of RFC after immunization are thymus-dependent antigen sensitive cells, and Storb *et al.* (1969) have also found rosette-forming macrophages in immunized mice.

Table V

Classification of rosette-forming cells

	Normal mice	Immunized mice (day 6)
Number of RFC per spleen	10^5	4×10^6
Macrophages (morphology)	25% (Bach)	25% (Storb)
Antigen sensitive cells (θ antigen)	60% (Bach)	25% (Greaves)
Antibody forming cells (and PFC progenitors?)	15%	50%

The work of Biozzi *et al.* (1966, 1968) demonstrates that most RFC produce the receptors responsible for rosette formation, but not that all of them are antibody-forming cells. Table V shows that the normal animal represents the best situation for studying antigen sensitive cells. RFC found in normal animals are much more sensitive *in vivo* and *in vitro* to ALS (Bach, 1969) and antimetabolites (Bach *et al.*, 1969a) than spleen RFC in immune animals and bone marrow RFC. Therefore they may well be the target of these immunosuppressive agents. This hypothesis is supported by the correlation between *in vitro* immunosuppressive potency of ALS and antimetabolites and their rosette inhibiting activity (Bach *et al.*, 1969a, b, c).

REFERENCES

Ada, G. L. and Byrt, P. (1969). *Nature, Lond. 222*, 1291-1292.
Bach, J.-F. (1969). *C.r. Acad. Sci Paris 268*, 883-884.
Bach, J.-F. (1970). Submitted for publication.
Bach, J.-F. and Antoine, B. (1968). *Nature, Lond. 217*, 658-659.
Bach, J.-F. and Dardenne, M. (1969). *C.r. Acad. Sci. Paris 269*, 751-754.
Bach, J.-F., Dardenne, M. and Fournier, C. (1969a). *Nature, Lond. 222*, 998-999.
Bach, J.-F., Dormont, J., Dardenne, M. and Antoine, B. (1969b). *Transplant. Proc. 1*, 403.
Bach, J.-F., Dormont, J., Dardenne, M. and Balner, H. (1969c). *Transplantation 8*, 265-280.
Bach, J.-F., Dardenne, M. and Muller, J.-Y. (1970a). *Nature, Lond.* 1970, *227*, 1251-1252.
Bach, J.-F., Dardenne, M. and Muller, J.-Y. (1970b). *C.r. Acad. Sci. Paris 270*, 2142-2144.
Biozzi, G., Stiffel, C., Mouton, D., Liacopoulos-Briot, M., Decreusefond, C. and Bouthillier, Y. (1966). *Annls Inst. Pasteur Paris 110*, 7-32.
Biozzi, G., Stiffel, C. and Mouton, D. (1967). *In* "Immunity, Cancer and Chemotherapy" (E. Mihich, ed.), pp. 103-139. Academic Press, New York.
Biozzi, G., Stiffel, C., Mouton, D., Bouthillier, Y. and Decreusefond, C. (1968). *Immunology 14*, 7-20.
Burnet, F. M. (1959). "The Clonal Secretion Theory of Acquired Immunity". Cambridge University Press.
Byrt, P. and Ada, G. L. (1969). *Immunology 17*, 503-516.
Cheng, V. and Trentin, J. J. (1967). *Proc. Soc. exp. Biol. Med. 126*, 467-470.
Cunningham, A. J. (1969). *Immunology 16*, 621-632.
Davies, A. J. S. (1969). *Transplant. Rev. 1*, 43-91.
Greaves, M. F. and Hogg, J. (1971). In this Symposium.
Greaves, M. F. and Möller, E. (1970a). "Cellular Immunology". (In press.)
Greaves, M. F. and Möller, E. (1970b). In press.
Greaves, M. F., Torrigiani, G. and Roitt, I. M. (1969). *Nature Lond. 222*, 885-886.
Harris, R. and Ukaejiofo, E. O. (1969). *Lancet ii*, 327.
Humphrey, J. H. and Keller, M. V. (1969). *In* "Developmental Aspects of Antibody Formation". In press.

Laskov, R. (1968). *Nature, Lond. 219,* 973-975.
Liacopoulos, P., Amstutz, H. and Gille, F. (1970). In press.
Mandel, T. Byrt, P. and Ada, G. L. (1969). *Expl. Cell Res. 58,* 179-182.
Mason, S. and Warner, N. L. (1970). *J. Immun. 104,* 762-765.
Raff, M. C. (1969). *Nature, Lond. 224,* 378-379.
Reyes, F. and Bach, J.-F. (1970). *C.r. Acad. Sci. Paris 270,* 1702-1704.
Schlesinger, M. and Yron, I. (1970). *J. Immun. 104,* 798-804.
Šterzl, J. (1967). *Cold Spring Harb. Symp. quant. Biol. 32,* 493-506.
Storb, U., Bauer, W., Storb, R., Fliedner, T. M. and Weiser, R. S. (1969). *J. Immun. 102,* 1474-1485.
Warner, N. L., Byrt, P. and Ada, G. L. (1970). *Nature, Lond. 226,* 942-943.
Wigzell, H. and Mäkelä, O. (1970). *J. exp. Med 132,* 110-126.

Specific Lethal Radioactive Antigens

J. H. HUMPHREY, G. ROELANTS and N. WILLCOX

National Institute for Medical Research, Mill Hill
London, England

It has sometimes been suggested in discussions among immunologists that it would be interesting to develop "lethal" antigens, which would kill rather than stimulate the immunologically competent cells with which they interact. Such antigens would provide a means of testing the validity of clonal selection theories and might conceivably be useful for inducing specific immunological tolerance. An obvious theoretical possibility for making a lethal antigen would be to introduce enough atoms of a soft β-ray emitting radio-isotope into the antigen molecules, so that any cell with which they interacted in sufficient numbers would receive a lethal dose of radiation. Having unintentionally achieved this aim, as described briefly below, we have considered retrospectively what might be the necessary properties of such an antigen. It is a fair assumption that it would be held by receptors at the surface of lymphocytes and possibly also be ingested. If the cells were to be selectively irradiated, without significant damage to neighbouring cells, the radiation would need to be largely absorbed within about 5 μm, but also to penetrate for at least 1 μm so as to traverse the narrow layer of cytoplasm of a lymphocyte and reach the nucleus. Gamma rays are in general too penetrating, but the β-rays of tritium (33% <4 keV, 48% 4-10 keV, 19% 10-14 keV; Perry, 1962) or of [125]I (77% <4 keV, 23% 22-34 keV; Myers and Vanderleeden, 1960) have about the right characteristics, which are very much the same as those needed for high resolution autoradiography (see, e.g. Ada *et al.*, 1966). Because tritium has a long (12 year) half life compared with [125]I (60 days), really high specific activities are only obtainable with radio iodine, and this happens also to be the more convenient atom to introduce chemically. [125]I would thus be the most suitable choice on general grounds. If we assume conservatively that one-third of the electrons emitted from antigen bound at the surface will be captured effectively within the nucleus, and if there are N molecules of antigen, molecular weight M and specific radioactivity (in terms of γ radiation) R μCi/μg, the number of electrons captured in 24 hr would be 2.5 NMR x 10^{-9}. If for the purposes of illustration,

$N = 10^4$, $M = 10^5$ and $R = 10^3$ this would give 2500 electrons. There seems to be little hard information about the number of electrons which need to be captured within a nucleus to kill mouse cells. However, Kisieleski *et al.* (1964) using ^{3}H-thymidine incorporation into the nucleus of mouse spermatocytes, calculated that about 400 released electrons were enough to kill half these cells. Lymphocytes are, however, well recognized as being peculiarly sensitive to radiation damage. Schrek and Stefani (1964) examined the effect of X-rays on suspensions of purified suspensions of human peripheral blood lymphocytes *in vitro*, and concluded that 90% of resting lymphocytes were killed by a dose of 75 r and 95% by 150 r, whereas after stimulation with PHA about 30% survived 1200 r. It is difficult to relate such figures directly to electrons. The effect of 75 r would be to produce about 12,500 ion pairs in a mass of air equal to 10^{-10} g, about equal to the mass of a lymphocyte nucleus, but unfortunately we do not know what proportion of these ions would be effectively captured. Perhaps the most that can be stated on existing evidence is that the findings discussed below were not wholly unpredictable.

At the Prague Symposium on "Developmental Aspects of Antibody Formation and Structure" in 1969, Humphrey and Keller (1970) described experiments in mice using a synthetic polypeptide antigen (TIGAL) labelled with ^{125}I at very high specific radioactivities.

TIGAL was made by iodination of the tyrosine residues of the multichain polypeptide (T,G) -A-L 509 (Sela *et al.*, 1962) so as to introduce 200-320 atoms of iodine per molecule. (M.W. about 200,000, containing about 200 tyrosine residues). ^{125}I was introduced by direct iodination of (T,G) -A-L using the chloramine-T method on a micro scale, or by exchange of ^{125}I in TIGAL. This antigen was chosen because it has a large number of similar (though not all identical) antigenic determinants containing on average 2T and 4G residues at the end of its 100 side chains; it is a reasonably good immunogen in mice; and in mice (but not in rabbits) the antibody elicited is almost completely specific for the iodinated material. Since ^{125}I-TIGAL could be made with specific activities in excess of 2500 μCi/μg (sufficient for one molecule in one half life to yield about 50 silver grains by autoradiography), and the radioactive label is an essential part of the antigenic determinant groups, it offered the possibility of testing unequivocally whether antibody containing cells might contain the antigen or even antigenic fragments.

This purpose was frustrated because highly radioactive ^{125}I-TIGAL failed to elicit either primary or secondary antibody responses, although corresponding amounts of non-radioactive-TIGAL gave regular and reproducible responses. This failure was specific for the radioactive antigen, and was not attributable to general radiation damage since the antibody responses to other antigens introduced by the same route at the same time were unaffected. On subsequent further injection of non-radioactive TIGAL, the mice responded as though they had never received ^{125}I-TIGAL, i.e. they were neither made tolerant nor primed by the radioactive antigen. Humphrey and Keller (1970) concluded that the most probable explanation was that highly radioactive ^{125}I-TIGAL acted as a lethal antigen, and killed selectively those cells that came into contact with it and had sufficiently avid receptors to bind immunologically significant amounts. However, since the mice failed to respond not only to ^{125}I-TIGAL 0.2 μg at 850 μCi/μg, to 2 μg at 170 μCi/μg, to 4 μg at 60 μCi/μg or to 40 μg at 10 μCi/μg (though they gave small responses to 2 μg at 10 μCi/μg) they considered that a macrophage concentrating mechanism might be required to provide sufficient irradiation of adherent lymphocytes at the lower specific activities. A possibility which could not be completely discounted, though unlikely on the evidence, was that the highly radioactive TIGAL was not immunogenic because it destroyed itself too rapidly by self-irradiation.

At the same Symposium, Ada *et al.* (1970) (see also Ada and Byrt, 1969) reported experiments in which they treated normal mouse spleen cells *in vitro* with ^{125}I labelled polymerized flagellin from one or other of two different strains of *Salmonella*, transferred the cells to 750 r irradiated mice, and tested the ability of the recipients to make primary antibody responses to a mixture of the unlabelled flagellins. These experiments showed very elegantly that exposure of the cells to one radioactive antigen *in vitro* for 16-20 hr, under conditions where active uptake of the antigen was precluded, specifically abolished or significantly diminished the subsequent response to that antigen without affecting the response to the other. Ada and his colleagues concluded that these results implied that at least some of the lymphocytes that have been shown by autoradiography specifically to bind radioactive antigens (Naor and Sulitzeanu, 1967; Ada *et al.*, 1970; Humphrey and Keller, 1970) are cells that take part in an antibody response; that at any one time most or all of the cells in the mouse spleen that can be stimulated to produce antibody within eight days are able to react with antigen *in vitro*; that the specificity of inactivation is difficult to reconcile with any theory other than one requiring cell populations with a very restricted potential for antigenic stimulation; and that the ability to inactivate selectively particular cells in mixed populations should prove useful for assessing the contribution of such cells in the immune response.

Partly because we agree with these conclusions, and partly because Byrt and Ada's system offered a means of eliminating the possibility that ^{125}I-TIGAL was non-immunogenic *in vivo* because it was destroyed by self-irradiation, we have extended their studies with the antigens TIGAL and Keyhole limpet haemocyanin (KLH) employed in earlier studies. Spleen cells from (CBA x C57) F1 hybrid mice primed and boosted with TIGAL and KLH 8-13 weeks previously were incubated with various combinations of radioactive or non-radioactive TIGAL (2 μg/ml) or KLH (4 μg/ml) in medium 199 containing 10% fetal calf serum for 60 min at 0°C. The cells were separated from unattached antigen by centrifugation through a fetal calf serum gradient as described by Byrt and Ada (1969), and left for about 20 hr in 50% fetal calf serum at 4°C. Viability counts were made (usually 65-75% were viable by the trypan blue exclusion test) and known numbers of viable cells were injected i.v. or i.p. (in different experiments) into syngeneic mice irradiated one day beforehand with 700 r from a Co^{60} source. One day after receiving the donor spleen cells the recipients were injected i.p. with a mixture of 10 μg TIGAL plus 10 μg KLH. Blood samples taken 7 and 13 days later were assayed for anti-KLH by

TABLE I

Transfer of spleen cells after contact in vitro *followed by boost* in vivo

Cells transferred	Treatment in vitro	Boost in vivo	Mean responses to			
			TIGAL[a]		KLH[b]	
			7-day	13-day	7-day	13-day
3×10^7 primed	TIGAL* + KLH	TIGAL + KLH	1.0	1.1	4,000	34,000
3×10^7 primed	TIGAL + KLH*	TIGAL + KLH	3.0	25.0	45	830
3×10^7 primed	TIGAL + KLH	TIGAL + KLH	12.0	31.5	7,500	52,000
3×10^7 primed	nil	TIGAL + KLH	10	30	14,000	71,000
3×10^7 primed	nil	nil	3.3	6	10	160
no cells	nil	TIGAL + KLH	0	0	0	0

[a] per cent of antigen bound.
[b] haemagglutination titre.
Nine or more mice were used in each group. Asterisks indicate radioactive antigen.

haemagglutination of KLH-coated tanned formolized sheep erythrocytes, and for anti-TIGAL by an indirect antigen binding test (Janeway and Sela, 1967). The results were quite clear cut. When the specific activities of the radioactive TIGAL and KLH were 1000 and 420 μCi/μg respectively, and 30 million cells were transferred, pretreatment of the cells *in vitro* with radioactive antigen specifically prevented them from responding to the same non-radioactive antigen after transfer (Table I). There was some delay in the response to the other antigen, detectable in the seven-day responses in this and similar experiments, which is probably attributable to overall irradiation of the cell suspension before the excess of radioactive antigen was removed, but it was no longer significant by 13 days. These results confirm that Ada and Byrt's findings with normal spleen cells apply also to primed cells. They also indicate that TIGAL labelled at 1000 μCi/μg had retained its capacity to react with specifically responsive cells, and that the earlier *in vivo* findings of Humphrey and Keller were unlikely to be due to self destruction of the antigen.

We examined the distribution of antibodies between 19S and 7S by density gradient centrifugation of serum pools from each group. In the seven-day bleeds there were small but definite 19S antibody peaks against both antigens, but the great bulk of the antibody at seven days and all the antibody at 13 days was 7S. The cells inactivated by radioactive antigen were therefore probably potential 7S antibody producers.

We examined the effect of varying the specific radioactivity of TIGAL and KLH used for preincubation with primed spleen cells in a similar experiment to that outlined in Table I. The results (Table II) indicated that at the concentrations used the specific activity required to cause practically complete

TABLE II

Relation of specific radioactivity to suppressive effect of ^{125}I *antigens on response of primed mouse spleen cells*

Specific activity		% suppression of response at 14 days
TIGAL (2 μg/ml)	1000 μCi/μg	100
	310	>99
	75	70
	2	0
KLH (4 μg/ml)	420	100
	185	99
	38	90
	1.5	0

At least six mice were used as recipients for each group.

suppression of the antibody response 14 days after boosting lay between 100 and 200 μCi/μg for either antigen. In these experiments autoradiographs of smears of the cells exposed to ^{125}I antigens *in vitro* were made before injecting them into the irradiated recipients. About 0.6% of the cells exposed to TIGAL and 0.2-0.3% of those exposed to KLH were so heavily labelled as to be classified as "hedgehogs" (Humphrey and Keller, 1970) and probably represent those which would have been killed by the radiation.

Byrt and Ada (1969) showed that *in vitro* binding of radioactive flagellin and haemocyanin by normal mouse spleen lymphocytes could be inhibited by pretreatment with a polyvalent anti-Ig serum. Warner *et al.* (1970) have also shown that such pretreatment largely blocks the ability of normal mouse spleen cells adoptively to transfer the capacity to produce a primary response in syngeneic recipients against a variety of antigens. Anti-L chain and anti-μ chain antisera were effective, but not anti-α, anti-γ or anti-γ 2 heavy chains, suggesting that cells with IgM receptors were predominantly involved in these primary responses. It seemed to us possible that pretreatment with anti-immunoglobulin serum, by preventing attachment of radioactive antigens, might protect the cells against specific radiation damage, and that on subsequent transfer to irradiated hosts the cells might lose their coating of anti-Ig and recover their capacity to respond to the unlabelled antigen. We tested this in an experiment in which donor spleen cells from mice primed with TIGAL and KLH 12 weeks beforehand were pretreated at $0°C$ for 1 hr with normal rabbit serum or polyvalent rabbit anti-Ig or specific rabbit anti-μ chain (provided by Dr Michael Parkhouse), washed and then treated with various combinations of radioactive and non-radioactive TIGAL and KLH. The specific activities of the ^{125}I-TIGAL and KLH were about 150 and 75 μCi/μg respectively. Two days after transfer of 15 million viable cells to 700 r irradiated recipients they were boosted with 10 μg each of TIGAL and KLH, and blood samples were taken 7 and 18 days later. At 21 days they were boosted again with similar amounts of antigens and bled once more after 11 days (Table III). Pretreatment with anti-μ alone caused a six-fold diminution in the antibody responses to both antigens at seven days, but these had largely recovered by 18 days and completely by 32 days (11 days after the second boost); anti-Ig alone caused a 50-fold drop at seven days, and the subsequent anti-KLH levels never reached more than a quarter of the controls; ^{125}I-TIGAL alone almost completely suppressed the response to TIGAL up to 18 days, but it had recovered to about a quarter of the controls by 32 days; ^{125}I-KLH alone in this experiment caused only a 10-fold suppression of the response to KLH at seven days, which recovered to one quarter by 18 days. Successive treatments with anti-immunoglobulin and radioactive antigens not only gave no protection against the effects of the latter but caused a very definite increase over the suppressive effects of either treatment alone and greatly diminished the extent of recovery after 18 days. Since it was of interest to know whether 19S and 7S antibody responses would be differentially affected by pretreatment with anti-μ chain, all the 32-day and some of the 18- and 7-day sera containing sufficient antibody were pooled in groups and examined after density gradient ultracentrifugation. The detectable antibody at seven days was all 7S; by 18 days some 10% was 19S, except in the groups pretreated with anti-μ in which 19S antibody content was hardly detectable; but after the second boost, by 32 days all the groups had quite large 19S antibody contents, which in the two groups pretreated with polyvalent anti-Ig almost equalled the 7S antibody titres. This last observation requires further investigation. In general our findings in relation to already primed cells confirm those of Ada and his colleagues with adoptive primary responses made by normal spleen cells, although we have not yet tested the effect of specific anti-γ sera. *

* Two attempts have since been made to reproduce the effect of pretreatment of spleen cells from primed mice with anti-μ and anti-Ig in suppressing the antibody response to booster injections of TIGAL and KLH. In neither was significant suppression observed, even though the original antisera were used and the experiments included groups (other than those involving radioactive antigens) as similar as possible to those described in the preceding paragraph. We have no reason to suspect the experimental results recorded in Table III, but have at present no explanation for our failure to repeat them.

TABLE III

Effect of combined treatment with anti-Ig and radioactive antigen [a]

Group	Pretreatment in vitro	Treatment in vitro	Boost in vivo	anti-KLH			anti-TIGAL [b]		
				7d.	18d.	32d.	7d.	18d.	32d.
1	—	—	—	<25	35	600	1.6	2.3	12.4
2	—	no cells	TIGAL + KLH	<25	50	25	1.7	1.5	0.3
3	—	TIGAL + KLH	TIGAL + KLH	4600	200,000	58,000	23	41	42
4	Normal rabbit serum	TIGAL + KLH	TIGAL + KLH	4800	500,000	47,000	22	48	49
5		TIGAL + KLH*	TIGAL + KLH	390	44,000	12,000	14	37	41
6	—	TIGAL* + KLH	TIGAL + KLH	1200	75,000	15,000	1.6	4	34
7	anti-μ	TIGAL + KLH	TIGAL + KLH	800	120,000	47,000	7	37	46
8	anti-μ	TIGAL* + KLH*	TIGAL + KLH	34	9,000	3,400	0	1.3	25
9	anti-Ig	TIGAL + KLH	TIGAL + KLH	100	31,000	16,000	1.7	32	46
10	anti-Ig	TIGAL* + KLH*	TIGAL + KLH	<50	250	2,200	0	1.2	31

* 125I TIGAL 150 μCi/μg; 125I KLH 75 μCi/μg. 15 million cells were transferred. Each group contained six recipients irradiated with 700 r. They were boosted with antigen 2 and 21 days after cell transfer.

a Part of these results have not proved repeatable. See footnote p. 128.

b A 10% difference in antigen binding corresponded to about a four-fold difference in antibody content.

It is interesting, and at first sight puzzling that specific anti-μ chain antibody should suppress what would have been largely a 7S response. However, in view of the evidence of Greaves and Hogg (in this Symposium) that receptors on helper cells react with anti-μ or anti-L sera, it would not be surprising if the helper function had been suppressed in this experiment. Assuming, as seems likely, that co-operation can occur in respect of different determinants on the same complex antigen, inhibition of helper cells would be expected to diminish the antibody response even though the antibody-forming cells would make 7S antibody. The fact that pretreatment with anti-μ not only caused an initial inhibition of the whole response but also a subsequent relative inhibition of the 19S component after recovery can be interpreted as indicating that potential 19S producing cells had also been affected.

The results reported so far have only a speculative bearing on the problems of cell co-operation, since they provide no indication of whether the cells which react with and are inactivated by radioactive antigens are derived from thymus or bone marrow. An experiment to test this has indicated that the spleen cells from primed mice that bind large amounts of antigen *in vitro,* in the case of one antigen but not of another, are probably not thymus-dependent. Spleen cell suspensions from mice primed against TIGAL and spider crab haemocyanin (MSH) were treated with anti-θ serum (prepared by Dr M. Raff in AKR mice) and complement or with normal AKR serum and complement. Anti-θ lysed the θ positive (and presumably thymus-dependent) cells, and the proportion destroyed was estimated by comparison with the controls. The cell suspensions were then incubated in the cold with highly radioactive TIGAL or MSH, washed through fetal calf serum gradients, and autoradiographs were done on smears. The results of counting the proportion of very heavily labelled cells are given in Table IV. The calculated proportion of "hedgehogs" in the case of TIGAL was just that expected if θ positive cells were not involved; however, in the case of

TABLE IV

Antigen uptake by spleen cells of primed mice

Antigen	Cell treatment	"Hedgehogs"	"Hedgehogs" + "not quite Hedgehogs"
TIGAL	Normal AKR serum	70/40,000	122/40,000
	anti-θ serum	117/40,000	200/40,000
MSH	Normal AKR serum	58/18,000	90/18,000
	anti-θ serum	30/18,000	47/18,000

Calculated incidence, assuming that heavily labelled cells are unaffected by anti-θ

TIGAL		118/40,000	204/40,000
MSH		71/18,000	108/18,000

MSH it appeared that the antigen binding cells include more than half that are θ positive. This may be significant in view of the findings relating to the carrier effect with MSH discussed below. Some additional evidence is provided by a brief report by Naor et al. (1970) of autoradiographic studies indicating that peripheral blood leucocytes from a child with congenital agammaglobulinaemia contained far fewer, and from an adult with acquired agammaglobulinaemia significantly fewer, cells capable of binding ^{125}I-labelled BSA and/or guinea pig serum albumin than were present in the blood of corresponding normal controls. Since thymus-dependent lymphocytes are little if at all impaired in agammaglobulinaemia, this would suggest that lymphocytes capable of reacting to a detectable extent with these two antigens are largely bone marrow-derived.

We have also done some experiments in collaboration with Dr Brigitte Askonas to test the effect of preincubation with highly radioactive ^{125}I-MSH on the capacity of cells from mice primed with MSH to co-operate with cells from mice primed with DNP-ovalbumin, so as to produce a secondary response to the hapten following a booster injection of DNP-MSH. Although these experiments are only preliminary, they are sufficiently germane to the theme of this Symposium to report them. The system used was essentially that devised by Mitchison (1968). Donor C3H mice were primed by a single injection i.p. of 100 μg alum precipitated MSH or DNP-ovalbumin with 10^9 killed H. pertussis organisms. Two months later spleen cell suspensions from the donors, either separate or mixed were injected i.p. into irradiated (660 r) syngeneic recipients, together with either DNP-ovalbumin or DNP-MSH. The antibody responses of each mouse to DNP and to MSH were measured 10-21 days later. In some groups the MSH primed cells were pretreated with high specific activity ^{125}I-MSH in order to test whether their ability to act as co-operating cells and/or their ability to make anti-MSH antibody would be impaired by specific irradiation. The control for this group was to pretreat the DNP-ovalbumin primed cells with ^{125}I-MSH under the same conditions.

In the first experiment, involving 12 recipient mice in each group, the number of donor cells transferred was relatively small (5×10^6) and the amounts of antigen used for boosting were 1 μg DNP-MSH or 10 μg DNP-ovalbumin. Antibody against DNP was barely detectable by 14 days after transfer and boosting but readily detectable by 21 days. The 21-day levels are shown in Table V. They demonstrate the expected co-operation effect of MSH primed cells and DNP-ovalbumin primed cells in the DNP antibody response to DNP-MSH, but they also show (group 5) that preincubation of MSH primed cells with ^{125}I-MSH inhibited both co-operation in respect of formation of anti-DNP antibody and formation of antibody to MSH. When the DNP-ovalbumin primed cells were preincubated with ^{125}I-MSH (group 6) there was no such inhibition. The results of this experiment imply that both the co-operating cells that recognize the MSH and the cells that are involved in making antibody to MSH had receptors able to bind sufficient ^{125}I-MSH to inactivate them. However, an

TABLE V

Effect of pretreatment with ^{125}I MSH on co-operative effect of MSH primed cells in low cell dose transfers

Group	Priming of donor cells (5×10^6)	Treatment *in vitro*	Challenge	Antibody at 21 days anti-DNP[a]	anti-MSH[b]
1	DNP-OV	−	1 μg DNP-MSH	0	0
			10 μg DNP-OV	31	0
2	DNP-MSH	−	DNP-MSH	22	6
3	MSH	−	DNP-MSH	0	10
4	DNP-OV	−	DNP-MSH	18	14
	MSH	−	DNP-OV	13	0
5	DNP-OV	−	DNP-MSH	0	0
	MSH	MSH*	DNP-OV	11	0
6	DNP-OV	MSH*	DNP-MSH	35	24
	MSH	−	DNP-OV	19	0
7	Normal	−	DNP-MSH	0	

Twelve mice were used as recipients per group—each serum was assayed individually.
* ^{125} I MSH 210 μCi/μg, 5 μg/ml.
[a] Geometric mean of per cent DNP hapten bound by serum diluted 1/6.
[b] Geometric mean of μg MSH bound per ml of serum.

experiment similar in most respects, except that the number of cells transferred to the recipients was 20×10^6 and the amounts of antigen used for boosting were 10 μg both of DNP-MSH or DNP-ovalbumin, gave results which were interestingly different. These are shown in Table VI. With the larger number of cells antibody was already present at 10 days; production of antibody against MSH was, as expected, inhibited but this time there was no indication that pretreatment of MSH primed cells with ^{125}I-MSH had impaired their ability to co-operate with DNP-ovalbumin primed cells, since a good response to DNP was elicited by DNP-MSH (group 3). Three explanations, which are not mutually exclusive, can be offered for these apparently discrepant findings. One is that both helper and potential antibody-forming cells possess sufficient receptors to bind lethal amounts of ^{125}I-MSH, but that even dead or moribund helper cells can co-operate, provided that the anti-DNP response takes place sufficiently rapidly (as in the second experiment). A second is that there are more helper cells than potential antibody-forming cells, and that even if the great majority of both are killed the surviving helper cells from 20×10^6 cells, though not from 5×10^6 cells, are sufficient to co-operate effectively.* A third is that the helper

* The evidence that helper cells are normally present in excess is discussed in Mitchison's paper in this symposium.

TABLE VI

Effect of pretreatment with ^{125}I MSH on co-operative effect of MSH primed cells with DNP primed cells in high dose transfer

Group	Priming of donor cells (20 × 10⁶)	Treatment in vitro	Challenge	Antibody at 10 days anti-DNP	anti-MSH
1	DNP-OV	—	10 μg DNP-MSH	0	0
			10 μg DNP-OV	25	
2	DNP-OV	— ⎫	DNP-MSH	16	86
	MSH	— ⎭	DNP-OV	17	
3	DNP-OV	— ⎫	DNP-MSH	28	0
	MSH	MSH* ⎭	DNP-OV	15	
4	DNP-OV	MSH* ⎫	DNP-MSH	7	150
	MSH	— ⎭	DNP-OV	3	

* ^{125}I-MSH 400 μCi/μg, 5 μg/ml.
Twelve mice were used as recipients in each group and antibody was assayed as in experiment of Table V.

cells are less sensitive to ^{125}I-MSH, perhaps because they bind fewer antigen molecules, and that enough survive in the larger but not in the smaller quantity of cells to be effective. Further experiments will be needed to decide between these alternatives.

The antigens that we have employed may not be typical of all antigens, because TIGAL has a large number of similar determinant groups attached to its poly-lysine backbone, and both MSH and KLH are large polymeric molecules which must have several identical determinants. Such molecules are likely to bind to cells with numerous receptors, especially if these are multivalent, by a multi-point attachment which would not be readily reversed by washing the cells or by dilution of the antigen. Even if each receptor is not particularly avid the whole antigen will bind quite firmly (as suggested, e.g. by Pasanen and Mäkelä, 1969; Frank and Humphrey, 1969), whereas smaller antigens such as monomeric BSA, which may contain only one representative of each determinant per molecule (e.g. Humphrey, 1964), are likely to bind less firmly without some additional process, such as macrophage presentation, to align them.

The possible importance of such considerations was suggested by an experiment done in collaboration with Dr N. A. Mitchison, to determine whether cells capable of binding labelled BSA would be detectable by autoradiography in the spleens or lymph nodes of mice paralysed with BSA, according to various schedules ranging from low zone to high zone paralysis. Although it might be expected at first sight that detectable antigen binding cells would be absent from animals fully tolerant of that antigen, Ada *et al.* (1970)

reported that the spleen cells of rats made tolerant to flagellin or to haemocyanin contained the same proportion of cells able to bind labelled antigen in the cold (1 : 10,000 and 1 : 1100 respectively) as those of normal rats. Humphrey and Keller (1970) found that in mice made tolerant from birth to TIGAL or MSH (by frequent injection of relatively large amounts of antigen) the proportion of very heavily labelled cells was one-third to one-fifth that in normal mice. Naor and Sulitzeanu (1969) found in one experiment that BSA binding cells were absent from 10,000 cells examined from BSA tolerant mice, but this was not a constant finding. In the experiment to be described, spleen or lymph node cells from normal mice, or adult mice made tolerant by 600 r followed by 3 x weekly injections of 10,100, 1000 or 10,000 μg BSA (Mitchison in this Symposium) were washed, incubated in the cold with precentrifuged ^{125}I-BSA (about 40 μCi/μg) washed thoroughly through a normal rat serum gradient and spread on slides for autoradiography. After exposure for one half life 5000-10,000 cells were scanned for grains with an oil immersion lens. Only "live" cells (about 50%, both by trypan blue exclusion and by staining and morphological characteristics) were scored. The findings are given in Table VII, from which it appears that the paralysed mice contained numbers of antigen binding cells similar to normal mice and possibly more cells with >50 grains, though in view of the small numbers involved this may not be significant. These results could mean that paralysis, whether by high dose or low dose régimes, does not involve elimination of antigen binding cells (whether thymus-dependent

TABLE VII

Effect of different paralysis régimes on the number of mouse cells binding detectable amounts of precentrifuged ^{125}I BSA (40 μCi/μg)

Paralysis régime		Cells per 1000 with	
(μg BSA 3 x weekly)		10-50 grams	51-150 grams
0	L. Node	0.22	−
	Spleen	0.6	−
10	L. Node	<0.13	−
	Spleen	0.5	0.33
100	L. Node	0.33	−
	Spleen	1.8	0.27
1,000	L. Node	1	−
	Spleen	0.22	0.32
10,000	L. Node	0.16	−
	Spleen	0.7	0.2

5-10,000 "live" cells were scored under the high power lens after exposure for 62 days.

or bone marrow-derived, or both). They could also mean that the labelled cells, which never resembled what Humphrey and Keller termed "hedgehogs", were irrelevant as far as the immune response to BSA is concerned and that the truly relevant ones were not observed. From calculations based on the autoradiographic efficiency of ^{125}I (Ada *et al.*, 1966) even a cell with 120 grains would only have about 400 attached molecules of BSA. This number is much smaller than the 40,000 or more molecules of flagellin which Byrt and Ada (1969) calculate were attached to their antigen reactive cells, or the more than 200,000 β-galactosidase molecules which Sercarz and his colleagues (this Symposium) observed on their antigen binding cells. Either the attachment of BSA to cell receptors is reversible during the washing process, and consequently the autoradiographic method underestimates the extent of binding when in contact with BSA, or immunologically relevant cells with large numbers of receptors were too rare in any of the cell suspensions described in Table VII for them to be observed in scanning 10,000 cells. The first possibility might be tested by using polymerized ^{125}I-BSA instead of monomeric BSA, and the second by scanning many more cells.

ACKNOWLEDGEMENTS

Some of these experiments are the result of many helpful discussions with Dr B. A. Askonas, and we are greatly indebted to her. We are also very grateful for the excellent technical assistance of Mrs R. M. deRossi, Mr G. F. Kenny, and Misses M. J. Bilbie and J. A. Gorham.

SUMMARY

1. Spleen cells from mice primed with two antigens, keyhole limpet haemocyanin (KLH) and fully iodinated (T,G) -A-L (TIGAL), were treated *in vitro* with the antigens labelled with ^{125}I at very high specific radioactivity, and tested for their capacity to transfer adoptive antibody responses in irradiated syngeneic mice. Pretreatment with a radioactive antigen specifically inhibited the expected response to challenge with that antigen, which was largely 7S antibody. Complete inhibition was obtained when antigen-specific radioactivities were 200 $\mu Ci/\mu g$ or more.

2. Pretreatment *in vitro* of primed spleen cells with specific anti-μ or polyvalent anti-Ig caused temporary suppression of both 7S and 19S antibody responses by the transferred cells, suggesting that cells with μ chain receptors may co-operate in 7S antibody production. Additional pretreatment with radioactive antigens increased and prolonged the inhibition. *

* Part of these results have not proved repeatable. See footnote p. 128.

3. Spleen cells from mice primed with TIGAL and *Maia squinada* haemocyanin (MSH) were treated with normal AKR serum or AKR anti-θ serum and complement, and the proportion of cells that bound large amounts of labelled TIGAL or labelled MSH was examined by autoradiography. Virtually all cells binding TIGAL but less than half those binding MSH were found to be θ negative.

4. Co-operation between spleen cells from mice primed with DNP-ovalbumin and from mice primed with MSH was tested by transfer to irradiated syngeneic recipients, which were challenged with DNP-MSH, and testing for antibody production against DNP and against MSH. Pretreatment *in vitro* of the MSH primed cells with highly radioactive MSH suppressed the formation of antibody to MSH; the co-operative anti-DNP response was also suppressed, but only when the number of cells transferred was small. Some possible explanations of the differential susceptibility of antibody producing and co-operating MSH-sensitive cells are suggested.

5. An autoradiographic study was made of cells able to bind radioactive BSA *in vitro* from lymph nodes and spleens of normal mice and mice made tolerant of BSA by various regimens. Only cells binding small amounts of BSA were observed, and these were present in all groups. The significance of this finding is questioned.

REFERENCES

Ada, G. L. and Byrt, P. (1969). *Nature, Lond. 222*, 1291-1292.
Ada, G. L., Humphrey, J. H., Askonas, B. A., McDevitt, H. O. and Nossal, G. J. V. (1966). *Expl. Cell Res. 41*, 557-572.
Ada, G. L., Byrt, P., Mandel, T. and Warner, N. L. (1970). *In* "Developmental Aspects of Antibody Formation and Structure" (J. Šterzl and I. Říha, eds). Academic Press, New York.
Byrt, P. and Ada, G. L. (1969). *Immunology 17*, 503-516.
Frank, M. M. and Humphrey, J. H. (1969). *Immunology 17*, 237-247.
Humphrey, J. H. (1964). *Immunology 7*, 462-473.
Humphrey, J. H. and Keller, H. U. (1970). *In* "Developmental Aspects of Antibody Formation and Structure" (J. Šterzl and I. Říha, eds). Academic Press, New York.
Janeway, C. A., Jr. and Sela, M. (1967). *Immunology 13*, 29-38.
Kisieleski, W. E., Samuels, L. D. and Hiley, P. C. (1964). *Nature, Lond. 202*, 458-459.
Mitchison, N. A. (1968). *In* "Recognition of Antigen in Differentiation and Immunology" (K. B. Warren, ed.). Academic Press, New York.
Myers, W. G. and Vanderleeden, J. C. (1960). *J. Nuclear Med. 1*, 149.
Naor, D. and Sulitzeanu, D. (1967). *Nature, Lond. 214*, 687-688.
Naor, D. and Sulitzeanu, D. (1969). *Int. Archs Allergy appl. Immun. 36*, 112-113.
Naor, D., Bentwich, Z. and Cividalli, G. (1970). Submitted for publication.
Pasanen, V. J. and Mäkelä, O. (1969). *Immunology 16*, 399-407.

Perry, R. P. (1962). *In* "Methods in Cell Physiology" (D. M. Prescott, ed.),
    Vol. 1, p. 324. Academic Press, New York.
Schrek, R. and Stefani, S. S. (1964). *J. natn. Cancer Inst.* 32, 507-521.
Sela, M., Fuchs, S. and Arnon, R. (1962). *Biochem. J.* 85, 223-235.
Warner, N. L., Byrt, P. and Ada, G. L. (1970). *Nature, Lond.* 226, 942-943.

Origin of Antigen Binding Cells in Mice Tolerant to *E. coli* Polysaccharide

OLOF SJÖBERG

Division of Immunobiology
Karolinska Institutet
Wallenberglaboratory
S-10405 Stockholm 50, Sweden

INTRODUCTION

An immune response involves the appearance of both antibody secreting and antigen binding cells. Antibody secreting cells can be counted *in vitro* using the local haemolysis in gel technique (Jerne and Nordin, 1963), and are then called plaque-forming cells (PFC). Antigen binding cells can be visualized by their ability to bind particulate antigens, such as sheep red cells (SRBC) to their surfaces, and have been named rosette-forming cells (RFC) (Nota *et al.*, 1964; Zaalberg, 1964). It is likely that RFC are a mixture of antibody secreting and non-antibody secreting cells.

Howard *et al.* (1969) studied the RFC response in the spleens of mice tolerant to pneumococcal polysaccharide. Although serum antibodies could not be detected in these animals a considerable number of RFC was found. In fact there were more RFC in "tolerant" mice than in optimally immunized animals. They explained these findings on the basis that antibody secreting cells existed, but that the humoral antibodies were neutralized extracellularly by persisting undegraded antigen. The cellular basis for tolerance towards another poly-saccharide, of *E. coli* origin, has been examined in this study.

EXPERIMENTAL DESIGN AND RESULTS

Two to five months old CBA mice were used. The antigen was alkali detoxified polysaccharide from *E. coli* 055 : B5 (CPS) (Britton, 1969). Tolerance was induced by five 3 mg doses of CPS given over 14 days, and maintained by weekly injections of 3 mg. The animals tolerant to CPS had a normal PFC response to SRBC.

The number of PFC in the spleen was tested by the assay of Jerne and Nordin (1963) as modified by Mishell and Dutton (1967). CPS was conjugated to SRBC as described by Möller (1965). Since CPS stimulates 19S antibody synthesis exclusively for a prolonged time period only direct PFC were studied (Britton and Möller, 1968).

RFC were detected by a slight modification of the centrifugation-resuspension method of McConnell *et al.* (1969). The CPS was adsorbed to CBA mouse erythrocytes by the same method as to SRBC (Möller 1965).

Anti-theta serum was obtained by retro-orbital bleeding of AKR mice that had been repeatedly injected with CBA thymocytes. Spleen cells were treated with anti-theta serum as described by Greaves and Möller (1970).

Spleen cells from normal, immune and tolerant mice were assayed for the presence of PFC and RFC. Immune animals were assayed five days after an injection of 0.01 mg of CPS, at the peak of the PFC response (Britton, 1969). One of the tolerant groups, "tolerant mice A" was assayed five days after the last tolerance maintaining dose of 3 mg of CPS. The other tolerant group, "tolerant mice B", received 0.01 mg of CPS seven days after the last tolerance injection, and was tested five days later.

In non-immune mice the mean number of PFC was $0.6/10^6$ spleen cells and of RFC $115/10^6$ spleen cells. Optimally immunized animals showed a mean of 154 $PFC/10^6$ spleen cells and of 382 $RFC/10^6$ spleen cells (Table 1).

TABLE I

Cellular immune response in spleens of mice immunized with E. coli *polysaccharide*

	$PFC/10^6$ spleen cells	$RFC/10^6$ spleen cells
Normal mice	-0.23 ± 0.23 (0.6)	2.06 ± 0.07 (115)
Tolerant mice A	-0.23 ± 0.22 (0.6)	2.21 ± 0.14 (164)
Tolerant mice B	0.66 ± 0.25 (4.5)	2.37 ± 0.04 (235)
Immune mice	2.19 ± 0.08 (154)	2.58 ± 0.06 (382)

Figures are given in $logs_{10}$ for the mean ± standard error. Antilogs of the means in parentheses. Tolerant mice A were tested five days after the last tolerance maintaining dose of 3 mg CPS. Tolerant mice B were given 0.01 mg of CPS seven days after the last injection of 3 mg of CPS and assayed five days later. Immune mice were tested five days after injection of 0.01 mg of CPS.

In tolerant mice A the appearance of PFC was completely inhibited, with the exception of one mouse in which a small response was detected. However, the RFC tended to increase in number as compared to the unimmunized mice, with a mean of 164 $RFC/10^6$ spleen cells, although the difference from normal mice was not significant.

Tolerant mice B had a mean number of 4.5 $PFC/10^6$ spleen cells and 235 $RFC/10^6$ spleen cells. The number of RFC in this group was significantly increased over background values ($P < 0.005$), but was not as high as in optimally immunized mice.

Treatment with anti-theta serum caused little or no inhibition of the RFC against CPS (Table II) in tolerant or in optimally immunized mice. In contrast to this, RFC against SRBC in mice injected with SRBC five days earlier were inhibited by 20-32% after the same treatment.

TABLE II

% Inhibition by anti-theta serum treatment of spleen cells

A mice immunized to CPS	8, −2, 3
B mice tolerant to CPS	−4, 18, −16, −2
C mice immunized to SRBC	27, 20, 32, 32, 26

A mice were given 0.01 mg CPS five days before assay.

B mice were injected with 0.01 mg CPS seven days after the last tolerance maintaining dose and assayed five days later.

C mice were immunized with 4×10^8 SRBC five days before assay.

DISCUSSION

These findings cannot be explained on the basis suggested by Howard *et al.* (1969), i.e. that the tolerant state is due to neutralization of antibodies by excess antigen. Thus tolerance to endotoxin appears to inhibit the steps leading to the appearance of antibody secreting cells, but does not affect those resulting in RFC to the same extent. In fact Howard has recently reinvestigated the immune status in mice tolerant to pneumococcal polysaccharide, and has also found antigen binding cells in situations where the antibody secreting cells are paralyzed by very high antigen doses (personal communication). A number of different alternatives could explain the findings.

1. The receptors on the RFC in tolerant mice may be directed against different antigenic determinants from the antibodies secreted by the PFC in immune animals. These determinants may occur in such a low density on the coated SRBC that antibodies against them do not cause plaques. This possibility cannot be ruled out, but seems unlikely since Humphrey and Dourmashkin (1965) found that a single hit by a 19S antibody molecule can lyse a red cell.

2. Tolerant animals may have cells secreting antibody directed against the same determinants as the antibodies producing plaques in immune animals, but of very low affinity (Theis and Siskind, 1968). Cells producing antibody of such low affinity might be detected as RFC but not as PFC. This explanation seems unlikely, however, if one considers that receptors on the RFC must have an affinity sufficiently high to bind the antigen during the suspension of the cell pellet. It is assumed in this discussion that the antibodies produced have the same properties as the receptor on the precursor cells for antibody production

(Mäkelä *et al.*, 1969). Furthermore, the finding of Möller (1968) that 50-100% of a spleen cell population may occasionally be detected as PFC, suggests that most of the antibody secreting cells are in fact detected as PFC.

3. Tolerance induction leads to blockage of antibody secretion without affecting division and differentiation of precursor cells for antibody production. It can be argued against this possibility that RFC are fewer in tolerant than in optimally immunized mice. This would not be the case if tolerance meant simply inhibition of secretion of antibodies.

4. Tolerance may be caused by inhibition of division and differentiation of precursor cells for antibody production. In that case the RFC in tolerant mice may be formed by thymus-dependent lymphocytes (T cells) or by bone marrow-derived lymphocytes (B cells) that do not give rise to PFC. Recent findings have shown that T cells and B cells co-operate in the antibody response to certain antigens (Mitchell and Miller, 1968). The lipopolysaccharide antigen of *E. coli* does not need T cells to induce a humoral antibody response (Andersson and Möller, personal communication). This does not necessarily imply that there are no receptors for this antigen among the T cells. The thymus-independence of an adequate immune response to CPS may be related to the fact that the structure of CPS is characterized by repeating units of identical determinants. Such an antigen may bind well to bone marrow precursor cells with multivalent receptors and stimulate them to antibody production without the presence of T cells. Then T cells might be stimulated to respond in tolerant animals. Such cells would be RFC but not PFC, and might therefore be the explanation for the finding of RFC in tolerant animals.

Recently it has been found that a high proportion of the RFC in mice immune to SRBC are T cells, using isoantigenic markers (H-2, theta) on T cells (Greaves and Möller, 1970). As shown above anti-theta serum inhibited few or none of the RFC against CPS. The RFC against CPS are therefore likely to be B cells not T cells. It cannot be decided at this stage whether the lack of theta-positive cells in immune animals is due to lack of receptors for CPS on T cells, or to induction of tolerance or absence of stimulation of T cells.

It is possible that a high dose of antigen will only stop division of B cells whose receptors for the antigen are above a certain level of affinity. Such cells would give rise to PFC after an optimal immunization dose. B cells with receptors of lower affinity would be stimulated to divide, but not to produce antibodies, after optimal stimulation, since the energy of binding is not sufficiently high. In tolerant animals these cells are not prevented from dividing because they escape tolerance due to the low affinity of their receptors. Such cells could therefore be the explanation for the RFC in tolerant animals. If this alternative is correct it would also imply that a level of affinity sufficiently high to bind the antigen to the cell surface is not necessarily high enough to induce antibody production.

The lack of thymus-dependent RFC to CPS may explain some features of the response to this antigen. T cells are unable to increase the response. Low dose tolerance to CPS could not be obtained. This is to be expected if low dose tolerance is caused by inactivation of T cells, as suggested by Mitchison (personal communication). Long lasting memory, which generally is attributed to T cells, is poor for CPS. Finally the poor IgG production to CPS might be explained by the lack of thymus-dependent RFC, since the production of IgG has been shown to be more dependent on T cells than the production of IgM (Wortis et al., 1969).

ACKNOWLEDGEMENTS

I want to thank Miss Gun Stenman for excellent technical assistance. The work was supported by the Swedish Medical Research Council, the Swedish Cancer Society, the Anders Otto Swärds Stiftelse and the Damon Runyon Memorial Fund (DRG-1038).

SUMMARY

The numbers of antigen binding cells (rosette-forming cells, RFC) and antibody producing cells (plaque-forming cells, PFC) against polysaccharide from *E. coli* were studied in normal, immune and tolerant mice.

Optimal immunization increased the number of PFC from $0.6/10^6$ spleen cells in unimmunized mice to $154/10^6$ spleen cells, whereas the RFC increased only three times. The tolerant mice had no or only a marginal increase in PFC compared to normal mice. However, tolerant mice generally had more RFC than normal mice.

It was not possible to inhibit the formation of RFC to polysaccharide in immune or tolerant mice by treatment with anti-theta serum. Different alternative explanations for the increase in RFC in tolerant mice compared to unimmunized mice are discussed.

REFERENCES

Britton, S. (1969). *Immunology 16,* 513-526.
Britton, S. and Möller, G. (1968). *J. Immun. 100,* 1326-1334.
Greaves, M. F. and Möller, E. (1970). "Cellular Immunology". (In press.)
Howard, J. G., Elson, J., Christie, G. H. and Kinsky, R. G. (1969). *Clin. exp. Immun. 4,* 41-53.
Humphrey, J. H. and Dourmashkin, R. R. (1965). *In* "Complement" (G. E. W. Wolstenholme and J. Knight, eds), Ciba Fdn Symp., p. 175. Churchill, London.
Jerne, N. K. and Nordin, A. A. (1963). *Science, N.Y. 140,* 405.

Mäkelä, O., Cross, A. M. and Ruoslahti, E. (1969). *In* "Cellular Recognition" (R. T. Smith and R. A. Good, eds), pp. 287-294. Appleton-Century-Crofts, New York.

McConnell, I., Munro, A., Gurner, B. W. and Coombs, R. R. A. (1969). *Int. Archs Allergy appl. Immun. 35*, 209-227.

Mishell, R. and Dutton, R. (1967). *J. exp. Med. 123*, 423-442.

Mitchell, G. F. and Miller, J. F. A. P. (1968). *Proc. natn. Acad. Sci. U.S.A. 59*, 296-303.

Möller, G. (1965). *Nature, Lond. 207*, 1166-1168.

Möller, G. (1968). *J. exp. Med. 127*, 291-306.

Nota, N. R., Liacopoulos-Briot, M., Stiffel, C. and Biozzi, G. (1964). *C.r. Acad. Sci. Paris 259*, 1277.

Theis, G. A. and Siskind, G. W. (1968). *J. Immun. 100*, 138-141.

Wortis, H. H., Dresser, D. W. and Anderson, H. R. (1969). *Immunology 17*, 93-110.

Zaalberg, O. B. (1964). *Nature, Lond. 202*, 1231.

DISCUSSION

NOSSAL. I want to say that I consider by far the most likely hypothesis the one which states that tolerance has only been induced for cells with high affinity receptors. We all know that the induction of tolerance is dose dependent, and that in partial tolerance, low avidity clones are the ones which survive. It seems to me very clear that in what we call complete tolerance, there may be immunization of very low avidity cells, which, in a simplistic sense, "see" the antigen dose as a lower, immunogenic one. Such cells may be secreting immunoglobulin, but of insufficient quality to give a plaque. Essentially, it does not matter for this argument whether there are "T" cells, not secreting; or "M" cells, secreting. In either case, we could have rosettes but not plaque formation.

Antigen Binding Sites on Mouse Lymphoid Cells

M. F. GREAVES and N. M. HOGG

National Institute for Medical Research
Mill Hill, London

Recent evidence suggests (see *Transplant Rev.*, 1969) that the humoral antibody response to many antigens involves collaboration between different cell types. Although the nature of these cellular interactions is uncertain, it seems likely that one function is to serve as a concentrating device to present antigen to the precursors of antibody secreting cells that would otherwise react sub-optimally.

Cellular co-operation has been analysed almost exclusively in mice using heterologous red cells, albumins and hapten-protein conjugates as antigens. Experiments with chromosomal and allogeneic antigen markers on immunoglobulins and cell surfaces have established that secreted antibody is the product of cells derived from bone marrow precursors (B cells) (Mitchell and Miller, 1968; Nossal *et al.*, 1968; Taylor *et al.*, 1966; Davies *et al.*, 1967). The degree of response of these cells is, however, dependent upon a "helper" function of other cells, namely macrophages and thymus-derived/processed lymphocytes (T cells). Whereas the macrophage probably functions in a non-specific way with respect to antigen, the T cells are almost certainly specifically antigen reactive, although the evidence for this is equivocal. It is also likely that the role of the T cell in collaboration is linked to the capacity of these cells to initiate cellular hypersensitivity and transplantation reactions (Roitt *et al.*, 1969). The hypothetical receptor or antigen binding site(s) on T cells has been termed IgX, where X stands for an *unknown* class of immunoglobulin (Mitchison, 1969).

Both untreated "normal" mice and mice injected with sheep red blood cells (SRBC) have antigen binding cells (ABC) that can be identified by the rosette technique (Nota *et al.*, 1964; Zaalberg, 1964). While many of the reactive cells in immunized mice are undoubtedly antibody *secreting* cells and a small proportion are macrophages, a considerable number appear not to be actively secreting antibody. Thus when the rosette test is performed at 37°C some reactive cells bind multiple layers of SRBC, whereas others bind only a single layer and are presumably either minimally secreting or not secreting at all (Duffus and Allan,

1969; Greaves *et al.*, 1970). The latter group are primarily lymphocytes and "blast" cells and may be reacting with SRBC via antibody-like receptors associated with the plasma membrane.

With the use of antisera to cell surface allo-antigens (H-2, theta) it can be demonstrated that both T and B lymphoid cells bind SRBC (Greaves and Möller, 1970). The effect of pretreating mouse spleen cell suspensions with anti-theta (θ) serum plus complement prior to carrying out the rosette test is shown in Table I. The results indicate that a considerable proportion of ABC carry the

TABLE I

Effect of AKR anti-CBA thymus (anti-θC3H) serum on ABC and direct PFC in the spleens of CBA mice immunized with SRBC

Days after SRBC immunization	ABC/10^6 cells[a]			D.PFC/10^6 cells Mean $\pm$ SE		
	Anti-θ	NS	% inhibition	Anti-θ	NS	% inhibition
0	528	889	41	NT		
0	376	723	48	NT		
3	1111	1852	40	10.1 $\pm$ 1.8	9.6 $\pm$ 1.3	−5.2
5	5343	8628	38[b]	114.8 $\pm$ 17.1	97.3 $\pm$ 3.4	−17.9
6	4282	5613	24	NT		
25	2199	3125	29	NT		

[a] Between 20,000 and 100,000 spleen cells were scanned in each sample.

[b] Eleven experiments were carried out with spleen cells from animals injected 5-6 days earlier, the mean inhibition of ABC with anti-θ serum was 36 $\pm$ 2%. Taken from Greaves and Möller (1970) NS = normal AKR serum.

θ-antigen, in contrast to antibody secreting cells (plaque forming cells), which were unaffected by the antiserum and are presumably θ negative. A similar result was obtained with anti-H-2 sera in a system in which irradiated animals were reconstituted with semi-syngeneic thymus and syngeneic bone marrow mixtures (Greaves and Möller, 1970).

Rosette-forming cells can be inhibited by pretreatment with anti-immuno-globulin sera (Biozzi *et al.*, 1967; McConnell *et al.*, 1969). This system thus affords a means of analysing antigen binding sites on the two major categories of mouse lymphoid cells.

The anti-immunoglobulin sera used in this study are listed in Table II. Several methods have been used to assay specificity (Table III), and all results discussed in this paper were obtained using sera that were assayed in two or more test systems and shown to be specific. Full details of the methods for raising anti-immunoglobulin sera and evidence for their specificity will be presented elsewhere (Greaves, 1971).

TABLE II

Anti-immunoglobulin sera used to inhibit ABC

Antisera[a]		Immunogen
Directed against	Raised in	
IgA	Rabbits	47A
IgM	Rabbits	MOPC-104E
IgG_1	Rabbits	RPC-23
IgG_{2a}	Rabbits	5563
IgG_{2b}	Rabbits	MOPC-141
Fab	Rabbits	Fab 5563
$L_{(\kappa)}$	Rabbits	L chains from serum IgG
$L_{(\lambda)}$	Rabbits	L chains from MOPC-104E
Ig-1[a]	C57B1	CBA anti-pertussis antibodies
Ig-1[b]	Balb/c	C57B1 anti-pertussis antibodies

(In the Immunogen column, the first five rows — 47A, MOPC-104E, RPC-23, 5563, MOPC-141 — are bracketed together as **Balb/c myelomas**.)

All rabbit sera tested were absorbed with insoluble Fab and/or myeloma proteins of other classes. With some sera soluble myeloma proteins were used for absorption. These were added in a stepwise fashion until no further precipitation occurred. Control experiments indicate that soluble antigen-antibody complexes do not inhibit ABC.

All sera were absorbed with sheep and mouse erythrocytes.

[a] All rabbit sera tested were from pools derived from 2 to 12 animals.

TABLE III

Specificity controls for anti-immunoglobulin sera

1. Immunoprecipitation in agar
2. Immunofluorescence/tumour cells
3. Co-precipitation of labelled antigen
4. Additivity of combinations of antisera in PFC assays
5. Selective absorptions

In initial experiments we have studied the effect of anti-Fab and anti-light chain sera on spleen cells from both uninjected mice and mice immunized with SRBC. In both situations virtually 100% inhibition of ABC could be obtained with these reagents. It is interesting to note that a considerable degree of inhibition was obtained with antisera specific for λ light chains. This was rather surprising since immunoglobulins with λ light chains constitute less than 5% of the serum immunoglobulins (McIntyre and Rouse, 1970), but might be explained by analogy with results in other species which have indicated that some antigens may select particular immunoglobulin types (Nussenzweig and Benacerraf, 1966; Weiler and Weiler, 1968).

The effect of absorbing these sera with various immunoglobulin preparations is shown in Table IV. The inhibiting effect of both anti-Fab and anti-light chain sera could be removed by absorption with IgG in either a soluble or insoluble form. After such absorption the anti-light chain sera still showed some residual antibody activity in agar gel precipitation tests against free light chains (but not against Fab or IgG). This was to be expected as the sera were raised against free light chains, and presumably contained antibodies against the so-called "hidden" determinants that are not detectable on Fab (Takahashi *et al.*, 1969).

TABLE IV

Inhibition of ABC with anti-Fab and anti-light chain sera

Directed against	Antisera Absorbed with	Maximum % ABC inhibition
Fab	—	99
	Insoluble 5563[a]	2
	Insoluble IgG	−2
	Insoluble BSA	98
	Soluble IgG[b]	11
	$L_{(\kappa)}$ chains[c]	98
$L_{(\kappa)}$ chains	Soluble MOPC-315 (λIgA myeloma)	85
$L_{(\lambda)}$ chains	Insoluble Fab 5563	26
	Insoluble Fab$_\gamma$	26
L chains	Insoluble IgG	0
L chains	Insoluble BSA	92

Spleen cells obtained from mice 5 to 20 days after a single injection of 10^8 SRBC i.p. Proteins were made insoluble using glutaraldehyde (Avramaes, 1969).

[a] 0.5 ml of antiserum passed twice through a 5 × 1 cm column of Sepharose-linked 5563 (κ IgG$_{2a}$ myeloma) (Porath *et al.*, 1967).

[b] 50% ammonium sulphate precipitate of mouse serum passed through DEAE cellulose.

[c] Prepared from 5563.

Absorption with light chains did not appear to reduce the maximal degree of inhibition obtained with the anti-Fab serum.

The results can be interpreted to suggest that effectively all ABC have Fab determinants on their surface. It is therefore probable that θ positive (T cell) ABC have immunoglobulin-like antigen binding sites. This is in accord with evidence from other experimental systems, which has been reviewed recently (Greaves, 1970).

A considerable proportion of ABC can also be inhibited by pretreatment with antisera to class-specific heavy chains. Up to 75% inhibition can be obtained with individual sera depending on the time after immunization (Greaves, 1971). However, since θ positive ABC are almost always a minority it is difficult in these situations to find out whether the θ positive cells do in fact have receptors that are similar to any of the five serum immunoglobulin heavy chain classes. The rosette system is also inappropriate for analysing the immunoglobulin class of antigen binding sites, since many reactive cells are probably actively secreting antibody and so one may not be detecting "receptors" as such, but antibody molecules in the process of being secreted.

To overcome these difficulties we have attempted to select for non-secreting ABC by the following two methods: filtration of reactive cells through cotton wool (i.e. removal of adhesive cells) and inhibition of antibody secreting cells by injection of anti-SRBC antibody. These techniques are known not to inhibit antigen sensitive cells (Wigzell, 1967; Uhr and Möller, 1968).

As Table V shows both of these techniques produce suspensions of ABC that are almost completely free of secreting cells (plaque forming cells). In the cotton wool filtration experiments macrophages are also eliminated. It is interesting to note that both of these methods produce some selection in favour of θ positive ABC, although it is clear that many minimally or non-secreting θ negative (B?) ABC are reactive. A similar selection of non-secreting antigen binding populations using "passive" antibody treatment has been previously reported (Horibata and Uhr, 1967).

TABLE V

Selection of antigen binding cells

Treatment	PFC[a] % reduction/ 10^6 cells	ABC		
		% reduction/ 10^6 cells	% $\theta+$[b] remaining	% non-secreting[c] remaining
Control	0	0	32	64
Plus hyperimmune anti-SRBC serum (day 3)	92	41	69	91
Cotton filtered	99	50	85	>90

All cells harvested 7-30 days after injection of SRBC. Results are pooled data of five experiments.

[a] Direct (19S) and indirect (7S) plaque forming cells (PFC) were similarly affected.

[b] Calculated as described previously (Greaves and Möller, 1970).

[c] Single and multiple layered ABC enumerated using the technique of Duffus and Allan (1969).

Results of inhibition studies on non-selected cell suspensions using anti-heavy chain sera are shown in Table VI. The spleens of "normal" mice contain both θ positive and θ negative ABC. These cells can be inhibited by antisera to Fab and light chains but very little suppression is observed with antisera to IgA, IgG_1, IgG_{2a} and IgG_{2b}. This is in contrast to the degree of inhibition obtained with these sera when tested against immune cells harvested 3-6 days after injection of

TABLE VI

Inhibition of ABC with anti-heavy chain (class-specific) sera

Days post SRBC	% inhibition with antisera directed towards:							
	IgA	IgG_1	IgG_{2a}	IgG_{2b}	Ig-1[a]	Fab	κ	λ
0 (i.e. "normal")	<10	<10	<10	<10	<10	>90	85	25
3	66	26	26	22	21	>90	–	–
5/6	33	48	47	41	45	>90	85	26

The maximum degree of inhibition was obtained when rabbit antisera were diluted 1 in 20 and anti-allotype sera 1 in 5.

SRBC. It is also clear that with immune cells the sum of the inhibitions is greater than 100. In additivity studies with purified anti-IgG_1 and anti-IgG_{2a} antibodies and in anti-allotype inhibition assays on Ig-1[a/b] heterozygotes, there is also a clear lack of summation of the effects of individual antibodies (Greaves, 1970). We are currently investigating the significance of these observations. Our results so far indicate that the high degree of inhibition obtained with individual sera, and the lack of summation, cannot be ascribed to macrophages or cross reactions of sera, but is probably due to lymphoid cells with more than one class or allotype of immunoglobulin on their surface. We have not yet eliminated the possibility that such cells have absorbed antibody secreted by other cells.

These results also illustrate a danger inherent in this type of experiment. One might anticipate that a cell with more than one class of antibody on its surface would *not* be prevented from binding antigen by antisera to any individual class, since the remaining antibodies/receptors would still be free to react. This appears not to be the case, and it would therefore seem that antisera to immunoglobulins on the cell surface can block adjacent antibody molecules by steric hindrance (compare Boyse *et al.,* 1968). The problems this raises with respect to receptor identification are self-evident.

When anti-IgA and anti-IgG sera were applied to *selected* ABC only small amounts of inhibition were obtained (Table VII), suggesting that the majority of these cells including reactive T cells do not have antigen binding sites with exposed determinants in common with IgA or IgG.

TABLE VII

Inhibition of selected ABC with anti-heavy chain (class-specific) sera

Selection method	% inhibition with antisera directed towards:							
	IgA	IgG$_1$	IgG$_{2a}$	IgG$_{2b}$	Ig-1[a]	Fab	κ	λ
Passive antibody suppressed mice (day 7-14)	17	6	12	2	15	>90	81	29
Cotton filtered cells (day 11-30)	12	<5	0	<5	–	>90	83	26

Both anti-κ and anti-λ sera gave some inhibition although the majority of selected ABC would appear to have antigen binding sites with κ light chains.

The results with anti-IgM have been variable, in so far as three sera showed considerable inhibiting capacity against various ABC populations while another was considerably less active. Comparative results with two of these antisera are given in Table VIII. Only a small proportion of ABC in "normal" mice could be inhibited by anti-IgM, although the degree of suppression was greater than that obtained with anti-IgA or anti-IgG.

TABLE VIII

Effect of anti-IgM sera on ABC

Spleen cells	% inhibition of ABC with anti-IgM sera	
	AS.3	AS.5
Normal	15	5
Immune day 3	60	18
Immune day 5	76	53
Immune day 40	55	0
Immune passive antibody-suppressed	97	56
Immune cotton filtered	85	26

A high percentage of ABC could be inhibited by both of the anti-IgM sera when the cells were obtained from mice injected 3-5 days previously with SRBC. This period coincides with the peak of IgM secreting (plaque forming) cells. Later in the immune response when the IgM secreting cell level has declined there was a discrepancy in the results with the two sera, only one of which was inhibitory. When these sera were tested against selected ABC almost complete inhibition was observed with AS.3. A similar degree of suppression has been observed with three of the sera so far tested. The inhibitory capacity of AS.3 was removed by absorption with MOPC-104E (the λ light chain IgM myeloma protein used to raise the antiserum), but not with MOPC-315 (a λ chain IgA myeloma), α_2 macroglobulin, Fab_γ or λ light chains from MOPC-104E.

As θ positive cells constitute the majority of selected ABC it would seem that many T cells may have antigen binding sites on their surface with κ light chains and μ heavy chains. The majority of antigen binding B cells (θ negative) selected in this way may also have IgM-SRBC receptors.

One might consider why θ positive ABC from "normal" mice are not inhibited by anti-IgM. One possibility that we cannot yet exclude is that the IgM on selected immune θ positive ABC is in fact passively absorbed antibody, that has been synthesized and secreted by other (B?) cells. It is important in this context to remember that early cytophilic antibody in mice is IgM (Tizard, 1969). We have performed a number of control experiments involving attempts to induce "passive" ABC by *in vitro* or *in vivo* treatment with anti-SRBC serum. These results have suggested that although such passively reactive cells may exist in immunized animals they probably do not constitute more than 10% of the total ABC population. Clearly however this is a critical point that requires further investigation.

If we assume that IgM detected on the surface of ABC does represent a product of synthesis of those same individual cells, then an alternative explanation for the difference between θ positive ABC in "normal" and immune mice can be proposed (see Fig. 1). It might be that "resting" T cells in "normal" mice have cell surface antigen binding sites so aligned in the plasma membrane that most, if not all, of the Fc piece of the molecule is buried. Antigen binding reactions by such cells would only be inhibited by anti-Fab antibodies. If, following activation by antigen (SRBC), changes occurred at the cell surface that resulted in a greater exposure of receptor molecules then antigen binding might be inhibited, not only by anti-Fab, but also by antibodies to Fc determinants, particularly if the latter were in the "hinge"-region.

It is interesting to consider this possibility in the light of the results of Warner *et al.* (1970). They found that virtually all cells from "normal" mice that bound iodinated flagellin or haemocyanin could be inhibited by both anti-light chain *and* anti-IgM sera. In our studies only 15% of ABC in "normal" mice could be inhibited by anti-IgM. These differences might be explained in terms of cell populations. Flagellin and haemocyanin are relatively thymus-independent antigens in mice as compared to SRBC. It seems possible that a greater proportion of ABC reacting against the former antigens might be B cells. It is interesting to note in this context, that anti-θ serum has practically no inhibiting effect on antigen binding reactions of cells from mice immunized against endotoxin—another relatively thymus-independent antigen (E. Möller and O. Sjöberg, personal communication). If B cells represent the majority of ABC in

Antigen Binding Cells

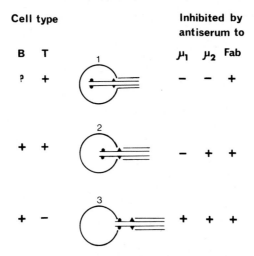

Fig. 1. Inhibition of ABC by anti-immunoglobulin sera. A scheme to represent the effects of three variables.

A. Cell type involved: (i) T (Thymus-derived/processed); (ii) B (Bone marrow-derived/non-processed).

B. State of activation: (1) Resting; (2) Activated; (3) Secreting.

C. Specificities of antiserum used, i.e. antibodies to different parts of the immunoglobulin molecule: (i) Fab determinants (light chain and/or Fd); (ii) ▲ "Hinge" region determinants on the H chain (i.e. destroyed by papain) (μ_2); (iii) ◢ C terminal end determinants on H chain (μ_1).

the study of Warner *et al.* (1970), then their inhibition by anti-IgM might reflect a greater exposure of antigen binding sites as compared to "resting" T cells. In this respect these reactive cells might resemble activated T cells (see Fig. 1).

It should also be considered that antisera to any one immunoglobulin class might vary in their content of antibodies to different determinants, and that this might affect their capacity to inhibit ABC. For example, in light of the above suggestions, a serum with a high content of antibodies to "hinge"-region determinants might be more inhibitory than another antiserum raised against the same class, but which contained fewer of such antibodies. Antibodies to heavy chain determinants near the C terminal end of Fc might be inhibitory only when assayed against *secreting* cells.

It seems likely that cell surface immunoglobulin receptors for antigen will be associated with the plasma membrane via the Fc piece. The degree of exposure of the molecules may depend on the cell type involved (**T** or **B**) and its state of activation. There may also be differences in the expression of receptors of different immunoglobulin classes (Pernis, 1969).

ACKNOWLEDGEMENTS

This work was supported by the Medical Research Council. We should particularly like to thank Dr G. Torrigiani for providing and testing many anti-Ig sera.

SUMMARY

Both "normal" and immunized mice have spleen cells that bind sheep red cells to their surface (= antigen binding cells, ABC). Both T and B cells are reactive.

Non- or minimally-secreting ABC can be selected for by filtering reactive spleen cell suspensions through cotton wool, or by "passive" antibody suppression of the humoral response.

The majority of both T and B ABC appear to have μ chain antigen binding sites.

It is suggested that activation of T cells may involve increased exposure of Ig molecules on the cell surface.

REFERENCES

Avramaes, S. (1969). *Immunochemistry 6*, 43-52.
Biozzi, G., Stiffel, C. and Mouton, D. (1967). *In* "Immunity, Cancer and Chemotherapy" (E. Mihich, ed.), pp. 103-131. Academic Press, New York.
Boyse, E. A., Old, L. J. and Stockert, E. (1968). *Proc. natn. Acad. Sci. U.S.A. 60*, 886-893.
Davies, A. J. S., Leuchars, E., Wallis, V., Marchant, R. and Elliott, E. V. (1967). *Transplantation 5*, 222-231.
Duffus, W. P. H. and Allan, D. (1969). *Immunology 16*, 337-347.
Greaves, M. F. (1970). *Transplant. Rev. 5*, 45-75.
Greaves, M. F. (1971). To be published.
Greaves, M. F. and Möller, E. (1970). *Cellular Immunology 1*, 372-385.
Greaves, M. F., Möller, E. and Möller, G. (1970). *Cellular Immunology 1*, 386-405.
Horibata, K. and Uhr, J. W. (1967). *J. Immun. 98*, 972-978.
McConnell, I., Munro, A., Gurner, B. W. and Coombs, R. R. A. (1969). *Int. Archs Allergy appl. Immun. 35*, 209-227.
McIntyre, K. R. and Rouse, A. M. (1970). *Fedn. Proc. 29*, 704.
Mitchell, G. F. and Miller, J. F. A. P. (1968). *J. exp. Med. 128*, 821-837.
Mitchison, N. A. (1969). *In* "Immunological Tolerance" (M. Landy and W. Braun, eds). Academic Press, New York.
Nossal, G. J. V., Cunningham, A., Mitchell, G. F. and Miller, J. F. A. P. (1968). *J. exp. Med. 128*, 839-853.
Nota, N. R., Liacopoulos-Briot, M., Stiffel, C. and Biozzi, G. (1964). *C.r. Acad. Sci. Paris 259*, 1277.
Nussenzweig, V. and Benacerraf, B. (1966). *J. exp. Med. 124*, 805-818.

Pernis, B. (1969). Presented at the Anniv. 500th meeting of the Biochemical Society, London.

Porath, J., Axén, R. and Ernback, S. (1967). *Nature, Lond. 215*, 1491-1492.

Roitt, I. M., Greaves, M. F., Torrigiani, G., Brostoff, J. and Playfair, J. H. L. (1969). *Lancet ii*, 367-371.

Takahashi, M., Yagi, Y. and Pressman, D. (1969). *J. Immun. 102*, 1268-1273.

Taylor, R. B., Wortis, H. H. and Dresser, D. W. (1966). *In* "The Lymphocyte in Immunology and Haemopoiesis" (J. M. Yoffey, ed.), pp. 242-244. Edward Arnold (Publishers) Ltd., London.

Tizard, I. R. (1969). *Int. Archs Allergy appl. Immun. 36*, 332-346.

Transplant Rev. (1969). *1.*

Uhr, J. W. and Möller, G. (1968). *Adv. Immun. 8*, 81.

Warner N. L., Byrt, P. and Ada, G. L. (1970). *Nature, Lond. 226*, 942-943.

Weiler, E. and Weiler, I. J. (1968). *J. Immun. 101*, 1044-1058.

Wigzell, H. (1967). *Cold Spring Harb. Symp. quant. Biol. 32*, 507-516.

Zaalberg, O. B. (1964). *Nature, Lond. 202*, 1231.

Beta-galactosidase Binding by Thymus and Marrow Cells: Relationship to the Immune Response

E. SERCARZ, J. DECKER, D. De LUCA
R. EVANS, A. MILLER and F. MODABBER

Department of Bacteriology, University of California
Los Angeles, California 90024, U.S.A.

INTRODUCTION

The advantages and necessities of cell co-operation have been pointed out already at this symposium. In imagining an ideal partitioning of functions, it may be instructive to consider the desirable features of each partner in performing the two functions of antigen presentation and antibody production.

For the helper function of antigen presentation the organism must have sufficient cells available for reactivity with the majority of antigens, and must then present any particular antigen in an appropriate topological matrix to the relevant cell. Otherwise the risk of tolerance might be assumed, since free antigen may act as a competitive inhibitor of appropriately presented antigen in the induction process. No real need exists for whole antibody molecules to be synthesized by these cells.

The associated function of antibody production demands an adequate synthetic machinery and some means of recognizing antigen (receptors). Whereas it would seem of importance for presentation cells to bear their antigen tightly, it is more essential for the synthetic precursor to be able to couple antigen-receptor contact to a trigger, which activates cell division and the antibody production machinery.

Each partner should be able to amplify its signal considerably. Clonal expansion of antibody-forming cell precursors should accomplish this on the synthetic limb; a similar expansion might affect the density of antigen and matrix formation on presentation, or helper cells. Perhaps a redistribution of receptors to other cells would be a more rapid means of amplification for presenting cells.

Five years ago our laboratory began to study the binding of an enzyme by the immune system. This potentially could be measured at the molecular level. Boris

Rotman (1961) showed that microdrops containing single beta-galactosidase (BGz) molecules could be detected with a fluorogenic substrate, fluorescein-di-beta-galactoside (FDBG). Since antibody did not interfere with the enzymatic activity of BGz, it seemed to us that antibody molecules should also be detectable with great sensitivity.

MATERIALS AND METHODS

BGz from *E. coli* strain ML308 has been provided generously by I. Zabin and A. Fowler, UCLA. We are grateful to the following people for gifts of serum: anti-mouse Ig (class specific)–J. Fahey, NIH; anti-mouse, $Ig_{2a\text{-}b}$–L. Herzenberg, Stanford; anti-mouse kappa-chain–E. Lennox, Salk Institute; anti-rabbit class and allotype sera–C. Todd, City of Hope Hospital.

What we measure is the ability of cells from various organs to bind the BGz molecule. Cells from the appropriate organ are washed in TPF-tris-phosphate buffer, pH 7.2, with 5% fetal calf serum (GIBCO; Oakland, Calif.). They are then mixed with a saturating amount of enzyme, 50 μg/ml, for 1 hr at ice-bath temperature. After incubation, the cells are washed extensively. FDBG is added at a final concentration of 2×10^{-5} M. The appearance of fluorescence is followed with time as an indication of the activity of the preparation.

Usually we distribute cells in droplets of 0.2 μl in slide chambers under mineral oil (Nujol) and measure the fluorescein produced by cell-bound enzyme with a photomultiplier microphotometer (American Instrument Co.) after passage through the optical system of a Zeiss fluorescence microscope (Sercarz and Modabber, 1968). The droplet size can easily be varied for greater sensitivity, or when a mass assay is desirable, when we measure the reaction in tubes with a Turner Model 111 fluorometer.

In the determination of frequency of antigen-binding cells, a visual scanning technique is used in which a very few positive droplets (containing 10-25 cells each) are readily seen among thousands of inactive droplets, when a Kodak Wrattan No. 15 filter is used in goggles as the secondary filter (Modabber *et al.,* 1970a).

EXPERIMENTAL

In this report we would like to focus on several questions relating to the binding of BGz by thymus and bone marrow cells of unimmunized mice and rabbits, and also to consider the change in distribution of the antigen-binding units after antigen injection.

1. How many cells have antigen-binding activity? A larger number of cells can be comfortably scanned by the use of the above technique than has been attempted with other techniques (Sulitzeanu and Naor, 1969; Byrt and Ada,

1969). Table I shows the rather low frequency of BGz-binding thymus and marrow cells in the A/J mouse. Cells that were *not* incubated with BGz were also examined to look for intrinsic enzyme activity, presumably due to mouse beta-galactosidase. Under our conditions intrinsic activity is not a complicating factor. The values for enzyme-incubated cells represent pools from at least four different experiments. They clearly indicate that unimmunized thymus and bone marrow both contain cells capable of binding *E. coli* BGz.

TABLE I

BGz-binding by unfixed thymus and marrow cells from uninjected A/J mice

Cell source	Incubation with BGz (50 µg/ml)	No. of cells assayed	No. of ABC	ABC/10^6 nucleated cells
Thymus	+	42×10^5	21	5
	−	5×10^5	0	0
Bone marrow	+	15×10^5	15	10
	−	5×10^5	0	0

Antigen-binding cells (ABC) determined by the A-H assay-enzymatic fluorogenic group hydrolysis with fluorescein-di-beta-galactoside as substrate.

2. How many receptors are present on active cells? Droplets containing putatively active cells from an initial screening were followed at intervals for several hours. Readings can be plotted as in Fig. 1, which shows four thymus droplets and four marrow droplets each containing 10 cells, and presumably one active cell. Microphotometer units are readily convertible to numbers of antigen molecules. It should be noted that our limit of detection using the relatively large 0.2µl droplets is about 10,000 BGz molecules per cell. The final concentration of fluorescein in the droplets containing a cell with lower levels of bound enzyme is not great enough to permit the visual discrimination over background that is required by the scanning test.

Although the values for enzyme molecules bound per cell (Table II) are distributed over a broad range, it is evident that the median marrow cell bears four times as many receptor molecules as the median thymus cell. Some cells in each group bind nearly one million molecules of BGz. (The activity in background droplets is very low.) This is a large number, which exceeds the values reported by Naor and Sulitzeanu (1968) for BSA, or Byrt and Ada (1969) for flagellin. The number of TIGAL molecules bound by the "hedgehogs" (exhibiting the highest binding in the terminology of Humphrey and Keller (1970)) is not known to us. We feel that our higher value can be attributed to

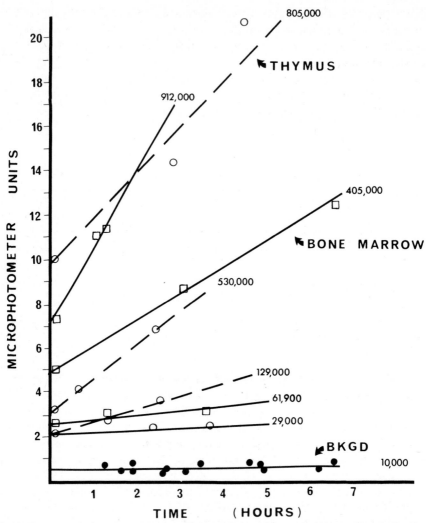

Fig. 1. Activities of individual unfixed thymus and bone marrow cells from uninjected mice. Activities are plotted as arbitrary microphotometer units against time. The number of BGz molecules bound, calculated from the slopes of the curves, are shown on the right. ○--○ thymus, □--□ bone marrow, ●—● background.

working under conditions of saturation, where further increases in antigen concentration bring about no increase in the number of cells with binding ability (De Luca *et al.*, 1970).

The cell surface of our most active cells would appear to be half-covered by bound BGz. Such an estimate is based on the arbitrary assumptions that a lymphocyte of 8 μ diameter has a billiard ball surface, and that an equal number

TABLE II

Number of BGz molecules bound by unfixed cells from uninjected mice

Cell	BGz molecules ($\times 10^3$)	
	Thymus	Marrow
1	29	62
2	38	131
3	80	238
4	95	316
5	98	372
6	117	378
7	129	405
8	530	510
9	805	610
10		912
Median	98	375
Mean background	0.31	0.47

of BGz molecules (125 Å x 125 Å x 50 Å) are on edge and are resting flat on the cell surface. Such a cell should have room for 2×10^6 BGz molecules and 2×10^7 Fab binding sites. (The unreasonableness of the lymphocyte as a billiard ball is evidenced by immunofluorescence studies (Modabber *et al.*, 1970b) showing substantial uncovered areas on the surface of antigen-binding normal thymocytes.) Thus in very active cells a steric factor may be implicated in denying access of BGz to all potential binding sites. It is perhaps of interest that we have never found a cell with more than 2×10^6 BGz molecules attached (Modabber and Sercarz, 1970b).

3. Can we detect all cells able to bind BGz? We had originally studied antibody-containing cells using this method, and had fixed the cells to increase the availability of internal antibody to BGz (Sercarz and Modabber, 1968; Modabber and Sercarz, 1967, 1970a, b). To determine whether all cells with binding sites were being detected in unimmunized mouse thymus and marrow, we used the fixation method shown in Fig. 2, essentially that of de Petris *et al.* (1963).

The increase in the number of active cells in the thymus was dramatic as can be seen in Table III. The top line represents a summation of all the experiments before 1970. Each other entry is an individual experiment and demonstrates the reproducibility of the results.

Values obtained by the A-H assay (using fluorogenic FDBG) and the BIG assay (in which 5-bromo-4-chloroindol-3-yl-β-D-galactoside substrate gives rise to a precipitable blue product) were similar. Thus 100 times more cells could bind BGz under these conditions than in the unfixed state.

Cell suspension: 1. Incubate at 37°C for 30 min.
2. Wash three times in TPF.
3. Fix in 4% formalin for 15 min.
4. Enzyme binding—freeze, thaw into BGz, incubate for 1 hr at 4°C.
5. Wash four times in TPF.
6. Cell count.
7. Assay—A-H or BIG.

Fig. 2. Method of fixation of cells before testing for BGz-binding by A-H or BIG assays. TPF = Tris-phosphate buffer, pH 7.2, with 5% fetal calf serum. The initial incubation at 37°C was designed to allow detachment of cytophilic antibody in immunized animals, and may be omitted.

TABLE III

Frequency of antigen-binding cells in the fixed mouse thymus

Assay	No. of cells assayed	No. of ABC	$ABC/10^6$ nucleated cells
BIG	243×10^5	16,400	674
	10.7×10^5	638	596
	9.0×10^5	569	632
	10.0×10^5	1,039	1039
	2.7×10^5	66	248
	10.3×10^5	809	785
	2.5×10^5	195	780
A-H	3.7×10^5	207	552
	1.25×10^5	36	288
	2.5×10^5	156	624

A-H assay see Table I. BIG assay—using 5-bromo-4-chloroindol-3-yl-β-D-galactoside, which gives a precipitable blue product.

The results are even more striking in the marrow. Table IV gives frequencies of positive cells in the marrow, spleen and lymph nodes. In these experiments 0.3-2% of all marrow nucleated cells were positive. We were troubled by this unusual number of binding cells, especially when one considers that probably only a quarter of the nucleated cells were lymphoid. It should be emphasized that intrinsic enzyme was negative in all fixed cells.

4. Are the BGz-binding cells immunologically specific? Specificity does not ensure relevance. Nevertheless we have been able to show that the binding of antigen by our cells can be prevented with (a) immunologically cross reactive, but enzymatically inactive proteins (Cz) and (b) appropriate anti-immunoglobulin reagents.

TABLE IV

BGz-binding by fixed cells from uninjected mice

Assay	Thymus	ABC/10^6 nucleated cells		
		Marrow	Spleen	Lymph node
BIG	598	3,100	7200	610
A-H	624		1800	
BIG	780	4,200		
A-H	288	3,900		
BIG	632	22,000	1200	680

Intrinsic activities were zero in all cases. Assays, see Tables I and III. ABC = antigen-binding cells.

Inhibition experiments were performed with fixed cells by the non-quantitative BIG assay, which only gives the reduction in cell numbers binding detectable antibody, and sometimes also by the quantitative A-H tube assay. Inhibitor was added at the same time as enzyme, at the concentrations shown in the tables. Table V shows, in several experiments with Cz, that the binding of BGz can be inhibited effectively even at rather low Cz/BGz ratios. Much higher excesses of cold flagellin were required by Byrt and Ada (1969) to inhibit the binding of radio-labelled flagellin, again suggesting non-saturation. Inhibition has also been attempted with an extract from an amber mutant (Uz) with a mutation very early in the Z gene. Such an extract gave minimal inhibition. The experiments with Cz merely show that the binding entity finds Cz, but not other proteins, similar to BGz.

That the binding entity is immunoglobulin in nature and presumably

TABLE V

Inhibition of BGz-binding by immunologically related materials

Cell source	ABC/10^6 nucleated cells: BIG assay		
	Control	Cz	Uz
Thymus	785	648	935
Thymus	527	14	210
Thymus	487	137	467
Marrow	1,130	520	
Marrow	25,800	425	

Inhibitors (Cz and Uz, see text) were added together with BGz at a protein concentration 10 times that of control, normal enzyme. BIG assay see Table III. ABC = antigen-binding cells.

antibody-like is shown in Table VI. An extensive study of this problem will be found elsewhere (Decker *et al.*, 1970). Here we present data with the best of our inhibitory antisera. An anti-mouse kappa serum essentially inhibited all the marrow binding activity, in an experiment where the highest number of active marrow cells were evident. A polyvalent antiserum is also shown. Inhibition with pure anti-Ig_{2a-b} serum and with other class-specific antisera was not significant or reproducible.

TABLE VI
Inhibition of BGz-binding rabbit by anti-immunoglobulin sera

| Inhibitor | $ABC/10^6$ nucleated cell: BIG assay | |
	Thymus	Marrow
None	487	25,800
Polyvalent anti-mouse gamma globulin	212	0
Anti-mouse kappa chain	161	529
Normal rabbit serum	450	28,700

Fixed cells. Antisera and BGz added together. BIG assay see Table III. ABC = antigen-bindiing cells.

5. Are these binding cells immunologically relevant? The crucial problem after the demonstration of specific antigen-binding, is relating this fact to subsequent immune reactivity.

Earlier experiments using the A-H system (Modabber and Sercarz, 1970a) had shown a rapid doubling or tripling (before 12 hr) in the BGz-binding activity of A/J mouse spleen, which remained at this plateau until two days after primary antigen injection. Then it started climbing to a peak achieved at about the sixth day.

In experiments being carried out for the most part in Boston (Modabber, Morikawa and Coons, personal communication) the thymus binding activity has been followed after BGz injection. An experiment carried out in Los Angeles will illustrate the type of result obtained (Table VII). The typical 12 hr rise

TABLE VII
Change in distribution of BGz-binding cells after immunization in mice

| Treatment | $ABC/10^6$ nucleated cells: BIG assay | | |
	Thymus	Marrow	Spleen
None	632	21,800	1160
12 hr after 100 μg BGz i.p.	185	10,400	5300

Pools of three mice were used. BIG assay see Table III. ABC = antigen-binding cells.

occurs in the spleen; this is concomitant with a decrease in thymus and marrow antigen-binding cells (ABC). Modabber and Coons (1970b) have carried out thymectomy experiments which point to the thymus as the agent responsible for the splenic rise. If a mouse is thymectomized a day before antigen injection, the 12 hr rise in the immunized spleen does not occur. Since there is approximate parity in the number of lymphoid cells in thymus and spleen, the thymectomy experiment says that the departure of 45,000 cells from the thymus leads to an increase of about 415,000 ABC in the spleen. A critical point for understanding this result will be the determination of the binding activity per cell in the 12 hr spleen. However, disregarding the quantitative aspects, which will be discussed later, these experiments can be taken to show that even the cells that can only bind antigen after fixation (the vast majority) are in some way related to immunization.

The departure of the antigen-reactive cells from their resting place upon immunization has previously been shown in the rabbit (Singhal and Richter, 1968). *Does the rabbit show the same pattern of BGz-binding in thymus and marrow as the mouse?* We have only looked at about a dozen rabbits to date, but several results are worth summarizing. First, thymus cells of the adult rabbit, either fixed or unfixed, do not bind BGz. The marrow cells bind at a level comparable to the mouse. Second, in the spleen, anti-b4 light chain allotype serum did not inhibit in experiments where polyvalent serum or anti-heavy chain allotype serum inhibited well. Third, experiments similar to Richter's have been carried out in four pairs of rabbits (see Table VIII) in which marrow binding was examined three to five days post-immunization. At such a time, there was a reduction (mean 56%) in the marrow ABC of injected animals relative to normals, again establishing a connection between immunization and antigen-binding.

TABLE VIII

Reduction in the number of BGz-binding cells in rabbit marrow 3-5 days after immunization

ABC/10^6 nucleated cells: BIG assay	
Normal	After immunization
12,400	7500
8,930	4400[a]
6,700	2430
10,960	2950
9,750 Mean	4320 Reduction 56%

All rabbits received 10 mg BGz i.v. except [a] which received 1 mg. Rabbits were tested in pairs. ABC = antigen-binding cells.

Discussion

We would like to address ourselves to two general points in this discussion: (1) the large numbers of BGz-binding cells in the marrow of mice and rabbits; (2) the question of the amplification of the presentation function.

At first sight it is an alarming number of cells that can bind BGz. We were worried about the possibility of a grossly artifactual situation. The arguments against this idea can be listed: (a) The binding activity is localized in relatively few cells and not distributed homogeneously over the entire population. (b) The results with two different substrates, FDBG and BIG, using different assays are in quantitative agreement. (c) The binding of BGz is inhibited by Cz, a cross reacting enzymatically inactive protein, but not by other proteins in *E. coli* extracts or fetal calf serum. (d) The binding is inhibited by anti-immunoglobulins, with only particular specificities of sera being effective. (e) The number of antigen-binding cells is related to antigenic experience. A large number of the binding cells leave the central lymphoid organs upon injection of antigen. (f) The functional absence of antigen-reactive cells to several antigens from rabbit thymus (Abdou and Richter, 1969) is apparently true also for our BGz by direct binding tests. Thus thymus cells of all sorts don't possess an unusual (carbohydrate?) BGz binding site. (g) The thought that all rabbits and mice are hyperimmunized to BGz is belied by the very characteristic primary response. There is a definite increase in binding activity in the spleen over the first 12 hr, a plateau of about 36 hr, and finally a second increase. All of our uninjected mice show such a response, whereas primed mice give a steady increase in binding activity from day 0 to a peak at about day 2. Nevertheless we are still studying this point. (h) Several other remote possibilities will not be explored here, such as a class of cells in the marrow that accumulates specific antibodies, or a non-specific attachment caused by fetal calf serum (but see (c) and (d) above).

If we can accept that somewhere between 1 and 10% of all marrow lymphoid cells have binding sites for BGz, the possibility must be considered that the antibody-forming cell precursors need not be unispecific. We are starting to look, in normal animals, for cells that show immunologically related binding of other enzymes, such as glucose oxidase and horseradish peroxidase, to look for some doubles. These might be present at a low but detectable frequency.

In general there is homogeneity of the globulin produced by a clone of normal immunocytes (Mäkelä and Cross, 1970). Actually there is little theoretical value in unispecificity for precursor cells. More important, there are few completely convincing arguments or experiments to exclude random oligospecificity, especially of identical allotype, within the precursor cells. This would never be noticed in depletion experiments; it would also be rather difficult to find in selected populations (e.g. from antigen-coated columns). The tactical value of having a *large* pool of precursors for any one antigen is undisputed, and for the cell-bound activities, due to thymus-dependent immunocytes, there are several cases in which an unexpectedly large number of normal cells can respond to alloantigens (summarized in Mäkelä and Cross, 1970). More relevant is that there are a few exceptions in systems involving immunoglobulin secretion (e.g. Takahashi *et al.,* 1968; Liacopoulos *et al.,* 1970; Bussard and Lurie, 1967). If the precursor is multipotent and the mature cell unspecific, all that would be necessary is a means of restricting the specificity after antigen contact. It seems premature to exclude such a possibility in the face of the available evidence.

With regard to the presentation function, let us return to the mention in the Introduction of rapid amplification. The evidence of our later experiments can be used to support the following model: antigen combines initially with very high affinity receptor (VHAR) sites on thymus cells. A highly energetic reaction between the large BGz molecule and VHAR sites perturbs the complex enough to release it from the cell surface. Antigen-VHAR complexes are extremely cytophilic, and the thymus cell is most efficiently globulinophilic (Modabber and Coons, 1970a). Nearest neighbor thymus cells, probably of various specificities, could thus passively acquire antigen-VHAR. What would be accomplished is receptor spreading and quite rapid amplification.

The large increase in the number of spleen cells binding BGz apparently was at the expense of thymus cells, from the thymectomy evidence of Modabber and Coons (1970b). However, complete balance sheet experiments on total receptor content of thymus, marrow, and spleen must be performed before the receptor distribution idea can be strongly defended.

The two ideas of receptor distribution on the T cells and oligospecificity of the B cells can be comfortably accommodated within an episomal theory of antibody variability (Gally and Edelman, 1970; Sercarz, 1970), but that is a story for some other evening.

ACKNOWLEDGEMENTS

We would like to thank Donna Finley and Robert Ezzell for inspired technical assistance. This work was supported by American Heart Association grant 69-1000, NSF grant GB-5410, and a grant from the Academic Senate, UCLA.

SUMMARY

The binding of *E. coli* beta-galactosidase (BGz) by thymus and marrow cells was studied using both fixed and unfixed cells. With fixed cells enzymatic activity was determined with two different substrates—fluorogenic fluorescein-di-β-galactoside (FDBG), and precipitinogenic 5-bromo-4-chloroindol-3-yl-β-D-galactoside (BIG).

About five binding cells per million normal thymus cells were found in the mouse by droplet distribution, whereas none were visible in the rabbit thymus. Formalin fixation increased the number of mouse positives 100-fold; rabbit thymus remained negative. Binding activity could be partially or completely inhibited by cross reacting material (Cz), from extracts of bacteria with missense mutations. Mutant extracts without Cz were not inhibitory. Inhibition of binding with anti-immunoglobulin sera was successful to varying degrees.

The number of active cells was 10-20 times higher in fixed bone marrow than in thymus, and could reach 2% of the total population of nucleated cells. Mouse and rabbit marrow had similar activities. (Fixed marrow cells not incubated with BGz were totally inactive.)

By measuring the increase in fluorescence with time in active droplets the number of BGz molecules bound per cell could be calculated. Active thymus cells bound a median number of 98,000 molecules of BGz, and marrow cells 375,000 molecules.

The binding activity of mouse spleen cells increased rapidly within 12 hr of BGz injection. There was a concomitant loss of activity in the thymus. Rabbit marrow was partially depleted of active cells after antigen injection.

Thus a large number of antigen-binding molecules can be found on the surface of some thymus and bone marrow cells. However, a much larger number of cells appear to bind BGz (specifically) only after fixation. Oligospecificity of the precursors of antibody-producing cells was discussed, and a possible mechanism for rapid amplification of helper activity by antigen-reactive cells.

REFERENCES

Abdou, N. I. and Richter, M. (1969). *J. exp. Med. 129,* 757-774.
Bussard, A. E. and Lurie, M. (1967). *J. exp. Med. 125,* 873-891.
Byrt, P. and Ada, G. L. (1969). *Immunology 17,* 503-516.
Decker, J., De Luca, D., Miller, A. and Sercarz, E. (1970). To be published.
Gally, J. M. and Edelman, G. (1970). *Nature, Lond. 227,* 341-348.
Humphrey, J. H. and Keller, H. U. (1970). *In* "Developmental Aspects of Antibody Formation and Structure" (J. Šterzl and I. Říha, eds). (In press.) Academic Press, New York.
Liacopoulos, P., Gille, F. and Amstutz, H. (1970). *C.r. Acad. Sci., Paris 270,* 1049-1052.
De Luca, D., Decker, J., Miller, A. and Sercarz, E. (1970). In preparation.
Mäkelä, O. and Cross, A. M. (1970). *Prog. in Allergy 14,* 145-207.
Modabber, F. and Coons, A. H. (1970a). *Bact. Proc. 70.* (In press.)
Modabber, F. and Coons, A. H. (1970b). *Fedn Proc. 29,* 697.
Modabber, F. and Sercarz, E. (1967). *Bact. Proc. 67,* 90.
Modabber, F. and Sercarz, E. (1970a). *J. Immun.* (In press.)

Modabber, F. and Sercarz, E. (1970b). *Proc. Soc. exp. Biol. Med.* (In press.)
Modabber, F., Mäkelä, O., De Luca, D., Rotman, B. and Sercarz, E. (1970a). Manuscript in preparation.
Modabber, F., Morikawa, S. and Coons, A. H. (1970b). *Science, N.Y.* Submitted.
Naor, D. and Sulitzeanu, D. (1968). *Life Sci.* 7, 377.
de Petris, S., Karlsbad, G. and Pernis, B. (1963). *J. exp. Med.* 117, 849-862.
Rotman, B. (1961). *Proc. natn. Acad. Sci. U.S.A.* 47, 1981-1991.
Sercarz, E. (1970). *Folia Biol.* To be submitted.
Sercarz, E. and Modabber, F. (1968). *Science, N.Y.* 159, 884-885.
Singhal, S. K. and Richter, M. (1968). *J. exp. Med.* 128, 1099-1125.
Sulitzeanu, D. and Naor, D. (1969). *Int. Archs Allergy appl. Immun.* 35, 564-578.
Takahashi, M., Tanigaki, N., Yagi, Y., Moore, G. E. and Pressman, D. (1968). *J. Immun.* 100, 1176-1183.

The Use of Affinity-labelling in the Search for Antigen Recognition Sites

PAUL PLOTZ

Division of Experimental Biology,
National Institute for Medical Research,
Mill Hill, London, England

Over the past two years I have tried to adapt the method of affinity-labelling, developed by Wofsy *et al.* (1962) for locating the antigen-binding site of antibodies, to the study of the anatomy and physiology of cellular antigen recognition sites. My starting point has been the belief that antigen recognition takes place at cell surface molecules. These have the same specificity for antigen that the antibody subsequently produced will have; much evidence points to the Fab portion of immunoglobulin as the recognizing molecule (Greaves, 1970).

The early studies were designed to test whether or not compounds with affinity-labelling properties for antibody are able to affect irreversibly the function of receptors. In no other way can one be certain that apparently specific binding is in fact at the receptor. I have used a cell transfer system in which spleen cells from mice primed to NIP-chicken gamma globulin (NIP-CGG) are treated *in vitro* with an affinity-labelling compound and transferred to irradiated recipients, which are then boosted with NIP-CGG. I have shown that NIP-azide (Fig. 1) irreversibly inhibits the ability of primed spleen cells to respond to the hapten but not to the carrier, and that this inhibition can in part be prevented by simultaneous exposure of the cells to a reversibly-bound hapten, NIP-caproate (Plotz, 1969). I will now report subsequent studies with this system, document the affinity-labelling properties of NIP-azide for anti-NIP antibody, and describe studies with other affinity-labelling haptens.

NIP-azide is a very reactive compound. When it reacts with a spleen cell suspension, a large number of molecules are bound irreversibly to primed and unprimed populations alike (Table I). NIP-caproate has no significant effect on this binding, as one would expect, since the overwhelming majority of molecules are not bound in NIP-specific sites. Despite this large number of molecules bound, treated cells show unchanged viability by dye exclusion and form antibody, except anti-NIP, normally after transfer (Plotz, 1969). To rule out the

171

NIP – AZIDE

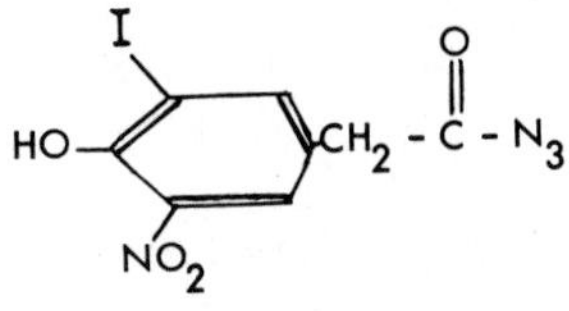

BANIT

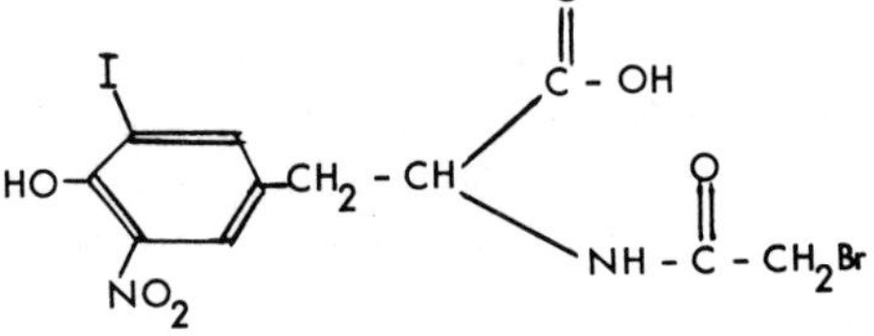

NIP – CAP – NPE

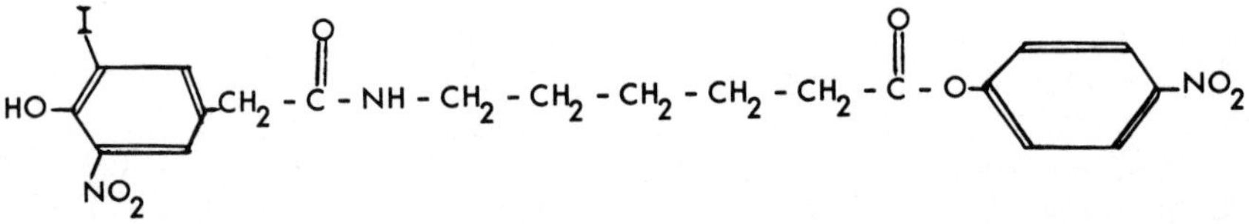

Fig. 1. NIP-azide was prepared according to Brownstone *et al.* (1966). BANIT (α-N-bromacetamido-5-iodo-3-nitrotyrosine) was prepared by bromacetylation of 3-nitrotyrosine by Paul Straussbach at the Weizmann Institute of Science (Weinstein *et al.*, 1969), and was iodinated with ICl in glacial acetic acid. When it was made with radioactive iodine, it was purified by chromatography on silica gel plates (E. Merck) with chloroform 400: methanol 100: glacial acetic acid 1, and eluted from the dry powdered gel with cold buffer just prior to use. NIP-CAP-NPE (p-nitrophenyl-6-(4-hydroxy-3-iodo-5-nitro-phenacetamido) caproate) was prepared by coupling NIP-caproate and p-nitrophenol with dicyclohexylcarbodiimide in dichloromethane, and the product was crystallized from glacial acetic acid by the addition of cold water. Alternatively, the coupling was done with NP-caproate (Brownstone *et al.*, 1966), and the NP-CAP-NPE was iodinated with ICl in glacial acetic acid.

possibility that the low levels of anti-NIP antibody were due to an immuno-adsorbent action of the non-specifically bound NIP, I established that anti-NIP titres in passively immunized recipients were not affected by simultaneous transfer of NIP-azide treated cells (Table II).

In order to establish whether the NIP-sensitive cells were alive, but with blocked receptors, or had been preferentially killed by NIP azide, I followed a lead from Goetzl and Metzger (1970). They showed that an affinity-labelled antibody had not utterly lost its antigen-binding sites; rather, the binding

TABLE I

Effect of NIP-caproate on the uptake of N^{125} IP-azide by normal and NIP-CGG-primed spleen cells

Source of cells	N^{125} IP-azide 10^{-5} M	NIP-caproate 10^{-3} M	Molecules/cell after fourth wash $\times\ 10^{-6}$
Normal	+	0	4.3
Normal	+	+	4.9
NIP-CGG-primed	+	0	5.5
NIP-CGG-primed	+	+	5.4

Normal or NIP-CGG-primed cells from CBA mice were incubated with N^{125}IP-azide in balanced salt solution at $0°$C for 15 min. They were then washed four times with cold BSS, and the radioactivity and number of cells in the resuspended pellet was measured.

TABLE II

Effect of NIP-azide-treated cells on the serum levels of passively-transferred anti-NIP antibody

Days after transfer of cells and serum	Anti-NIP $\log_{10}$ ABC	
	Control cells	NIP-azide treated cells
1	2.15 (0.15)	1.87 (0.18)
2	1.88 (0.17)	1.65 (0.01)
4	1.73 (0.17)	1.78 (0.4)
7	1.41	1.43 (0.06)

Irradiated (600 r) CBA mice were injected i.p. with 0.6 ml of serum from CBA mice bled 10 days after secondary stimulation with NIP-CGG, and with 1/3 spleen equivalent of a cell suspension of normal spleen treated with NIP-azide or DMSO for 15 min in the cold. They were bled at intervals, and the sera tested as described (Plotz, 1969). The results are the average of three mice in each group (SE).

constant had been reduced by a factor of 50. If that were true of the cells as well, raising the boosting dose of antigen should overcome the blockage, and anti-NIP antibody should be synthesized. In fact, as the boosting dose was raised, anti-NIP antibody did appear, reaching normal levels at high doses (Fig. 2(A)). Thus the cells appear to be living and responsive. Such antibody should have the same affinity as antibody from control cells, whereas the affinity should be lower if NIP-azide simply eliminates cells with high-affinity receptors. In a rough test of that prediction, I found no difference in the binding properties, using a modified Farr test, of sera from treated and untreated

C.I.—7

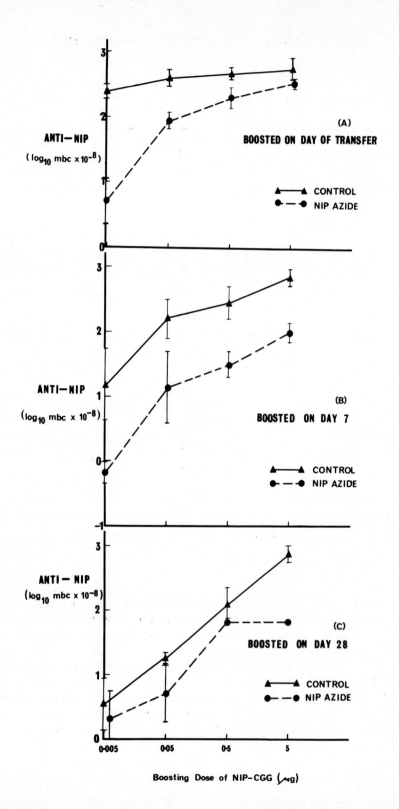

recipients boosted with high doses, but I have not compared antibody raised by high and low doses. These results do not prove that NIP-azide occludes recognition sites in a manner similar to that described by Goetzl and Metzger; that is a plausible but by no means unique interpretation of the results.

If NIP-azide does interfere with receptor function without killing NIP-sensitive cells, do the cells recover their sensitivity, perhaps by making new receptors during membrane turnover? To study that question I delayed the boosting injection as long as a month after treatment and transfer of the cells. With boosting seven days after transfer (Fig. 2(B)) both groups had diminished responsiveness, and there was still a differential between them. By a month, the responses of treated and untreated cells were almost equal, but this seems due primarily to the diminished responsiveness of the control cells; the treated cells did not decline proportionately as much. The evidence suggests that memory cells have no rapid turnover of receptors under these circumstances, though it is strange that the receptors should escape normal membrane turnover (Warren and Glick, 1968).

In view of the extensive non-specific binding of NIP-azide and the fact that NIP-caproate does not fully prevent the NIP-azide inhibition, it seemed possible that the inhibition might be due in part to NIP bound elsewhere than the NIP recognition site. To test that possibility, I took normal unprimed spleen cells, treated them with NIP-azide, washed them extensively, and transferred them along with untreated primed cells. In some experiments the treated normal cells specifically inhibited the subsequent response of the untreated primed cells, thus mimicking the effect of treating the receptor-bearing population directly, though the extent of inhibition was less (Table III). Such inhibition has occurred only with spleen cells, not with lymph node cells, and it has occurred inconstantly, for reasons I don't understand.

To conclude the NIP-azide story, I have established that NIP-azide does affinity-label purified rabbit anti-NIP antibody under conditions similar to those used with cells, and that NIP-caproate partially prevents the specific binding (Table IV).

In summary, I believe that the NIP-azide inhibition is due to two effects: NIP-azide binds by affinity-labelling to NIP recognition sites on sensitive cells, blocking the subsequent response to low doses of antigen without destroying the cells. It also binds non-specifically elsewhere, and such non-specifically bound NIP may specifically block the subsequent response by a peculiar kind of paralysis.

Fig. 2. Spleen cells from NIP-CGG primed mice were treated with NIP-azide or with DMSO and transferred to groups of six recipients irradiated the previous day with 600 r. The recipients were boosted on the same day (A), on day 7 (B), or on day 28 (C) with NIP–CGG and were bled 10 days after boosting. Anti-CGG titres, not shown, were virtually the same in the treated and untreated groups.

TABLE III

Effect of normal cells treated with NIP-azide on the response of NIP-CGG-primed cells

| Treatment | | After transfer | Antibody on Day 10 | |
| *In vitro* | | NIP-CGG | NIP | CGG |
NIP-CGG-primed spleen cells	Normal spleen cells	0.005 μg		
DMSO	DMSO	+	1.58	3.54
NIP-azide	DMSO	+	−0.31	3.13
DMSO	NIP-azide	+	0.69	4.81
DMSO	NIP-azide	0	−1.00	0.00

Spleen cells from normal and NIP-CGG-primed mice were incubated for 15 min with NIP-azide or with dimethyl sulphoxide (DMSO). They were then washed three times with balanced salt solution, and injected into CBA recipients irradiated (600 r) the previous day. Recipients were boosted within 4 hr with NIP-CGG and bled at day 10. Antibody titrated as described by Plotz (1969).

TABLE IV

NIP-azide affinity-labelling: pH curve

| pH | Moles NIP-azide/moles γ-globulin | |
	RAN-13	RGG
7.0	0.195 (0.098)	0.029 (0.023)
7.3	0.251 (0.111)	0.039 (0.029)
7.6	0.249 (0.126)	0.055 (0.038)
8.0	0.603 (0.329)	0.183 (0.121)
8.3	0.993 (0.574)	0.314 (0.260)

Rabbit anti-NIP antibody (RAN-13) was purified by adsorption onto Sepharose coupled with ε-N-NIP-lysine by the cyanogen bromide method (Porath *et al.*, 1967) and elution with 1 N propionic acid. Fluorescence quenching (performed by N. M. Green) and a Farr test with NIP-caproate gave $K_a \cong 3\text{-}5 \times 10^7$ and 1.3-1.6 binding sites per mole of gamma globulin. Rabbit gamma globulin (RGG) was purchased from Pentex. The concentration of protein was 2.8×10^{-6} M and of $N^{125}IP$-azide was 3.2×10^{-5} M. The reaction was stopped after 10 min at $0°C$ with five volumes of cold 5% trichloracetic acid in acetone with 0.02 M ε-aminocaproic acid. The resulting precipitate was washed once with 5% trichloracetic acid in acetone, and counted in a Nuclear Enterprises automatic gamma counter. The figures in parentheses represent NIP-azide bound in the presence of NIP-caproate at 1.1×10^{-3} M.

Faced with this complex picture and with the knowledge that NIP-azide binds non-specifically in amounts so large that it could not possibly be used to label receptors specifically enough for isolation, I decided to leave NIP-azide and search for other affinity-labelling reagents. I have stuck to reagents based on the NIP nucleus since, even with the most generous estimates of the number of receptors on a sensitive cell and good cell purification schemes, radioactive iodine seems to offer the best chance of finding receptors.

The first reagent, BANIT (Fig. 1), is analogous to David Givol's bromacetamido derivatives of DNP (Weinstein *et al.*, 1969; Givol *et al.*, 1969) and was made with the help of Paul Straussbach and Givol at the Weizmann Institute. It is an excellent affinity-labelling reagent for purified anti-NIP antibody, inhibitable by NIP-caproate, though it reacts very slowly (Table V). So far, however, I have been unable to block specifically cellular responses to NIP. Concentrations of the compound that block the anti-NIP response block the anti-carrier response too.

TABLE V

BANIT affinity-labelling: pH curve

	Moles BANIT/moles γ-globulin	
pH	RAN-13	RGG
7.3	0.32 (0.051)	0.025 (0.037)
7.6	0.38 (0.050)	0.040 (0.041)
8.0	0.40 (0.080)	0.051 (0.063)
8.3	0.54 (0.064)	0.057 (0.059)

Initial concentration of protein was 2.8×10^{-6} M; of BANIT 1.1×10^{-5} M; and of NIP-caproate 10^{-3} M. The reaction was stopped after 24 hr at $37°$C and the samples treated as described in Table IV.

The second reagent, NIP-CAP-NPE (Fig. 1) was made at the suggestion of N. M. Green of Mill Hill and with his help. It, too, is an excellent affinity-labelling reagent for purified antibody, though with a marked pH dependence so that it is poorly reactive under conditions that cells tolerate (Table VI). However, I have not yet been able to demonstrate specific inhibition of NIP-cell receptor function.

It is possible that both BANIT and NIP-CAP-NPE may attack groups on the cell surface more crucial for survival than the groups NIP-azide attacks, or that they may bind to receptors specifically, but in a way that does not measurably impair function. Thus there are two reagents with marked specificity for NIP-binding sites of antibody whose effect on receptor function I have been unable to establish. The possibility that the cellular antigen-binding site and the antibody active site differ markedly could account for these results but seems to me unlikely.

To continue to insist that impairment of receptor function is the only valid criterion of receptor binding would threaten to bring this promising approach to

TABLE VI

NIP-CAP-NPE affinity-labelling: pH curve

| pH | Moles NPE/moles γ-globulin | |
	GAN-3	GGG
7.0	0.030	0.004
7.3	0.040	0.003
7.6	0.062	0.005
8.0	0.144	0.009
8.3	0.250	0.018

Goat anti-NIP antibody (GAN-3) was purified in the same way as RAN-13. By the Farr test with NIP-caproate $K_a \cong$ 1-2 x 10^7 and there was 1.1 binding site per mole of gamma globulin. Goat gamma globulin was prepared by the method of Givol and Hurwitz (1969). The initial concentration of protein was 2.8 x 10^{-6} M, and of NIP-CAP-NPE 2.0 x 10^{-5} M. The reaction was stopped after 30 min at 0°C and the samples treated as described in Table IV.

an early grave. It may be more sensible to follow the lead of those who have successfully pursued the lactose transport protein in *E. coli* (Fox and Kennedy, 1965), or the binding site for hormones (Lefkowitz *et al.*, 1970), and seek in a population of antigen-sensitive cells membrane fractions that have preferentially bound affinity-labelling reagents. Such a search would be immeasurably advanced by a procedure for purifying sensitive cells.

ACKNOWLEDGEMENTS

I am grateful to Annabel Green and Marianne Watson for technical assistance and to N. A. Mitchison, R. Pitt-Rivers, A. Brownstone and N. M. Green for encouragement and advice. Parts of this work were presented at the British Society for Immunology in April 1970. I am a Fellow of the Helen Hay Whitney Foundation.

SUMMARY

Cells from mice immunized against NIP-chicken gamma globulin lose their ability to respond to NIP but not to chicken gamma globulin when they are treated *in vitro* with the affinity-labelling hapten, NIP-azide, prior to transfer and boosting. Subsequent studies have shown that larger boosting doses overcome this inhibition, implying that the cells have not been killed. There is no substantial recovery of responsiveness to low doses of antigen for several weeks after transfer, suggesting no rapid turnover of receptors. The effect of

NIP-azide on primed cells can be mimicked in part by mixing NIP-azide treated normal cells with untreated primed cells; thus NIP-azide inhibition is probably the sum of effects of receptor-bound NIP-azide and NIP-azide bound elsewhere.

NIP-azide is a specific affinity-labelling hapten for purified rabbit anti-NIP antibody. Two other compounds, a bromacetamido derivative of NIP and a nitrophenyl ester of NIP-caproate, are even better affinity-labelling haptens for purified antibody, but neither has yet been shown to block NIP cell receptor function specifically.

REFERENCES

Brownstone, A., Mitchison, N. A. and Pitt-Rivers, R. (1966). *Immunology 10*, 465-479.
Fox, C. F. and Kennedy, E. P. (1965). *Proc. natn. Acad. Sci. U.S.A. 54*, 891-899.
Givol, D. and Hurwitz, E. (1969). *Biochem. J. 115*, 371-375.
Givol, D., Haimovich, J., Segal, S., Globerson, A. and Feldman, M. (1969). *Nature, Lond. 223*, 1374-1375.
Goetzl, E. J. and Metzger, H. (1970). *Biochemistry 9*, 1267-1278.
Greaves, M. F. (1970). *Transplantation Rev.* (In press.)
Lefkowitz, R. J., Roth, J., Pricer, W. and Pastan, I. (1970). *Proc. natn. Acad. Sci. U.S.A. 65*, 745-752.
Plotz, P. (1969). *Nature, Lond. 223*, 1373-1374.
Porath, J., Axén, R. and Ernback, S. (1967). *Nature, Lond. 215*, 1491-1492.
Warren, L. and Glick, M. C. (1968). *J. Cell Biol. 37*, 729-746.
Weinstein, Y., Wilchek, M. and Givol, D. (1969). *Biochem. biophys. Res. Commun. 35*, 694-701.
Wofsy, L., Metzger, H. and Singer, S. J. (1962). *Biochemistry 1*, 1031.

Affinity-labelling of Cells with Dinitrophenyl Specificity

JOSEPH HAIMOVICH

Department of Microbiology, Washington University School of Medicine,
St. Louis, Missouri, U.S.A. *

Recognition of antigen by cells involved in an immune response is believed to be mediated by antibody-like receptor molecules, which are similar in their properties to the antibodies produced (Mitchison, 1967).

Affinity-labelling reagents are substances that can react specifically with the appropriate antibody and can bind covalently to certain amino acid residues

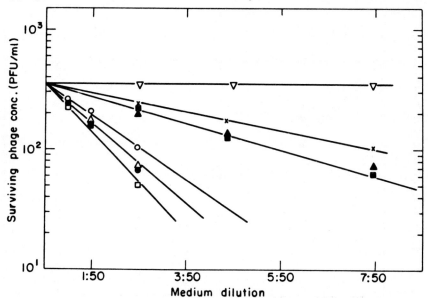

Fig. 1. Effect of BADL, BADE and DNP-lysine on antibody production *in vitro*. Inactivation of DNP-phage by incubation for 2 hr with medium from spleen explants. ▽ explants from normal animals; × explants from primed animals without *in vitro* stimulation with antigen (DNP-RSA); ▲ explants incubated with BADE before antigen; ■ explants incubated with BADL before antigen; ○ explants incubated with DNP-lysine before antigen; ● explants incubated with PBS before antigen; □ explants incubated with BADL three days after antigen; △ explants incubated with BADE three days after antigen (Segal *et al.*, 1969).

* On leave from the Department of Chemical Immunology, the Weizmann Institute of Science, Rehovot, Israel.

within the antibody combining site by virtue of a chemically reactive group (Singer *et al.*, 1967).

Bromoacetyl derivatives of dinitrophenyl ligands, namely the α,N-bromoacetyl, ε-N-(2,4-dinitrophenyl)-lysine (BADL) and N-bromoacetyl-N'-(2,4-dinitrophenyl)-ethylenediamine (BADE), were found to react specifically with anti-dinitrophenyl (DNP) antibodies and irreversibly block their activity (Weinstein *et al.*, 1969).

Recently it was shown (Segal *et al.*, 1969) that induction of a secondary response *in vitro* could be almost completely blocked by adding BADE or BADL to spleen explants from mice immunized with DNP-protein conjugates. The non-covalently binding ligand DNP-lysine had only a slight effect under the same conditions (Fig. 1). The effects of BADE and BADL on the secondary response *in vitro* could not be attributed to a non-specific toxic effect, because addition

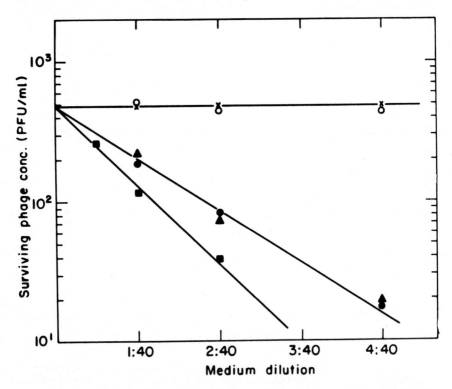

Fig. 2. Effect of BADL and BADE on antibody production to poly-DL-alanine *in vitro*. Inactivation of poly-DL-alanyl T4 phage by incubation for 4 hr with medium from spleen explants. x explants from normal animal; ○ explants from primed animal without *in vitro* stimulation with antigen (poly-DL-alanyl-RSA); ● explants incubated with PBS before antigen; ▲ explants incubated with BADE before antigen; ■ explants incubated with BADL before antigen (Segal *et al.*, 1969).

of these reagents to the spleen explants after the stimulatory antigen did not affect the production of antibodies. Moreover BADL and BADE had no effect whatsoever in a non-related system—spleen explants from mice immunized with a poly-DL-alanyl protein conjugate (Fig. 2).

A more direct approach to the detection and quantitation of antibody-like structures on cell surfaces has recently been developed (Sulica *et al.*, 1970). The assay for the receptors is based on the binding of chemically modified bacteriophages to specific cells followed by elution of the phage with a large excess of a specific monovalent ligand.

Spleen cells from guinea pigs immunized with DNP-bovine serum albumin $(5 \times 10^6$ cells) were incubated with DNP-bacteriophage T4 $(2 \times 10^7$ particles). The cells were then washed and the DNP-T4 was eluted with ϵ-DNP-amino-caproic acid $(10^{-3}$ M); 3.7×10^4 phage particles could be specifically eluted. Pretreatment of such spleen cells with BADL $(10^{-3}$ M for 3 hr at pH 7.6) almost completely inhibited the specific binding of the DNP-phage.

Affinity-labelling has been primarily used to determine the amino acid residues involved in the construction of the active sites of antibodies. These studies are hampered by the fact that antibody populations are heterogeneous

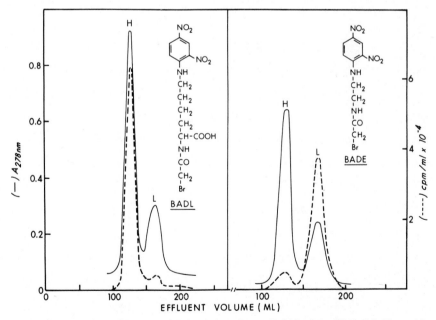

Fig. 3. Separation of heavy and light chains of protein 315 after affinity-labelling with ^{14}C-BADL (left) and ^{14}C-BADE (right). Reduced and carboxymethylated protein was applied to a Sephadex G-100 column (2.2 x 100 cm) equilibrated with 6 M urea-1 M acetic acid. Flow rate 5 ml per hour (Haimovich *et al.*, 1970).

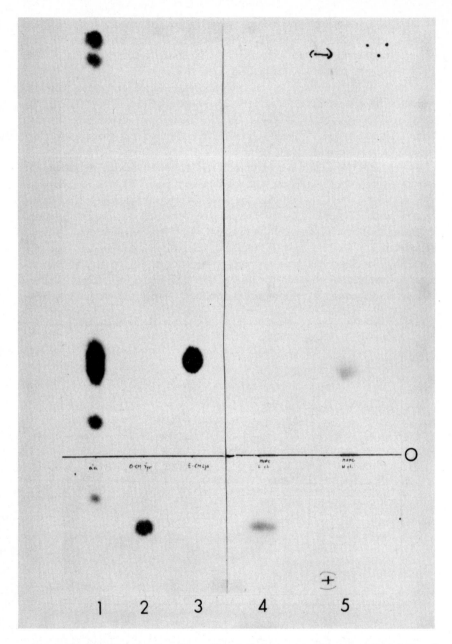

Fig. 4. Identification of radioactive residues in hydrolysates of heavy and light chains of affinity-labelled protein 315 by high voltage electrophoresis. An amount of hydrolysate containing about 20,000 cpm was electrophoresed at pH 3.5 (pyridine-acetate buffer) for

with respect to their primary structure. However, the mouse myeloma protein secreted by MOPC-315 tumour cells has a high anti-DNP activity. BADL labelled exclusively a lysyl residue of the heavy chain of this myeloma protein, whereas BADE labelled exclusively a tyrosyl residue of the light chain (Figs 3 and 4) (Haimovich *et al.*, 1970).

In attempting to isolate and characterize the antibody-like receptor molecules, it is of extreme importance to search for homogeneous populations of cells bearing such receptors.

REFERENCES

Haimovich, J., Givol, D. and Eisen, H. N. (1970). *Proc. natn. Acad. Sci. U.S.A.* 67, 1656-1661.

Mitchison, N. A. (1967). *Cold Spring Harb. Symp. quant. Biol. 32*, 431-438.

Segal, S., Globerson, A., Feldman, M., Haimovich, J. and Givol, D. (1969). *Nature, Lond. 223*, 1374-1375.

Singer, S. J., Slobin, L. I., Thorpe, N. O. and Fenton, J. W. (1967). *Cold Spring Harb. Symp. quant. Biol. 32*, 99-109.

Sulica, A., Haimovich, J. and Sela, M. (1970). *J. Immunol.* (In press.)

Weinstein, Y., Wilchek, M. and Givol, D. (1969). *Biochem. biophys. Res. Commun. 35*, 694-701.

---

30 min at 3.5 kv. After electrophoresis the paper was autoradiographed for 24 hr using Kodak BB-54 medical X-ray film. Columns 1-3 are ninhydrin-stained standards: (1) a mixture of amino acids; (2) 0-carboxymethyl tyrosine; (3) ϵ-carboxymethyl lysine. Columns 4 and 5 are autoradiographs of; (4) a hydrolysate of BADE-labelled light chain, showing only 0-^{14}C-carboxymethyl tyrosine; (5) a hydrolysate of BADL-labelled heavy chains showing only ϵ-^{14}C-carboxymethyl lysine (Haimovich *et al.*, 1970).

The Nature of the Cell Receptor in Delayed Hypersensitivity and Tolerance

SIDNEY LESKOWITZ

Department of Pathology, Tufts Medical School,
Boston, Mass. 02111, U.S.A.

It has become axiomatic for immunologists to consider the antigen recognizing receptor on immunocompetent cells to be an immunoglobulin molecule. While the biologically active portion corresponding to the Fc region may in fact be an as yet unrecognized type, as indicated by the often used designation IgX, it is reasonable to assume that the binding site is made up of light and heavy chain contributions, analogous to that found for all conventional antibodies. If these assumptions are correct one might then expect cell receptor immunoglobulins to behave in some respects like classic immunoglobulins.

The major distinguishing characteristic of antibody is specificity, the ability to recognize and react with the molecular configuration that initiated its production originally. Most studies of antibody specificity have made use of the enormous versatility and precise measurements possible with hapten conjugates, and these have provided an enormous amount of information about the binding characteristics and structural requirements of antibody. However, there has always existed, in a more or less recognized form, the dichotomy of behavior of haptens with respect to their ability to react with preformed antibody on one hand, and to initiate a biologic response (i.e. immunization, tolerance etc.) on the other.

The recognition that haptens alone, while capable of binding to antibodies with considerable specificity and high affinity, are unable in the absence of appropriate carriers to elicit responses, has required certain additional hypotheses concerning the mechanism of action of antigens with immuno-competent cells. One such suggests that the carrier plays a role by virtue of its contribution of additional determinants, which assist in bringing together several cell types whose co-operation is mandatory for an immune response.

While not denying the possible validity of a cell-cell co-operative mechanism in immunologic events, I would like to suggest an alternative interpretation based in part on some data from my own work.

The concept of specificity of antibody has relied heavily on measurements of binding affinity between antibodies and haptens. In many instances these have demonstrated that the hapten alone is the "immunodominant" portion, and provides the greatest contribution to the overall binding energy, especially when such hydrophobic substances as DNP are used.

However, many other studies (Kabat, 1968) indicate that the size of the combining site is considerably larger than the hapten alone. A number of estimates of combining site with various antigens are given in Table I. While studies with protein antigens are necessarily more difficult because of the great variety of different antigenic determinants on one molecule, it has still been possible in several instances to determine that the size of a particular determinant is not greatly different from the size of the antibody combining sites found with structurally simpler antigens.

TABLE I

Size of antibody combining sites

Antigen	Determinant	Size (Å)	Mol. Wt.
Dextran	Isomaltohexaose	34 x 12 x 7	990
Poly GLA	Hexaglutamic	36 x 10 x 6	792
Poly-lysyl-RSA	Pentalysine	27 x 17 x 6	659
α-DNP-lys$_7$	α-DNP-lys$_7$	30 x 17 x 6	1080
—	DNP	4 x 4	—

The other side of the coin concerns the nature of the immunogenic determinant which is capable of giving rise to immunologic responses. Since it is obviously larger than the hapten alone one may ask how large it must be. In studies involving hapten-protein conjugates, the specificity of the anamnestic response and delayed hypersensitivity reactions are considered to involve fairly exacting requirements for the carrier molecule, hence the designation "carrier" specificity. However, the question of how large a part of the carrier is required to produce the immunogen has never been satisfactorily answered. The work of Gell and Silverstein (1962) in which the hapten and carrier were kept constant and only the position of attachment of the hapten was varied, showed that specificity of delayed hypersensitivity was a function of the hapten and the carrier environment immediately adjacent to its site of attachment. In other words they did not act independently and additively, but functioned as a unit determinant. In the carrier the number of amino acids that functioned together with the hapten could not be determined. However, it should be kept in mind that with the varied, non-repeating amino acid sequences of proteins any particular stretch of 5-6 amino acids might well be a unique structure, and hence convey exquisite specificity.

A number of studies over the years have progressively decreased the size of the carrier required to induce a response to the hapten. In the case of a response to the DNP hapten, conjugates with bacitracin M.W. 1900 (Abuelo and Ovary, 1965), and heptalysine, M.W. 1080 (Schlossman et al., 1969); have proven to be immunogenic. In fact in some instances DNP coupled to a single amino acid has been shown to be immunogenic (Frey et al., 1969), although the question of whether the hapten transfers to another carrier is still open.

For arsanilic acid, the hapten with which I have most experience, the situation is not as exceptional as it may seem to some. Thus, p-azobenzene arsonate (ABA) conjugates of high molecular weight carriers are fully immunogenic. In addition Borek et al. (1967) have demonstrated that a conjugate of ABA and hexa-tyrosine (M.W. 1300) is fully immunogenic in the rabbit, eliciting both antibody and delayed hypersensitivity. Studies from my lab over the years have revealed that the ABA group coupled to a single tyrosine or histidine molecule is still immunogenic, albeit only with respect to delayed hypersensitivity (Leskowitz et al., 1966). Thus ABA-N-acetyltyrosine (M.W. 450) will sensitize guinea pigs and rabbits, will elicit delayed reactions, will produce in vitro blast cell transformation and inhibition of macrophage migration, and will induce tolerance (Leskowitz, Mills and Ferraresi, unpublished observations; Collotti and Leskowitz, 1970). From the standpoint of size therefore, these conjugates fall well within the size limits suggested for an antibody receptor site.

The requirements of immunogenicity cannot therefore be met by size alone since a considerable spectrum of immunogenic sizes exist. The question then becomes one of what other attributes must a molecule possess in order to become an immunogen?

One current hypothesis suggests that immunologic reactions are driven by thermodynamic forces favoring high affinity binding of antigenic determinants by antibody sites (Paul et al., 1968). Those cells competing favorably for the available antigen will be turned on, to yield an antibody representative of the high affinity receptor on their surface. In delayed sensitivity it is not possible to compare directly affinities of receptors on the cell surface as is done for purified antibody. Nevertheless some indirect comparisons were attempted in studies on delayed sensitivity to the ABA group.

Since the specificity of these reactions is almost entirely due to the ABA group, a series of studies was carried out to determine the effect of varying the carrier on the resulting antibody and delayed sensitivity reactions in vivo and in vitro. The results of a number of these experiments are summarized in Table II.

When comparisons are made amongst the various conjugates several points seem evident. ABA conjugates made with optically isomeric polymers, which are similar in other respects, behave identically in reactions with preformed

TABLE II

Comparison of immunologic reactivity of ABA-conjugates

| ABA-conjugate | React. with AB | | React. with del. sensitivity | | | | |
| | | | In vivo | | | In vitro | |
	In vitro C'fix.	In vivo PCA	Immuno-genicity for del. sens.	Elicit del. sens.	Tolerance induct.	Blast trans.	MIF
ABA-N-acet. tyr	(−)	(+)	+	+	+	+	+
ABA-n-propylphenol	(−)	N.D.	−	−	−	−	−
ABA-L-GAT	+	+	+	+	+	+	+
ABA-D-GAT	+	+	−	−	−	−	−
ABA-poly-L-tyr	+	+	+	+	+	+	+
ABA-poly-D-tyr	+	+	−	−	−	−	−
ABA-diazonum	N.D.	N.D.	N.D.	+	+	−	N.D.

antibody such as quantitative complement fixation and PCA reactions. There is, however, a very clear distinction between their ability to sensitize, elicit reactions, or produce tolerance to delayed sensitivity in guinea pigs.

Similarly a comparison of ABA-tyr with ABA-p-propyl phenol reveals a structural similarity except for the absence of an α-amino carboxyl function in the latter. This change results in a corresponding loss in ability to effect any of the biologic reactions studied. It was of interest that the unconjugated ABA diazonum salt could also effectively induce tolerance, although it could not be determined whether its effect was directly on the cell or via coupling to serum protein.

We have drawn several conclusions from these studies.

1. The antigenic determinant involved in this system is the ABA group plus part of the aromatic ring of the tyrosine residue. It is assumed that the receptor site on the sensitized lymphocytes is similar to the immunogen giving rise to it, and is therefore of the same order of size and conformation.

2. In reactions with preformed antibody, the optical isomerism of the carrier tyrosine plays no role in the reactivity of the conjugate, and both pairs are presumed to fit the antibody site equally well.

3. Since the immunogenic determinant in all cases is ABA-tyr and since this is present in all compounds, the inability of D-polymers to function as carriers is probably based on reasons other than affinity for the cell receptor.

Our general hypothesis to explain these observations in a single coherent way suggests that a biologic "processing" event must occur with an antigen, leading to a material altered in such a way that it now becomes immunogenic. To make it compatible with the general Fishman and Adler (1963) hypothesis, we would suggest tentatively that this biologic processing is in fact conjugation of the antigen to an RNA moiety by some sort of macrophage, with or without digestion to smaller units.

In this respect one may point to the general indigestibility of D-amino acid polymers compared to their L-isomers, as well as the absence on the non-immunogenic ABA-n-propyl phenol of an α-amino acid handle by which some enzymatic event might transpire. Also in support of this contention I might mention some continuing studies with Dr Arthur Gottlieb. Labelled poly-L or poly-D-GAT was fed to rat macrophages. These were then extracted and fractionated on cesium chloride gradients to test for small molecular weight RNA complexes thought to be involved in immunogenicity. Up to six-fold greater quantities of the labeled immunogenic L-polymer were found associated with this fraction (Gottlieb and Leskowitz, unpublished observations).

If productive combination with a cell receptor leading to sensitization or reaction requires an immunogen processed on or through a macrophage, can any effect of non-processed antigen be shown? Together with other investigators we have been applying the methods of hapten inhibition to this problem to see if effects on cells similar to those on antibody may be produced.

A number of years ago we reported some attempts at hapten inhibition of delayed sensitivity (Leskowitz and Jones, 1965). When sensitized guinea pigs were given large doses of non-immunogenic ABA conjugates (e.g. ABA-p-cresol) just prior to skin testing, the skin reactions were considerably depressed. Since reactivity returned within a few days, when the non-immunogen was presumably excreted, we interpreted this phenomenon as hapten inhibition of sensitized cells, to contrast it with the longer term desensitization following injection of immunogenic conjugates.

In a more recent experiment, efforts to block the induction of tolerance in guinea pigs by prior injections of the non-immunogenic ABA-p-propyl phenol were attempted. Guinea pigs given a single intravenous injection of 10^{-5} moles of ABA-tyr were shown to be unresponsive three weeks after immunization with 10^{-8} moles ABA-tyr in complete Freund's adjuvant. When the tolerizing injection was preceded and followed by injections of 10^{-5} moles ABA-p-propyl phenol, guinea pigs were still rendered unresponsive while control animals receiving the ABA-p-propyl phenol alone were responsive.

This apparent failure to produce an *in vivo* inhibition of tolerance induction does not represent a reasonable test. Because of the vagaries of differential excretion and circulation an intact animal does not provide a suitable milieu for achieving thermodynamic equilibrium conditions. We have therefore recently switched to *in vitro* systems.

Cells removed from the lymph nodes of guinea pigs sensitized with ABA-tyr in complete Freund's adjuvant were cultured *in vitro* for 3-4 days, and studied for incorporation of thymidine after addition of various ABA conjugates.

Our preliminary studies show that stimulation up to nine-fold over control tubes is readily achieved with the immunogen ABA-tyr. The non-immunogenic ABA-p-propyl phenol is unable to produce stimulation in doses ranging up to toxic levels. Combining these two we have attempted to show inhibition by the non-immunogen if it is incubated with the cells prior to addition of the immunogen. While the results in Table III indicate some slight specific suppression at non-immunogenic levels of 10^{-5} M, larger concentrations needed to test hapten inhibition proved toxic. It will be necessary, if this system is to prove useful, to decrease the amount of antigen required to give adequate stimulation.

TABLE III

Attempt at hapten inhibition of cell stimulation in vitro

		Addition to culture	
Pre-incubation	None	ABA-N-Acet. tyr. 2×10^{-4} M[a]	PPD 2.5 μg[a]
None	1	9	16
ABA-p-propyl phenol			
10^{-6} M	1	8	15
10^{-5} M	1	5	28
10^{-4} M	0.5	0.5	2.5

[a] Each value is stimulation in relation to untreated control cultures.

Finally we have begun to resort to the method of affinity labeling to test the reactivity of the receptor site on sensitized cells. As mentioned previously, when the ABA diazonum salt is given intravenously, animals can be rendered tolerant. However, it cannot be ascertained whether this is due to direct combination and conjugation at the lymphocyte receptor, or by way of prior conjugation to a serum protein. Our first and very preliminary results *in vitro* are mildly encouraging.

Table IV shows the results of prior incubation (with ABA diazonum salt) of lymph node cells from sensitized guinea pigs. After washing they were stimulated with ABA-tyr or PPD. Two tentative observations may be made. First, the diazonum salt itself appears to be only slightly stimulatory at doses that are not toxic. Secondly it appears to specifically inhibit the subsequent stimulation by ABA-tyr but not by PPD.

If these results are borne out in continuing studies it will provide valuable confirmatory evidence for the existence of an antibody-like cell receptor. Since an affinity labeling hapten is bound covalently at the receptor site its affinity

TABLE IV

Inhibition of cell stimulation in vitro *with an affinity labelling hapten*

| Pre-incubation | None | Addition to culture Stimulation index (Ag/control) | |
		90 µg ABA-tyr	5 µg PPD
Buffer	1	6	22
ABA-diazonum			
10^{-9} M	1.7	7	38
10^{-8} M	1.1	5.4	42
10^{-7} M	1.5	1.8	32

constant may be said to be infinitely large. If this does not produce an immune response it may be said that affinity alone is not sufficient to drive the immune response, but only serves in a selective capacity after a prior processing event has taken place.

SUMMARY

Using delayed sensitivity to arsanilic acid conjugates as a system in which little carrier specificity is evident, a study was begun of the nature of the cell receptor in this response. Attempts to achieve hapten inhibition of reactions *in vivo* and *in vitro* by non-immunogenic hapten conjugates had some success. Preliminary experiments were begun with the diazonum salt of arsanilic acid as an affinity label to specifically react with receptor sites on sensitized cells. So far results indicate that the affinity labeling reagent itself does not stimulate sensitized cells *in vitro*, but does block subsequent stimulation by antigen. If confirmed, these results will provide further evidence for an antibody-like receptor site on sensitized lymphocytes.

REFERENCES

Abuelo, J. G. and Ovary, Z. (1965). *J. Immun. 95,* 113-117.
Borek, F., Stupp, Y. and Sela, M. (1967). *J. Immun. 98,* 739-744.
Collotti, C. and Leskowitz, S. (1970). *J. exp. Med. 131,* 571-582.
Fishman, M. and Adler, F. L. (1963). *J. exp. Med. 117,* 595-602.
Frey, J. R., de Weck, A. L., Geleick, H. and Lergier, W. (1969). *J. exp. Med. 130,* 1123-1143.
Gell, P. G. H. and Silverstein, A. M. (1962). *J. exp. Med. 115,* 1037-1051.
Kabat, E. A. (1968). "Structural Concepts in Immunology and Immunochemistry". Holt, Rinehart and Winston.

Leskowitz, S. and Jones, V. E. (1965). *J. Immun. 95*, 331-335.
Leskowitz, S., Jones, V. E. and Zak, S. J. (1966). *J. exp. Med. 123*, 229-237.
Paul, W. E., Siskind, G. W. and Benacerraf, B. (1968). *J. exp. Med. 127*, 25-42.
Schossman, S. F., Herman, J. and Yaron, A. (1969). *J. exp. Med. 130*, 1031-1045.

DISCUSSION

MITCHISON: Are you not selling the pass when you postulate that ABA-tyrosine can be linked covalently by an enzyme to a macromolecule? After all for the last few years your critics have been suggesting that your hapten simply couples *in vivo* to a carrier molecule—presumably a protein—in much the same as do FDNB and the other standard skin-sensitizing agents. What made your system unique was that ABA-tyrosine seemed to lack a reactive group. Now that you suggest that ABA-tyrosine actually can be conjugated, it takes away much of the special interest of your case.

You can hardly suggest that a covalent bond can be made with RNA and not with protein—that would be just magic.

LESKOWITZ: Yes, it is certainly possible that arsanil-tyrosine may be coupled *in vivo* to a carrier protein to make a larger immunogen. While I have no direct evidence for or against this point, there is indirect evidence which makes this possibility less likely. 1. Other haptens such as sulfanilic acid or aminobenzoic acid coupled to tyrosine do not function as immunogens even though they could presumably be coupled to a protein via the tyrosine residue in the same way. Thus it is not the tyrosine residue alone or the arsanilic acid alone which leads to immunogenicity but rather the specific determinant produced by the combination. 2. Studies with α-DNP by Schlossman show that seven lysine units confer immunogenicity while 6 lysine units do not. It is difficult to see again why one could be bound to a protein and not the other.

What I would suggest is that "processing" as I envisage it may involve coupling of an immunogen, no matter how big or small, to an RNA moiety which acts to derepress a cell to which the determinant is bound specifically. This is somewhat different from the carrier function which you envisage as serving to bring several cells together by virtue of multiple determinants.

REACTIONS OF LYMPHOCYTES TO
ANTIGENIC STIMULATION

Function of Thymus-independent Immunocytes:
Some Properties of Antibody-secreting Cells as Judged by the Open Carboxymethyl-cellulose Haemolytic Plaque Technique

G. J. V. NOSSAL, HEATHER LEWIS and N. L. WARNER

The Walter and Eliza Hall Institute of Medical Research,
Melbourne, Victoria 3050, Australia

Though the haemolytic plaque technique (Jerne and Nordin, 1963; Ingraham and Bussard, 1964; Cunningham, 1965) has largely replaced the earlier microdroplet techniques (Nossal, 1958; Nossal and Mäkelä, 1962) as the method of choice for the analysis of antibody production at the single cell level, recent work (Nossal *et al.,* 1970a; Bussard *et al.,* 1970) has shown that the technology commonly employed to enumerate antibody-forming cells is still capable of considerable improvement. While studying the formation of antibodies to sheep red blood cells (SRBC) by normal mouse peritoneal cells, we developed an improved version of the Ingraham-Bussard (1964) carboxymethyl cellulose (CMC) haemolytic plaque technique. This involved the preparation of very thin microcultures containing SRBC, CMC, complement and mouse lymphoid cells, held at 37°C under liquid paraffin. This permitted access to the plaque-forming cells (PFC) by a micropipette. Using a microscope held at 37°C, one could continuously monitor plaque appearance and growth, and furthermore could transfer individual PFC to second or third culture monolayers and study the effects of various inhibitors. The basic technique was termed the open CMC plaque method.

In the present brief review, we will summarize three sets of experiments, which have used the open CMC method for the analysis of problems related to the *in vivo* response of mouse spleen cells to SRBC. Details of these will be reported elsewhere (Nossal *et al.,* 1971a, b; Nossal and Lewis, 1971). First, we show that the open CMC method exhibits a greatly increased sensitivity for the very early stages of an *in vivo* response, in fact being able to demonstrate eight times more "direct" and 13 times more "enhanced" PFC than the liquid monolayer technique of Cunningham (1965). Secondly, we present evidence for functionally symmetric divisions amongst antibody-forming blast cells. Thirdly, we report that a significant minority of single cells can form IgM and IgG antibodies simultaneously.

MATERIALS AND METHODS

Animals. CBA/Har/WEHI mice aged 10-12 weeks at first immunization were used.

Antigen. Immunization consisted of an intraperitoneal injection of 10^9 SRBC.

Haemolytic plaque and micromanipulation techniques. These were essentially as previously described (Nossal *et al.,* 1970a; Bussard *et al.,* 1970). For the open CMC method, thin spreads were prepared on coverslips. These contained the following constituents: mouse spleen cells (usually at a concentration of 5×10^6/ml, though the exact number varied with the expected plaque count); 10% guinea pig serum, previously absorbed with SRBC; SRBC (5×10^8/ml); CMC (15.4 mg/ml) and tris-buffered Eagle's medium. Each spread was about $12\,\mu$ thick and had a total volume of $3.6\,\mu l$; evaporation was prevented by overlying liquid paraffin. The very viscous sol proved relatively easy to micromanipulate, using conventional techniques. All micromanipulation of PFC was done with the microscope held in a translucent plastic incubator at $37°$ C.

The liquid monolayer haemolytic plaque technique of Cunningham was used as modified by Szenberg and Cunningham (1968).

Preparation of rabbit anti-mouse immunoglobulin sera. Great care is necessary in the preparation, absorption and testing of specific inhibitory and enhancing sera for use in plaque techniques. Table I summarizes the methods we have used

TABLE I

Specific antisera used for enhancement and inhibition of plaque formation

Rabbit no.	Immunizing Agent		Absorbing Agent	
	Number	Class	Number	Class
R16	HPC−8 myeloma	IgG$_{2b}$	MPC 76	K type BJP
			MOPC 104	IgM
R17	HPC−9 myeloma	IgG$_1$	HPC 1	IgA
R19	NZB mouse serum		HPC 1	IgA
	Sephadex G-200	IgM	HPC 32	IgG$_1$
	excluded peak		HPC 5	L chain BJP
			RPC 5	IgG$_{2A}$
			MPC 86	IgG$_{2B}$

Activity of sera after absorption in radioprecipitation assay: R16 + 17 pool; strong anti-γ_1 and γ_2. R19: strong anti-μ.

to prepare rabbit antisera specific for (a) mouse μ chains and (b) a mixture of mouse γ_1, γ_{2A} and γ_{2B} chains. These sera were tested for specificity by the sensitive technique of radioimmunoprecipitation (Herzenberg and Warner, 1968). When fully absorbed "enhancing" (anti-γ) and "inhibitory" (anti-μ) sera were mixed, the residual excess IgG myeloma protein used for absorptions of the inhibitory serum interfered with the enhancing effect and *vice versa*. This problem was overcome in some experiments by using sera that had been absorbed with myeloma proteins adherent to polyaminostyrene beads, but for the most part it was found satisfactory to mix unabsorbed antisera (respectively predominantly anti-γ and anti-μ) to achieve the desired effect of suppression of IgM plaques and facilitation of IgG plaques. In this respect, it was fortunate that anti-light chain activity, present in such unabsorbed sera, inhibited "direct" PFC but not "enhanced" PFC.

Testing of pairs of daughter cells arising in vitro *from PFC for antibody production.* In this series of experiments, immunized mice were given 150 μg of colcemid 2 hr before killing. Single PFC were identified using the micro-manipulation variant of the Cunningham technique (Cunningham *et al.,* 1966; Nossal *et al.,* 1968). Large PFC and smaller PFC believed by virtue of their morphology (Nossal *et al.,* 1968) to be metaphase-arrested mitoses were transferred to droplets containing tissue culture medium. After periods varying from 15 min to 2 hr, a proportion of these cells escaped from metaphase arrest, and divided. The two daughter cells were separated from each other by micromanipulation, and were placed, separately but close to each other, in a CMC monolayer for plaque detection. Antibody-secreting activity (plaque growth rate) was then carefully observed over the next 2 hr.

RESULTS AND DISCUSSION

Comparative sensitivities of open CMC and liquid monolayer techniques. Figure 1 gives the results of a study in which the numbers of PFC arising in mouse spleen after immunization with SRBC were assessed by two techniques, the open CMC method and the Cunningham method. The curves for direct plaques give the numbers of PFC/spleen when no rabbit sera were incorporated into the plaque test. The curves for "IgG" plaques are derived from a subtraction of the direct plaque count from the plaque number observed in monolayers containing a 1 : 100 dilution of pooled rabbit anti-mouse γ_1 and γ_2 sera. The curves using the Cunningham technique present no surprises, being in general similar to those reported with the Jerne technique by previous authors (Wortis *et al.,* 1969; Šterzl and Říha, 1965; Plotz *et al.,* 1968; Nordin *et al.,* 1970). The curves using the open CMC method differ from previous studies in several important respects. As regards the number of direct plaques, this is already significantly above background by 24 hr, having risen from 730/spleen to

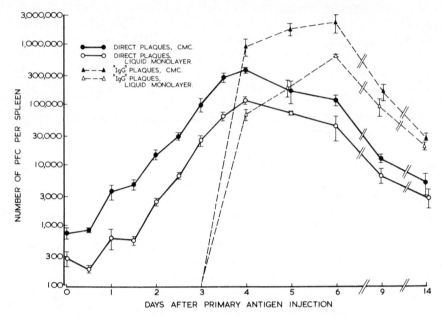

Fig. 1. Number of PFC/spleen using "direct" and "enhanced" monolayers, of either Cunningham or open CMC type. The number of IgG plaques as plotted represents the enhanced plaque number minus the direct plaque number. Vertical bars indicate standard errors of the mean.

3600/spleen. It rises little between 24 and 36 hr, and then resumes the expected exponential rise between 36 hr and 3.5 days, the doubling time over this interval being 8.1 hr. A peak value of 342,000 PFC/spleen is attained at four days, after which the plaque count falls. More obvious differences from reported results are evident for IgG plaques. At three days, there are no statistically significant differences between direct and enhanced counts, but by four days the number of IgG plaques is already 910,000/spleen, over twice the count of direct plaques. The IgG plaque count continues to rise, and by day 6 peaks at a value of 2,300,000/spleen, over 10 times the value reported by other authors. In other words, there is an explosive emergence of very large numbers of IgG PFC from day 3 to day 4, and a high proportion of the PFC that arise are simply not detected by conventional techniques.

In Fig. 2, we plot the mean ratios of the number of PFC detected by the open CMC method versus the Cunningham method at various stages of the immune response. The curves represent a different way of plotting the data of Fig. 1. It is evident that the open CMC technique has a special ability to spot PFC arising early in an immune response. For example, the ratio is 8.3 for direct plaques at 1.5 days, and then falls progressively to 1.97 at day 14. Similarly, when IgG PFC

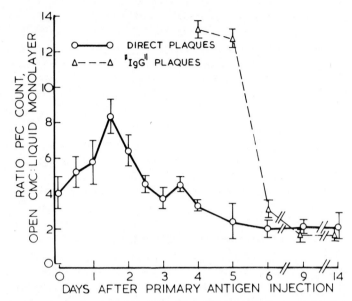

Fig. 2. Ratios of plaque numbers given by the open CMC versus the Cunningham techniques at various stages of the primary response of mouse spleen cells to SRBC.

first appear at day 4, the open CMC technique is 13.3 times more effective than the liquid monolayer method, and it maintains its lead on day 5, but thereafter there is a progressive fall in ratio, till at days 9 and 14, the open CMC technique is only 1.5-1.7 times as effective as the Cunningham method.

Obviously, with a method so much more sensitive than any previously reported, it is important to beware of possible artifacts and to ensure that the plaque technique is truly enumerating antibody-forming cells. We therefore performed the following experiments on IgG PFC to establish their validity. We showed that:

1. Plaque appearance was totally complement-dependent.

2. The number of plaques was proportional to the number of lymphoid cells in the monolayers.

3. Neither passive antibody given to a heterologously immunized animal nor *in vitro* incubation of lymphoid cells with antibody could cause cells to become PFC.

4. Cells removed by micromanipulation from the centre of IgG plaques reform plaques in an enhanced CMC monolayer in 98% of cases.

5. Plaques resulting from such a transfer grow in area proportionately to the time of incubation.

6. The PFC themselves resemble morphologically PFC detected by other methods, and are definitely lymphoid cells and not macrophages.

These considerations provide good evidence that the method is really measuring antibody-secreting cells and not cells that had adsorbed cytophilic antibody. We must now ask what special virtues the CMC method has. It is virtually a two-dimensional technique in contrast to the conventional Jerne method where the PFC must create a hemispherical zone of haemolysis; however, this is true of the Cunningham technique as well. The mechanical stability offered by the extreme viscosity of the medium is undoubtedly an important factor. For example, one can readily make a plaque-like hole in a Cunningham monolayer by jetting a fluid stream into it by micropipette. Brownian movement of erythrocytes fills this up again within a matter of minutes. In the CMC layer, such a drift of erythrocytes cannot occur. The delay in diffusion of antibody from the secreting cell occasioned by a viscous medium must be a second factor. Finally, the CMC may actually act chromatographically in reversibly binding secreted protein and thereby preventing escape. These factors may be especially important for the smaller IgG molecule.

This leaves us with the question of why cells appearing early in an immune response are so frequently missed by the conventional method. It seems clear that such cells must, on the average, be secreting either less or less avid antibody than the cells seen later in the response. We believe both considerations apply. Many of the cells seen at day 1-2 in the direct CMC plaques and at day 4 in the IgG plaques were very large, blast type cells that may well not yet have been secreting at an optimal rate. Also, it is probable that continuous selection occurs during an immune response for the sub-population producing the most avid antibody amongst the population of antibody-formers (Nussenzweig and Benacerraf, 1964). While usually experiments designed to show increasing avidity extend over a much longer time scale, it seems quite likely that clonal expansion might cease a day or two earlier amongst cells with receptors weakly reactive against an antigen than amongst the best-adapted cells. Both considerations would lead to the existence of a substantial population of cells, the secretory activity of which reaches the threshold of the CMC method, but not that of other conventional plaque techniques. It is, of course, quite likely that there exists another and perhaps even larger population of cells with very distant affinity for SRBC surface antigens, which was not detected by any of the methods. If this is so, it re-emphasizes the point that whether a globulin is considered an antibody to a particular antigen depends to a large extent on the sensitivity of the method used to ask the question.

The only method purporting to be one for the detection of antibody-forming cells against SRBC which reveals numbers of active cells comparable to those reported here is the rosette technique of Biozzi et al. (1966). However, preliminary and as yet unpublished work by Dr J. D. Wilson in my laboratory has suggested that the majority of IgG PFC are not typical "rosettes", and that the majority of rosettes fail to form IgG plaques. This leaves the nature of rosettes as a subject for further study.

Symmetry of division amongst PFC. The though many detailed models of the clonal proliferation resulting from the action of antigen on a lymphocyte have been proposed, we know neither the number of sequential divisions involved nor the exact pattern of differentiation in the expanding cell line. The issue has been further complicated by the finding that, for many immune responses, a collaboration must ensue between thymus-dependent and thymus-independent lymphocytes (Miller and Mitchell, 1969). Hence we thought it important to establish whether divisions amongst cells already forming antibody were functionally symmetric. In other words, did both daughter cells form antibody, and if so, in equal amounts? Alternatively, as has been proposed (Tannenberg and Malaviya, 1968), did one daughter cell form antibody and the other revert to the pool of antigen-reactive cells?

The special sensitivity of the CMC technique afforded an opportunity to study the point. Dividing, antibody-forming cells were identified as described in "Materials and Methods". Mitosis was allowed to proceed in microdrops, and the progeny cells were separated by micromanipulation and placed into CMC plaque-revealing monolayers for observation.

The first point noted was that, with rare exceptions, the daughter cells were closely similar in size and morphology under phase contrast microscopy. The functional results are given in Tables II to IV, which summarize our results using "direct" plaque formation. In every case, both daughter cells behaved symmetrically in that both either did or did not form antibody. We believe that the 13% of cells that failed to form plaques after division represented technical failures. Some cells proved extremely difficult to separate and may have been traumatized by forceful micromanipulation. Others took up to 2-3 hr *in vitro* to divide, and may have been in a suboptimal functional state, as plasma cells are known to be readily damaged in tissue culture. In fact, the 87% success rate (Table II) is quite satisfactory.

Table III shows that in the majority of cases, the rate of secretion, as judged

TABLE II

Symmetry of division amongst antibody-forming cells

Days after antigen	No. of experiments	No. of pairs of cells transferred	No. of cases with both cells forming plaques	No. of cases with neither cell forming plaques	No. of "asymmetric" results
2	4	10	7	3	0
3	6	31	28	3	0
4	4	27	24	3	0
Total	14	68	59	9	0

TABLE III

Equivalence of rate of antibody synthesis by daughter cells as measured by rate of increase of plaque diameter

Days after antigen	No. of cases with two plaques of equal size	Number of cases with plaques of unequal size	
		With one cell obviously larger than the other and forming a larger plaque	With no obvious discrepancy in cell diameter
2	7	0	0
3	23	3	2
4	22	1	1
Total	52	4	3

by rate of plaque growth, was the same in both daughter cells. In four of the seven exceptions, a size difference obvious at 200-fold magnification occurred in the division. Here, the larger cell made a larger plaque, though the differences never exceeded 20%. In three cases, a slight asymmetry of plaque size was not mirrored by an obvious cell size difference.

Another aspect that was noteworthy was the great similarity in plaque morphology. Thus, plaques can be clear or can exhibit various degrees of turbidity; they can have very sharp edges, or somewhat serrated edges, and so forth. These differences obviously reflect differences in the nature of the antibody secreted by the PFC. This seemed identical (as expected) in the two daughter cells. To document the point more clearly, we used CMC monolayers containing mixed goat and sheep erythrocytes. There is a known antigenic overlap here, and it is well documented that mouse PFC can make either clear or turbid plaques in such circumstances. Table IV gives the results of an experiment which shows that each member of a pair of daughter cells makes antibody of similar character.

TABLE IV

Nature of plaques formed by pairs of daughter cells in a mixed sheep and goat erythrocyte monolayer

No. of pairs of cells transferred[a]	No. of cases with both cells forming "clear" plaques	No. of cases with both cells forming "turbid" plaques	No. of cases with one clear and one turbid plaque	No. of cases with neither cell forming a plaque
13	4	7	0	2

[a] Mouse immunized three days previously with sheep erythrocytes.

These experiments would, no doubt, be of greater interest if it were feasible to follow two or more sequential divisions in antibody-forming cells. This has not proved technically possible as yet. As far as they go, they suggest that once an immunocyte of thymic-independent origin has reached the stage of detectable antibody secretion, it is committed to symmetric division into equi-potent daughter cells. It seems likely that such division would lead, eventually, to suicidal differentiation as occurs in the myeloid and erythroid cell lines. In that case, one would have to look elsewhere for the source of immunologic memory. Dr J. F. A. P. Miller and colleagues are currently actively exploring the question of whether memory in this system resides in the thymus-dependent cell lines.

Incidence of cells forming both IgM and IgG antibodies to SRBC. The open CMC detection system seems especially sensitive for the detection of IgG type PFC. This has provided us with an opportunity to study the question of the simultaneous synthesis of IgM and IgG antibody by single cells in a system far more refined than the one we used in 1964 to study the point (Nossal *et al.*, 1964). As single PFC can be serially transferred from monolayer to monolayer by micromanipulation, one can test cells first for their ability to produce IgM and can then transfer them to a new monolayer ("enhanced-inhibited") to test for the production of IgG.

In Table V we show how different types of PFC should behave on the basis of known or suspected characteristics of antibody of the various immunoglobulin classes. It can be seen that testing of PFC in an "enhanced" monolayer contributes nothing to our ability to detect PFC producing both IgM and IgG. Preliminary experiments on multiple micromanipulation transfers of single PFC showed us that while 93% of cells could make three plaques in succession, only 80% could make four, so that we decided to embark on a search for double (IgM-IgG) producers by transferring PFC through three CMC monolayers: either

TABLE V

Postulated behaviour of various cell categories on multiple transfer

Type of PFC	"Anti-immunoglobulin in monolayer"			
	Nil (direct)	Anti-γ (enhanced)	Anti-μ (inhibited)	Anti-γ + Anti-μ (enhanced-inhibited)
Typical IgM	++	++	−	−
Typical IgG	−	++	−	++
IgM-IgG double	++	++	−	++
"Lytic" IgG	+	+++	+	+++
Very strong IgM	+++	+++	+	+
"Enhanceable" IgM	−	−	++	++

C.I.−8

from "direct" to "enhanced-inhibited" to "inhibited"; or from "enhanced-inhibited" to "direct" to "inhibited". The theoretical requirements for a double-producer were that it would make a plaque in each of the first two monolayers but not (in contrast to a "lytic IgG" or "breakthrough-IgM"-producer) in the third monolayer.

Much pilot work had to be done before the main series of experiments could be undertaken. The results of this can be summarized as follows:

1. At day 3 of the primary response, enhancing sera failed to raise the observed number of PFC (Fig. 1), suggesting that IgG-formers were not yet in evidence. Titrations of anti-μ chain sera incorporated at various dilutions into the CMC monolayer were performed at that time point. It was shown that a 1 : 50 dilution of a particular, fully absorbed serum was 100% effective in eliminating plaque formation. This serum was used as the "inhibitory" serum throughout.

2. At day 4 of the primary response and thereafter, it was found that monolayers containing this inhibitory anti-μ serum did develop a few small, late-appearing plaques. Micromanipulation transfer readily allowed one to place these in two categories: "lytic IgG"-producers, and strong IgM-producers breaking through the inhibitory influence, as predicted from Table V. Lytic IgG producers formed small, late, very turbid, irregular plaques in the "direct" CMC monolayers and exactly the same type of plaque on transfer to the "inhibited" (anti-μ) monolayer, but when transferred to "enhanced" (anti-γ) monolayers, plaque formation was greatly accelerated. In contrast, "breakthrough" IgM-producers made small, late-appearing, round, clear plaques. On micromanipulation transfer to a direct layer, the PFC made greatly accelerated plaques and there was no rate change on further transfer to an "enhanced" monolayer. In fact, it was surprising how closely the observed behaviour fitted to the expected results.

3. No evidence for non-lytic ("enhanceable") IgM-producers, able to produce plaques facilitated by anti-μ sera, emerged in this work, either from plaque count data or from micromanipulation transfer.

4. Typical, early-appearing direct PFC were always inhibited in anti-μ sera, confirming their IgM nature. When the strongest of such cells did break through the inhibitory influence, the inevitable result was a plaque, appearing after a much longer latent period than in the direct layer, which was small and extremely slow-growing.

5. This inhibition was readily reversible. Cells held in anti-μ serum-containing CMC and retransferred to "direct" monolayers started plaque formation almost immediately.

6. The presence of "enhancing" (anti-γ) serum in a monolayer affected neither the latent period of plaque development nor the plaque growth rate when typical IgM PFC were transferred into such a layer.

7. The best conditions for the identification of IgG producers were when a monolayer was prepared containing a final concentration of 1 : 50 anti-μ 1 : 200 anti-γ_1 and 1 : 200 anti-γ_2 sera, each serum being unabsorbed.

8. Extensive experience both with the use of this mixture and through some thousands of individual micromanipulation transfers has allowed us to delineate with fair, though by no means total, accuracy the immunoglobulin class made by a PFC in an "enhanced" monolayer. With perhaps 20% of exceptions in each direction, IgM plaques tend to be round, sharp-edged and clear; they can frequently be identified within 5 min of incubation of the monolayer as a single, continuous ring of lysed SRBC around the PFC. IgG-producers are initially very turbid, and because of this are bigger and later when first identified 6-20 min after incubation of the reaction mixture. They grow just as rapidly as IgM plaques, and tend to clear later, but usually maintain rather ragged edges.

Details of these eight sets of experiments, and of other related ones will be presented elsewhere (Nossal *et al.,* 1971b). Some of the features emerge in Figs 3 and 4.

Tables VI and VII show the results of the investigation of 900 cells for the

TABLE VI

Incidence of cells producing both IgM and IgG against sheep erythrocytes

Day of primary response	Number of PFC transferred	Number[a] of cells producing			Per cent double producers
		IgM	IgG	IgM + IgG	
3 to 3.9	195	137	55	3	1.5
4 to 4.9	251	105	141	5	2
5 to 6	190	55	133	2	1
Total	636	297	329	10	1.5

[a] Non-random selection of cells; hence IgM : IgG ratios not representative.

TABLE VII

Incidence of cells producing both IgM and IgG against SRBC in the secondary response

Day of secondary response	Number of PFC transferred	Number of cells producing			Per cent double producers
		IgM	IgG	IgM + IgG	
2 to 2.9	34	19	14	1	3
3 to 3.9	160	68	90	2	1.25
4	70	23	46	1	1.4
Total	264	110	150	4	1.5

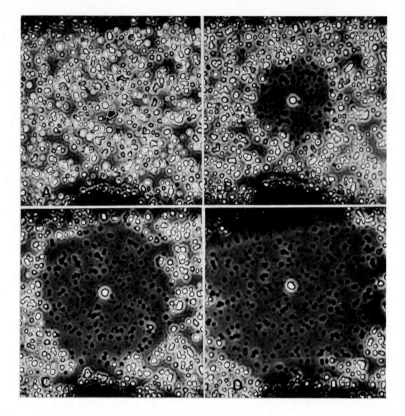

Fig. 3. Stages in the growth of a typical "direct" IgM-type plaque after micro-manipulation transfer of a PFC.
A. PFC immediately after transfer (arrowed).
B. Fifteen minutes later. A small, clear plaque has appeared.
C. After 30 min incubation.
D. After 60 min incubation.
The areas of blackening at the top and bottom of the photographs represent the Indian ink lines that mark the location of the transferred cell. Photographed at 200-fold magnification, phase contrast, and slightly enlarged.

simultaneous production of IgM and IgG. The actual number of double producers is small, about 1.5% of the total of cells examined. It could be a slight underestimate as typical plaque formation in both "direct" and "enhanced-inhibited" monolayers was demanded, and cells secreting predominantly one class with a trace of the second may have been missed. Furthermore, if a cell forming a plaque in both "direct" and "enhanced-inhibited" monolayers gave even a trace of lysis in the "inhibited" (anti-μ) monolayer, it was not termed a double producer, as it could then have been a lytic IgG or a "breakthrough"

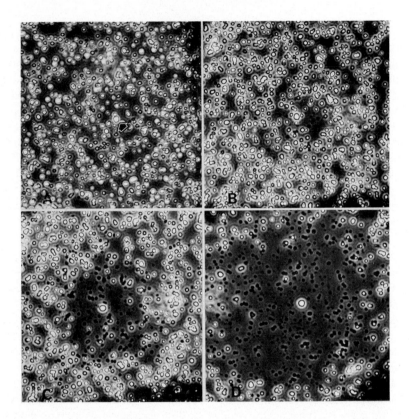

Fig. 4. Stages in the growth of an "IgG" type plaque in a monolayer containing anti-γ serum.

A. PFC immediate after transfer (arrowed).

B. At 10 min incubation. Ten or so ghosts are seen near the cell, as well as some more remote lysis.

C. At 20 min. A definite though very turbid plaque has appeared. The plaque morphology is highly characteristic.

D. At 60 min. The plaque is larger and clearer but still has an indefinite edge.

IgM. Thus a very few double producers may have escaped detection through the rigid criteria which we felt it necessary to set up.

The figure of 1.5% of double producers reported here compares with estimates in the literature varying from 0.5% (Cosenza and Nordin, 1970) to 1% (Pernis and Chiappino, 1964) or even 2-4% (Nordin *et al.*, 1970). It is lower than that found by us in an earlier, less satisfactory, test system (Nossal *et al.*, 1964). It is not possible with the present experimental approach to state whether an IgM to IgG transition is an obligatory event for all clones destined for IgG production. To determine this one would need a method for the continuous monitoring of antibody production in isolated, expanding clones. It may be wise to point out that we have never claimed such a transition as an obligatory event. Rather, we raise it as one possibility. What the present experiments do show is that a double IgM-IgG producer is a not very rare reality. This is consistent with the finding of Oudin and Michel (1969) of shared idiotypic determinants in IgM and IgG antibodies of individual rabbits. It is also consistent with the view of Gally and Edelman (1970) that V genes could be translocated as episomes to any one of a number of C immunoglobulin genes in a cell, and that occasionally an episome might "drift" from the C portion of the gene coding for μ chains to that of the gene coding for a γ chain, in the one cell or cell line. An alternative explanation for the occurrence of IgM-IgG double producers is that an occasional cell may break the rule of phenotypic restriction, and allow the simultaneous transcription of two heavy chain genes.

Summary and Conclusions

The open CMC plaque technique has been used to study a variety of problems in the immune response of mice to SRBC. The technique can discover at early stages of both the IgM and IgG responses, a large number of antibody-forming cells not revealed by standard methods. In fact, it has been shown that generally-quoted figures for numbers of IgG-producing plaques are underestimates by a factor of up to 13.

The open CMC method has been used to study the properties of daughter cells arising from mitoses in antibody-forming cells. It has been shown that such division is always functionally symmetric. Both daughter cells form antibody of similar quality usually at an identical rate.

The special sensitivity of the method has been useful for the study of cells capable of secreting both IgM and IgG antibodies against SRBC. Of 900 cells examined, 14 (1.5%) were double producers. Some IgG-producers capable of making "direct" plaques were also found. "Enhanceable" IgM-producers, that is cells making plaques only in the presence of anti-μ chain serum, were not seen in these experiments.

ACKNOWLEDGEMENTS

Our thanks are due to Miss Sue Hill for excellent technical assistance. This work was supported by the National Health and Medical Research Council, Canberra, Australia, the Australian Research Grants Committee, Canberra, the National Institutes of Health (AI-0-3958), and the United States Atomic Energy Commission (AT(30-D03695)).

This is publication 1424 from the Walter and Eliza Hall Institute.

REFERENCES

Biozzi, G., Stiffel, C., Mouton, D., Liacopoulos-Briot, M., Decreusefond, C. and Bouthillier, Y. (1966). *Annls Inst Pasteur Paris 110,* 7-32.

Bussard, A. E., Nossal, G. J. V., Mazie, J. C. and Lewis, H. (1970). *J. exp. Med. 131,* 917-935.

Cosenza, H. and Nordin, A. A. (1970). *Fedn Proc. 29,* 288.

Cunningham, A. J. (1965). *Nature, Lond. 207,* 1106-1107.

Cunningham, A. J., Smith, J. B. and Mercer, E. H. (1966). *J. exp. Med. 124,* 701-714.

Gally, J. A. and Edelman, G. M. (1970). *Nature, Lond, 227,* 341-348.

Herzenberg, L. A. and Warner, N. L. (1968). *In* "Regulation of the Antibody Response" (B. Cinader, ed.), pp. 322-348. Charles C. Thomas, Springfield, Illinois.

Ingraham, J. S. and Bussard, A. E. (1964). *J. exp. Med. 119,* 667-684.

Jerne, N. K. and Nordin, A. A. (1963). *Science, N.Y. 140,* 405.

Miller, J. F. A. P. and Mitchell, G. F. (1969). *Transplantation Rev. 1,* 3-42.

Nordin, A. A., Cosenza, H. and Sell, S. (1970). *J. Immun. 104,* 495-501.

Nossal, G. J. V. (1958). *Bri. J. exp. Path. 39,* 544-551.

Nossal, G. J. V. and Lewis, H. (1971). *Immunology.* (In press.)

Nossal, G. J. V. and Mäkelä, O. (1962). *A. Rev. Microbiol. 16,* 53-74.

Nossal, G. J. V., Szenberg, A., Ada, G. L. and Austin, C. M. (1964). *J. exp. Med. 119,* 485-502.

Nossal, G. J. V., Cunningham, A. J., Mitchell, G. F. and Miller, J. F. A. P. (1968). *J. exp. Med. 128,* 839-853.

Nossal, G. J. V., Bussard, A. E., Lewis, H. and Mazie, J. C. (1970). *J. exp. Med. 131,* 894-916.

Nossal, G. J. V., Lewis, H. and Warner, N. L. (1971a). "Cellular Immunology". (In press.)

Nossal, G. J. V., Warner, N. L. and Lewis, H. (1971b). "Cellular Immunology". (In press.)

Nussenzweig, V. and Benacerraf, B. (1964). *J. Immun. 93,* 1008.

Oudin, J. and Michel, M. (1969). *J. exp. Med. 130,* 619-642.

Pernis, B. and Chiappino, G. (1964). *Immunology, 7,* 500-506.

Plotz, P. H., Talal, N. and Asofsky, R. (1968). *J. Immun. 100,* 744-751.

Šterzl, J. and Říha, I. (1965). *Nature, Lond. 208,* 858-859.

Szenberg, A. and Cunningham, A. J. (1968). *Nature, Lond. 217,* 747-748.

Tannenberg, W. J. K. and Malaviya, A. N. (1968). *J. exp. Med. 128,* 895-921.

Wortis, H. H., Dresser, D. W. and Anderson, H. R. (1969). *Immunology 17,* 93-110.

The Common Cell Precursor for Cells Producing Different Immunoglobulins

J. ŠTERZL* and A. NORDIN

*Lobund Laboratory, Department of Microbiology, University of Notre Dame,
Notre Dame, Indiana 46556, U.S.A.*

The methods available for examining individual antibody producing cells have shown that they synthesize only one immunoglobulin class (Cebra *et al.*, 1966), one allotypic specificity (Pernis *et al.*, 1965), and one antibody specificity (Mäkelä, 1967). The synthetic capacity of these terminal and fully differentiated cells does not, however, permit any extrapolation to the potential of the original precursor cells.

Our experiments are concerned with the potentialities of the precursor cell. Does there exist an undifferentiated precursor cell with multipotential capacities with respect to the different immunoglobulin classes?

To determine the potential of the precursor cell we use the method of cloning transferred immunocompetent cells in lethally irradiated recipients (Makinodan *et al.*, 1960; Playfair *et al.*, 1965; Kennedy *et al.*, 1965). Precursors stimulated by antigen multiply and form clones. If precursor cells are already restricted as to class of immunoglobulin, the foci of antibody producing cells should contain cells producing only one type of immunoglobulin. However, if the precursor cell for the appropriate antigenic specificity can generate cells synthesizing more than one type of immunoglobulin the foci should contain both IgM- and IgG-producing cells. There is already some circumstantial evidence that changes occur at the level of activated cells (for nomenclature see Table I). There is not only the 100-1000 times increase in the number of the activated cells, but the slopes expressing the relation between antigen dose and number of cells in the primary response (competent cells) and in the secondary response (activated cells) suggest that activated cells have an increased antigen-binding capacity (Šterzl and Jílek, 1967).

Other studies (Šterzl *et al.*, 1970; Šterzl and Mandel, 1970) suggest that competent cells cannot be suppressed by an excess of antigen directly. In the true primary response (in germ-free animals) all doses of antigen (up to a certain

* Permanent address: Institute of Microbiology, Czechoslovak Academy of Sciences, Prague, Czechoslovakia.

TABLE I

Terms used for individual stages of cells differentiating towards antibody formation

Stem cell	Immunologically competent cell	Immunologically activated cell	Antibody producing cell
Primordial	Precursor cell	Primed, memory cell	Plaque forming cell
S-cell	X-cell	Y-cell	Z-cell
Germ cell	PC_1 cell	PC_2 cell	P_1-P_4 cell
	Progenitor cell Antigen sensitive cell		

threshold) stimulate antibody formation. Only at the level of immunologically activated cells were three zones observed: (a) low-dose tolerance, where minimum doses of antigen terminated the differentiation of memory cells without transforming them into antibody-forming cells; (b) secondary response, with active proliferation; and (c) high-dose tolerance, in which activated (memory) cells were transformed into antibody-forming cells and exhausted through terminal differentiation.

Further experiments demonstrate changes at the level of immunologically activated cells (Šterzl et al., 1969). In animals given 10^8 sheep red blood cells (SRBC), simultaneous injection of anti-SRBC antiserum completely inhibited the primary response and the secondary response. The same inhibitory dose of antiserum was injected 24 and 48 hr after antigen. In these groups the secondary response contained only cells producing IgM antibodies. Only if the antigen was removed by antiserum 72 hr or more after primary immunization did a typical secondary response develop with 10 times more IgG than IgM antibody-forming cells. This suggested that precursors for IgG production appeared only after several proliferation cycles of the immunologically activated cells. Similarly, restriction of proliferation by X-irradiation (Svehag and Mandel, 1964; Smith, 1964; Smith and Robbins, 1965) or 6-mercaptopurine (Smith and Robbins, 1965; Sahiar and Schwartz, 1964; Borel et al., 1965) significantly inhibited the appearance of IgG antibody-forming cells, but IgM antibody forming cells were not depressed. Thus, according to our hypothesis, we would expect that the precursor cell, activated by antigen, proliferates, that in the presence of antigen the first generations of activated cells give rise to IgM antibody-forming cells, and that only later do IgG-producing cells appear in the same foci that contain IgM producers (Fig. 1).

This paper demonstrates that on days 4 and 5 after antigen stimulation the foci are of IgM type but on days 6 to 8 some of the IgM foci also contain IgG-producing cells.

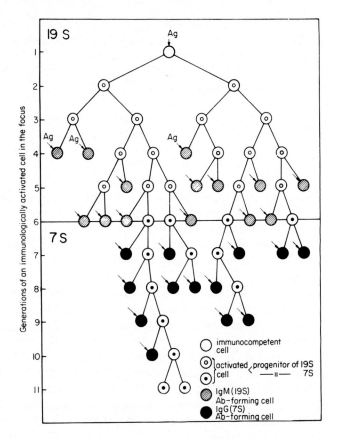

Fig. 1. Scheme of development of IgM- and IgG-producing cells from a common precursor (immunocompetent cell) in the focus.

METHOD

Animals. C3H/F inbred male mice, 8-10 weeks of age, were obtained from the Texas Inbred Mouse Company. They were kept under normal conditions and fed autoclaved Tek-Lab diet. To prevent post-irradiation infection they were routinely given HCl in their drinking water (McPherson, 1963).

Irradiation. Mice were irradiated with a 260-kvp clinical therapy X-ray machine, operated at 250 kvp and 15 ma, filtration 1.0 mm Al and 0.25 mm Cu (HVL = 1.05 mm Cu) at approximately 40R/min as measured in air with a Victoreen condenser R-meter.

Preparation of spleen cell suspension. Spleen cell suspensions were prepared by rubbing the spleen against a stainless steel wire mesh and filtering through two layers of surgical gauze. The cell suspension was washed by centrifuging at

1000 r.p.m. for 10 min and resuspending in Eagle's medium. An aliquot (0.1 ml) was added to 0.9 ml of a 1% solution of methylrosaniline chloride in 0.1 M citric acid and counted. The proper amount of Eagle's medium was then added to obtain the desired number of cells/ml. Viability of the cells was tested with trypan blue. For each experiment 2-3 spleens were pooled.

Cell transfer and immunization. Unless otherwise stated mice were injected i.v. with 0.2 ml of a mixture of equal parts of spleen cell suspension and 20% SRBC suspension 24 hr after irradiation. Control mice were also injected i.v. with either spleen cells or SRBC 24 hr after irradiation or left as uninjected controls.

Assay for foci. Spleens of irradiated recipient mice were assayed for foci of antibody producing cells by a method similar to that described by Kennedy *et al.* (1965) and Playfair *et al.* (1965). The spleen was removed either before perfusion or after perfusing anesthetized mice. Mice were perfused by injecting 10-20 ml of a saline-heparin solution into the heart after severing the major vessels of the abdomen. This was essential for removing preformed antibody in the vessels of the spleen.

The spleen was weighed and placed in a Petri dish with Eagle's medium and cut in half. Each half was placed on a tissue holder supplied with frozen Eagle's medium. The holder was put on the quick-freeze bar of the International Cryostat and covered with Eagle's medium during freezing. The temperature within the cryostat was $-20°$ C. Spleen sections 30 or 50 μ thick were placed on microscope slides previously coated with an agarose-SRBC mixture. To prepare these slides a thin layer of 0.5% agarose in 0.3% NaCl was first applied and thoroughly dried. This was essential to ensure strong adherence of the SRBC-agarose. The latter was prepared by placing 0.2 ml of 0.5% agarose in Eagle's medium in a prewarmed test tube followed by 0.02 ml of a 20% suspension of SRBC, mixing thoroughly and spreading on the slide with a glass rod. These slides were prepared immediately before use. Because the transfer of frozen cell sections often resulted in freezing, thawing and subsequent lysis of the SRBC at the point of contact, it was necessary to prewarm the slides. Approximately 8-10 sections were placed on a slide, which was then carefully covered with another slide containing a SRBC-agarose layer. To facilitate placing the slides together, frosted end slides were used and the frosted ends were not coated with SRBC-agarose (Fig. 2). A little clay was placed on the corners of the frosted ends of each slide. Four to five drops of Eagle's medium were added to the slide without spleen sections, and the slide with the sections was carefully positioned so that the frosted ends were at opposite ends of the replica pair. With some experience slides could be put together without disturbing the sections. Any air trapped between the slides will eventually cause lysis of the SRBC, and every effort should be made to avoid this since these areas could be

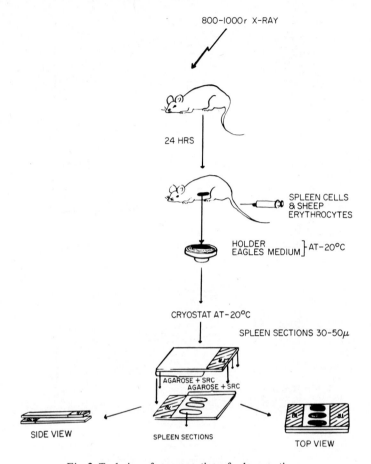

Fig. 2. Technique for preparation of spleen sections.

confused with true foci. The slides were then incubated at $37°$C in a CO_2 incubator for 1 hr.

After incubation the slides, still together, were placed on a projector which projected the image onto a parchment paper screen. The position of the slides could be exactly drawn because the outline of the slides and the margins created by the frosted ends were exactly traced along with the outline of each spleen section (Fig. 3). The slides were separated and appropriately marked as to bottom or top and left or right, depending on the orientation of the frosted end of the slide. Spleen sections were washed off and the slides immersed in Eagle's medium. The two slides were treated differently to detect IgM and IgG (see next

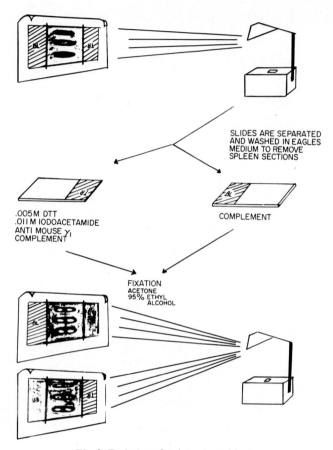

.005 M DTT
.011 M IODOACETAMIDE
ANTI MOUSE γ_1
COMPLEMENT

SLIDES ARE SEPARATED
AND WASHED IN EAGLES
MEDIUM TO REMOVE
SPLEEN SECTIONS

COMPLEMENT

FIXATION
ACETONE
95% ETHYL
ALCOHOL

Fig. 3. Technique for detection of foci.

paragraph), then fixed for 5 min in absolute acetone and 10 min in ethyl alcohol before being dried with a fan. To locate the foci the first slide of the replica pair was projected onto the corresponding previously drawn outline, the IgM foci were drawn in red, and then a second slide, showing IgG foci, was projected and the foci outlined with green on the same drawing. The number of matching and separate IgM and IgG foci was counted.

Detection of IgM- and IgG-producing cells in the foci. To detect IgM antibody one slide was treated with complement only (1/30 dilution of guinea pig complement, Fig. 4, procedures 1, 2 and 3). To detect IgG we did not use stimulation by specific antiserum, as used by Celada and Wigzell (1966) and Papermaster (1967), because both IgM and IgG are present on the same slide. Neither were we satisfied with separation of IgM and IgG foci by the specific

The methods for detection of IgM and IgG cells (foci)

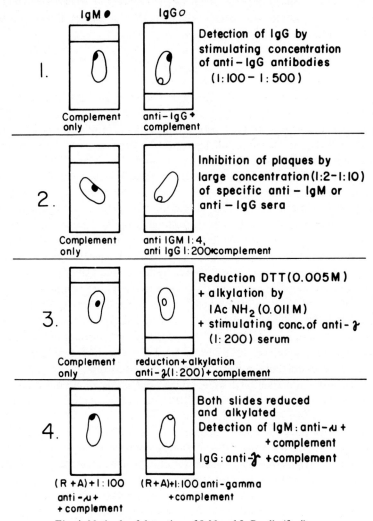

Fig. 4. Methods of detection of IgM and IgG cells (foci).

inhibitory effect of a large concentration of specific anti-IgM or anti-IgG serum (procedure 2). In most experiments reduction and alkylation and a subsequent treatment with specific anti-γ serum (for 1 hr before exposure to complement) were used to detect IgG antibodies (procedure 3). There could still, however, be the objection that foci detected by complement only could be the result of excess IgG antibodies or IgG with high hemolytic activity. Therefore both slides of the pair were reduced and alkylated, washed and incubated with specific anti-μ or anti-γ serum before adding complement (procedure 4). Experiments were performed to estimate how effectively reduced and alkylated IgM antibodies could be restored by anti-μ serum.

RESULTS

1. Restoration of IgM antibodies after reduction and alkylation. It is well established that IgM is more efficiently hemolytic than antibody of the IgG class (Humphrey and Dourmashkin, 1965). The less efficient IgG antibody, when present at levels below detection by routine procedures, can be enhanced by adding antiserum against IgG (Říha, 1964; Šterzl and Říha, 1965; Dresser and Wortis, 1965). However, if IgG antibody is present at high levels, hemolysis of red cells with complement will occur without the addition of anti-Ig serum (Říha, 1964; Humphrey, 1967). Therefore the method of enhancement by anti-γ serum is not sufficient to distinguish IgG from IgM hemolysin. To be sure that all IgM foci are specific we developed a method of restoring the activity of IgM antibody after reduction and alkylation by specific anti-μ sera. The hemolytic test was performed as described by Kabat and Mayer (1961).

Antiserum was obtained from C3H mice five days after immunization with 0.2 ml of a 20% suspension of SRBC. The antiserum was fractionated on a Sephadex G-200 column. Fractions I (IgM) and II (IgG) were concentrated to the original serum volume. By Ouchterlony gel precipitation the only immunoglobulin detected in fraction I was of the IgM type and the only immunoglobulin detected in fraction II was of the IgG type.

Whole serum and fraction I were reduced and alkylated either before or after the antibodies were absorbed onto SRBC. In both instances antibodies were reduced with 0.005 M dithiothreitol in 0.1 M Tris, 0.1 M NaCl at pH 8.6 and alkylation with 0.011 M iodoacetamide in the same buffer but at pH 7.3. After reduction and alkylation the SRBC-antibody complexes were washed five times with cold veronal buffer. The concentration of erythrocytes was then adjusted to 2% according to the calibration curve.

Anti-IgM was prepared in rabbits against μ heavy chains and anti-γ was prepared in sheep against the Fc fragment (Sell *et al.,* 1970). Anti-immunoglobulin sera were heat inactivated and absorbed with SRBC before use. Various dilutions of these antisera were added to the SRBC-antibody complexes and incubated at $4°C$ for 1 hr. After washing the cells, complement was added and the degree of hemolysis was measured.

The anti-SRBC serum produced 100% hemolysis in dilutions from 1 : 10 to 1 : 200 and 70% hemolysis at 1 : 400. The effect of anti-μ and anti-γ_1 sera on antibody, adsorbed onto SRBC and then reduced and alkylated, was estimated with two dilutions of anti-SRBC serum. The results (Fig. 5) demonstrate that a

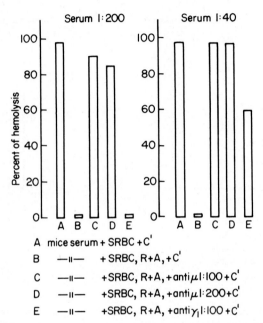

The restoration effect of anti-IgG sera on reduced + alkylated antibodies

A mice serum + SRBC +C'
B —"— + SRBC, R+A, +C'
C —"— +SRBC, R+A, +antiμI:100+C'
D —"— +SRBC, R+A, +antiμI:200+C'
E —"— +SRBC, R+A, +antiγ₁I:100+C'

Fig. 5. Restoration of reduced and alkylated (R + A) antibodies by anti-immunoglobulin sera.

1 : 100 dilution of anti-μ serum resulted in nearly complete recovery of the hemolytic activity of a 1 : 200 dilution of anti-SRBC serum. At an anti-SRBC dilution of 1 : 40, not only anti-μ but also anti-γ₁ effectively restored the hemolytic activity. These results indicate that the anti-SRBC serum contained IgG antibody, and further studies were done with the fractions obtained from Sephadex filtration of the anti-SRBC serum.

Fraction I (IgM) produced complete hemolysis at 1 : 50 and 30% hemolysis at 1 : 500 dilution (Fig. 6). Fraction II produced only weak hemolysis at 1 : 10 dilution. The activity of reduced and alkylated IgM antibody (fraction II) was partially restored by anti-μ (diluted 1 : 50 or 1 : 100) but not by anti-γ. When fraction I was reduced and alkylated before adsorption onto SRBC, the hemolytic activity could also be restored by specific anti-μ serum. Restoration by several antisera at three critical dilutions (1 : 20, 1 : 50, 1 : 500) of fraction I is shown in Fig. 7. By comparing the results of anti-μ (5) (diluted 1 : 100) on fraction I (diluted 1 : 50) in both Figs 6 and 7, it can be seen that recovery of the hemolytic activity of fraction I was slightly increased if the antibody was

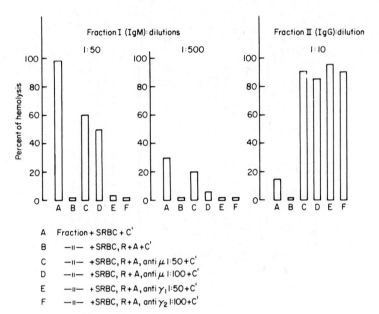

Fig. 6. Restoration of reduced and alkylated (R + A) serum fractions by anti-immunoglobulin sera.

reduced and alkylated after adsorption onto SRBC. On the other hand fraction II, in which we expected only IgG antibody, when diluted 1 : 10 was enhanced not only by anti-γ_1 but also by anti-μ. Additional experiments allowed us to conclude that: 1. The hemolytic activity of IgM antibody which is lost after reduction and alkylation can be reconstituted by anti-μ sera. 2. The Sephadex fraction of anti-SRBC serum, which characteristically contains immunoglobulins with sedimentation constants of 7S, was enhanced by anti-μ sera indicating the presence of naturally occurring subunits of IgM.

2. *Foci, containing IgM- and IgG-producing cells, detected by different methods.* It has been found that the quantity of antibodies (Makinodan *et al.*, 1960) and the number of foci (Kennedy *et al.*, 1965; Diener *et al.*, 1968; Miller and Mitchell, 1967; Armstrong and Diener, 1969) present in the spleens of irradiated mice is proportional to the number of spleen cells transferred. We examined this relation in our experimental conditions.

The average number of IgM foci per spleen section after the transfer of 10^6 spleen cells was approximately 1-4, after transfer of 5×10^6 spleen cells it was increased to 16-20. Both slides of the pair were developed by complement

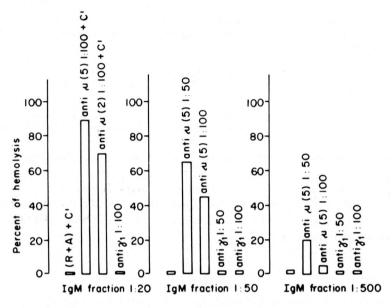

Fig. 7. Restoration of reduced and alkylated (R + A) IgM fraction by anti-immunoglobulin sera.

only to determine the efficiency for detecting the same foci on both slides. The results of several experiments have shown that 70-80% of the foci can be detected on both slides.

Three types of lethally irradiated controls were included in each experiment: some were injected with only spleen cells, some with only SRBC and others were uninjected. Control mice were assayed at various times after treatment. In mice receiving only spleen cells there were several foci of the IgM type 5-8 days after injection. The other two control groups usually had no detectable foci. We were never able to detect foci of the γ type in any of the controls. The possible explanations for individually occurring IgM foci will be discussed later.

The results of a typical experiment for detecting cells producing IgM and γ_1 in foci (using method 3, Fig. 4) at different days after the simultaneous injection of 10^6 spleen cells and 4×10^8 SRBC are presented in Table II. The foci detected at five days were all of IgM type. There was a slight increase in the number of IgM foci up to day 7 followed by a slight decline at day 8. Detectable γ_1 foci appeared at days 6, 7 and 8. The most critical point is that when γ_1 foci were first detected, almost all corresponded with the location of IgM foci on the

TABLE II

Number of foci at various times after injection of 10^6 spleen cells and 4×10^8 SRBC

Days after immunization	No. of foci IgM	No. of foci IgG	No. of matching foci
5	122	0	–
6	187	7	7
7	203	11	10
8	134	35	23

other slide. However, as expected, some of the γ_1 foci appeared separate and distinct as the development of the immune response progressed (day 8). In other experiments 5×10^6 cells were transferred, and with the higher number of cells larger numbers of foci were detected. Again 10% of IgM foci also contained γ_1 cells, and 95% of γ_1-producing foci matched with IgM. The results using method 3 are summarized in Table III.

TABLE III

Number of foci producing IgM and IgG antibodies at different days after transfer of cells (method 3)

Days after transfer	No. of cells transferred	No. of sections	No. of IgM colonies	No. of IgG colonies	No. of IgG identical with IgM	% IgM-IgG identity
4½	10^6	26	122	0	0	0
6	5×10^6	96	1135	108	102	95
6	10^6	41	187	6	6	100
7	10^6	36	163	1	1	100
7	10^6	34	203	11	10	90
8	10^6	48	349	10	8	80
8	10^6	48	134	35	23	66

To look for possible IgG hemolytic foci we used method 4, in which the activity of IgM antibody after reduction and alkylation is recovered by specific anti-μ serum. After reduction and alkylation, approximately one-third of the IgM foci could be recovered by anti-μ serum (Table IV, first row). Although the great majority of these foci were detectable on both slides, there were four foci that were not. This could be explained by the finding of Plotz et al. (1968) that there is an IgM antibody of low hemolytic efficiency, which can be enhanced by specific antiserum. Despite the low efficiency of recovery of IgM foci, both slides were reduced and alkylated and one slide treated with anti-μ serum and the other with anti-γ_1 before adding complement (Table IV, rows 2 and 3). In these experiments, again approximately one-third of the IgM foci were recovered (in one experiment 28, in the other 41). On the other slides 24 γ_1 foci were detected, of which 90 and 80% matched with IgM foci.

TABLE IV

Number of foci producing IgM and IgG antibodies six days after transfer of cells (method 4)

Complement only	R + A + anti-μ	No. of matching foci	% identity
63	22	18	81
R + A + anti-μ	R + A + anti-γ_1	No. of matching foci	
28	14	11	78.5
R + A + anti-μ	R + A + anti-γ_1	No. of matching foci	
41	10	9	90

R + A = reduced and alkylated.

DISCUSSION

Foci of antibody producing cells develop in the spleens of irradiated syngeneic recipient mice after the transfer of a spleen cells-SRBC mixture. The foci detected five days after transfer mainly contain cells secreting IgM hemolysin. From day 6 cells producing γ_1 antibody were found in the same foci as IgM-producing cells. We interpret the experimental data as evidence that both IgM- and IgG-producing cells appear in foci developing from single precursors. Of course, if antibody-forming cells are tested as isolated cells, most or all of them will contain only IgM or IgG antibodies.

Several points should be discussed: 1. Methods for detection of IgM and IgG antibody-forming cells in the foci. 2. The possibility of random association of IgM and IgG precursors. 3. The possible forming of clusters between different cells as the initial unit for the development of foci. 4. The character of cells used for the transfer.

1. Two methods (3 and 4) were used to detect IgG and IgM antibodies, both seem to produce convincing results. With method 3 the objection could be raised that treatment with complement alone might detect hemolysing IgG as well as IgM. As the concentration of the reducing agent does not affect IgG, we should detect some foci after reduction and alkylation with complement (no

addition of anti-γ serum). This never happened in our experiments. To be completely safe in this respect, a new method was developed. Both slides were reduced and alkylated: resistant IgG antibodies were detected with anti-γ_1 serum, IgM antibodies split into subunits were restored to serological activity by anti-μ serum.

There is further circumstantial evidence: when antiserum specific for γ_1 was applied to the slides (without reduction) no increase in numbers of foci was observed. However, using reduction and alkylation to destroy the activity of IgM and then treating the slide with anti-γ_1, these foci were readily detectable. The fact that they appeared in the areas of the IgM foci explains the lack of increase in numbers when reduction and alkylation procedures were not used.

2. Another point which should be discussed is the possibility of a random association of two immunocompetent cells, separate precursors for IgM and IgG. When a single cell suspension of 10^6 cells was injected intravenously approximately one immunocompetent cell reached the spleen and produced a clone. With this very small number of specific precursors for IgM and IgG only very strong forces favoring association of cells could cause agglutination in the suspension or accumulation in the same locality in the recipient's spleen. The chance of two precursor cells meeting at random would be of the order of 10^6. In our experiments, however, 10% of the foci contained both IgM- and IgG-producing cells. One could assume that foci were contaminated by randomly recirculating lymphocytes. This assumption seems to be excluded by experiments in which two antigens were injected. No foci contaminated mutually, i.e. being a mixture of cells producing two different specificities were detected (Celada and Wigzell, 1966; Papermaster, 1967).

3. All papers describing the use of lethally irradiated recipients colonized with donor cells show a linear relationship between the number of transferred cells and the numbers of foci producing antibodies or visible colonies of hematopoietic cells. This indicates that if the foci started from more than one cell, the cluster is determined by a functional cell. Otherwise there would not be this linear relationship (Armstrong and Diener, 1969). This raises the question of whether these experiments involve the co-operation of two cell types. One cannot exclude the possibility that the suspension of transferred cells is a mixture of competent cells and other cells helping them by presenting antigen or supplying metabolites for their proliferation process. If helper cells are present, then it is to be assumed that they are in excess, supporting the differentiation of transferred antibody-forming precursors. The role of helper cells does not interfere with the conclusion that the competent cell gives rise to the cell line in which subsequently gene regions for IgM and later for IgG are expressed.

An alternative possibility should be discussed: developing foci are limited by the number of helper cells, and the antibody-forming cell precursors are in non-limited quantities. This would result in associations between helper cell and antibody-forming cell precursors, which again are either (a) multipotential precursors, or (b) separate precursors for IgM and IgG. Transfer experiments prove that the helping effect is only relative and that the decisive role in clone formation is played by antigen sensitive bone marrow-derived cells (Shearer and Cudkowicz, 1969). Some antigens do not need the helper effect at all, but again the number of transferred cells is linearly related to the number of detected foci (Armstrong and Diener, 1969). Although a limiting number of helper cells is not supported by any of the data available, one could discuss what would happen if the clone of helper cells were seeded by two different IgM and IgG precursors. In such a case, the surplus of precursors must be so great that they could find every localized helper cell; only random contacts would not be followed by the linear cell-foci relationship. If helper cells are limiting, one would expect not only more random distribution of all three possibilities (isolated IgM, isolated IgG and mixed IgM-IgG foci), but particularly a random time sequence. However, our data show that if IgG-containing foci appear they are associated with IgM foci; IgG foci were never detected before the appearance of IgM.

4. In these experiments the percentage of foci in which IgG-forming cells developed was relatively low. In only approximately one out of 10 IgM foci were IgG antibody-forming cells detected. This fact can be explained by restricted multiplication of cells in the foci. Each antibody-forming clone contained only 16-100 cells, which indicates a limited number of proliferation cycles. So it is quite possible that foci in which antibody-forming cells develop are derived from cells already activated in the donor (immunologically activated, Y cells). This view is supported by the experiments of Papermaster (1967) who had to pre-immunize the donors of transferred cells in order to get both IgG- and IgM-containing foci.

In some papers the significance of the foci formed is discussed (Cunningham, 1969a). We agree that in some experiments the foci would be expected to be more numerous from the number of antibody-forming cells. However, with the method of pieces (Cunningham, 1969a) in which individual antibody-forming cells are detected by the plaque technique, the number of pieces containing both IgM- and IgG-producing cells is highly statistically significant. Results of this type were obtained by Cunningham (1969b) who concluded that they were in accordance with the idea of a switchover.

Summarizing the discussion, we have not found any reasons against interpreting the data to support the existence of a precursor cell, activated for multiplication by antigen and giving rise to both IgM and IgG cells. Such an interpretation is supported by the view of the regulation of immunoglobulin chains by two genes (Hood and Ein, 1968), by the finding of the same idiotype in IgM and IgG antibody (Oudin and Michel, 1969), and by the observation that malignant cells, producing immunoglobulins can express several heavy chains simultaneously (Fahey and Finegold, 1967; Takahashi et al., 1969).

SUMMARY

Limited number of syngeneic spleen cells were transferred to lethally irradiated recipients and immunized with sheep red blood cells (SRBC). Foci were detected on replica plates by two methods, which permitted detection of IgM- and IgG-producing cells. By both methods only foci containing IgM were found shortly after transfer. On day 6 after transfer IgG-producing cells were also detected in approximately 10% of the IgM foci. Almost all these IgG-containing foci matched with IgM foci (90% on day 6). These results are explained as resulting from a gene switch during the multiplication of immunologically activated cells, from a single precursor. The cells can be transformed by a second antigenic stimulus from activated cells to cells producing IgM and later, after approximately 6-7 multiplications, into IgG-forming cells.

REFERENCES

Armstrong, W. D. and Diener, E. (1969). *J. exp. Med. 129*, 371-391.
Borel, Y., Fauconnet, M. and Miescher, P. A. (1965). *J. exp. Med. 122*, 263-275.
Cebra, J. J., Colberg, J. E. and Dray, S. (1966). *J. exp. Med. 123*, 547-558.
Celada, F. and Wigzell, H. (1966). *Immunology 11*, 453-466.
Cunningham, A. J. (1969a). *Aust. J. exp. Biol. med. Sci. 47*, 485-492.
Cunningham, A. J. (1969b). *Aust. J. exp. Biol. med. Sci. 47*, 493-503.
Diener, E., Armstrong, W. D., Robinson, W. A. and Marbrook, J. (1968). *Biochem. Pharmac. Suppl.* p. 63.
Dresser, D. W. and Wortis, H. H. (1965). *Nature, Lond. 208*, 859-861.
Fahey, J. L. and Finegold, I. (1967). *Cold Spring Harb. Symp. quant. Biol. 32*, 283-288.
Hood, L. and Ein, D. (1968). *Nature, Lond. 220*, 764-767.
Humphrey, J. H. (1967). *Nature, Lond. 216*, 1295-1296.
Humphrey, J. H. and Dourmashkin, R. R. (1965). *In* "Complement" (G. E. W. Wolstenholme and J. Knight, eds), Ciba Fdn Symp., pp. 175-186. J. and A. Churchill, London.
Kabat, E. A. and Mayer, M. M. (1961). "Experimental Immunochemistry". Charles C. Thomas, Springfield, Ill.
Kennedy, J. C., Siminovich, L., Till, J. E. and McCulloch, E. A. (1965). *Proc. Soc. exp. Biol. Med. 120*, 868-873.
Mäkelä, O. (1967). *Cold Spring Harb. Symp. quant. Biol. 32*, 423-430.
Makinodan, T., Perkins, E. H., Shekarchi, I. C. and Gengozian, N. (1960). *In* "Mechanisms of Antibody Formation" (M. Holub and L. Jarošková, eds), pp. 182-189. Publishing House of Czech. Acad. Sci., Prague.
McPherson, C. W. (1963). *Lab. Animal Care 13*, 737-744.
Miller, J. F. A. P., Mitchell, G. F. and Weiss, N. S. (1967). *Nature, Lond. 214*, 992-997.
Oudin, J. and Michel, M. (1969). *C.r. Acad. Sci., Paris D 268*, 230-233.
Papermaster, B. W. (1967). *Cold Spring Harb. Symp. quant. Biol. 32*, 447-460.
Pernis, B., Chiappino, G., Kelus, A. S. and Gell, P. G. H. (1965). *J. exp. Med. 122*, 853-876.

Playfair, J. H. L., Papermaster, B. W. and Cole, L. J. (1965). *Science, N.Y. 149*, 998-1000.
Plotz, P. H., Colten, H. and Talal, N. (1968). *J. Immun. 100*, 752-755.
Říha, I. (1964). *Folia microbiol., Praha, 9*, 304-306.
Sahiar, K. and Schwartz, R. S. (1964). *Science, N.Y. 145*, 395-397.
Sell, S., Park, A. B. and Nordin, A. A. (1970). *J. Immun, 104*, 483-494.
Shearer, G. M. and Cudkowicz, G. (1969). *J. exp. Med. 129*, 935-951.
Smith, R. T. (1964). *Pediatrics 34*, 14.
Smith, R. T. and Robbins, J. B. (1965). *In* "Molecular and Cellular Basis of Antibody Formation" (J. Šterzl, ed.), p. 381. Publishing House Czech. Acad. Sci., Prague.
Šterzl, J. and Jílek, M. (1967). *Nature, Lond. 216*, 1233-1235.
Šterzl, J. and Mandel, L. (1970). *Europ. J. Immun.* (In press.)
Šterzl, J. and Říha, I. (1965). *Nature, Lond. 208*, 858-859.
Šterzl, J., Johanovská, D. and Milerová, J. (1969). *Folia microbiol., Praha, 14*, 351-358.
Šterzl, J., Šíma, P., Medlín, J., Tlaskalová, H., Mandel, L. and Nordin, A. A. (1970). *In* "Developmental Aspects of Antibody Formation and Structure". (J. Šterzl and I. Říha, eds), p. 865. Academia, Prague.
Svehag, S. E. and Mandel, B. (1964). *J. exp. Med. 119*, 21-39.
Takahashi, M., Takagi, N., Yagi, Y., Moore, G. E. and Pressman, D. (1969). *J. Immun. 102*, 1388-1393.

DISCUSSION

MAKELA. In my opinion you have not excluded the possibility that mixed IgM-IgG foci exist because specific helper cells were limiting, they determined the number of foci. One helper cell could well have helped one IgG-plasma cell precursor (PCP) and an IgM-PCP.

As the IgG response is later than the IgM response to an antigen, we must assume that if separate IgM and IgG clones exist the response of the later clones to an antigen is more long-lasting than of the former clones. It is conceivable that clones need several contacts with antigen to build up a full response as is suggested by the work of Möller and Wigzell. A virgin IgG-PCP (and perhaps IgM-PCP) is perhaps first helped by one helper cell to start dividing, but some of its daughter cells may move away to be helped for further divisions by other T cells. Now you would expect to find many mixed foci at the time between IgM and IgG peak responses and that is what you did.

ŠTERZL. I agree that the eventuality you mentioned should be discussed, i.e. that the number of foci is determined by the limiting number of specific helper cells, associated with immunocompetent cells. Such a situation is extensively considered in the discussion of our paper—point 3), where, among other objections, I also presented the statistical approach. To obtain a linear relationship between the number of transferred cells and the number of resulting foci, a great surplus of immunocompetent cells would be needed, in order to find and establish a clone around each helper cell.

Supposing that helper cells are limiting the number of foci, and since one focus develops after transfer of 10^6 spleen cells, one helper cell (H) is present among them. Assuming random contact of this H cell with transferred cells

(10^6), i.e. also with the cells engaged later in production of antibodies (immunocompetent cells, IC), the probability of the contact between H and IC cells is 10^{-6} . IC. Let us assume that the H cell randomly contacts a certain (A) number of cells from the transferred suspension. Then the probability that among the number (A) at least one IC cell would be present is:

$$1-(1-IC \cdot 10^{-6})^A \approx IC \cdot A \cdot 10^{-6} .$$

This formula expresses the simple fact that if a H cell contacts randomly 10 cells from the transferred suspension (10^6) there must be at least 10^5 immunocompetent cells (IC). Accepting that an H cell contacts 10^3 from among the cells transferred, at least 10^3 immunocompetent cells should be available. This is already a large number (compared with the experimentally estimated number of IC per 10^6 lymphoid cells), not assuming the contact of H cell with at least two different precursors in Ig classes.

It is difficult to suggest from experimental data, that helper cells are limiting in the suspension of spleen cells. Similar numbers of foci were produced by 5×10^6 spleen cells as well as by bone marrow cells mixed with unlimited amounts, 10^8, thymus cells (Shearer and Cudkowicz, 1969). In another system with *S. adelaide* flagellin antigen—not dependent on helper effect at all (Armstrong and Diener, 1969)—by transfer of 10^6 spleen cells approximately one focus was obtained, similarly as with SRBC antigen. So we are in favour of the conclusion that linear relationship between foci detected and the number of the spleen cells transferred results from the limiting numbers of immunocompetent cells in the suspension.

Characteristics of Surface-attached Antibodies as Analysed by Fractionation Through Antigen-coated Columns

H. WIGZELL,* B. ANDERSSON,* O. MÄKELÄ†
and C. S. WALTERS*

*Department of Tumour Biology, Karolinska Institutet,
104 01 Stockholm, Sweden
†Department of Serology and Bacteriology, Helsinki University,
Helsinki, Finland

This paper will deal with the characteristics of cell-attached antibodies as analysed by the filtration of immunocompetent cells through antigen-coated columns (Wigzell and Andersson, 1969a). After fractionation the immunological reactivity of passed, retained and control cells has been studied in different test systems (Wigzell and Andersson, 1969a, b; Wigzell and Mäkelä, 1970). Only a brief summary of the separation technique will be given (for details see Wigzell and Andersson, 1969a; Wigzell and Mäkelä, 1970). The column technique was designed on the assumption that immunocompetent cells express their potentialities by means of antigen-binding receptors on their outer surface. If the antigen-binding efficiency of these membrane-bound receptors is high enough these cells might be specifically retained during filtration through a column coated with the relevant antigen. In order to reduce production and possible shedding of receptors during separation, all fractionations were carried out at 4° C, and after separation all cells were washed twice before assaying for immune capacity. Most of the results will only compare passed to control cells. Due to "background" retention of cells for non-immunological reasons, retained cells have a lower factor of specific enrichment of immunocompetent cells than the factor of deprivation in the passed cells (Wigzell and Andersson, 1969a). New bead materials seemingly offer a better possibility of enriching cells than the beads used for these experiments (glass or polymetaacrylic plastic) (Truffa-Bachi and Wofsy, 1970).

The following cell types, classified according to function or test, have been found to express membrane-attached antibodies in such a way as to allow specific retention by the relevant antigen-coated columns.

High Rate Antibody-forming Cells

Cells producing and releasing antibodies at high rate as tested by an indirect plaque technique (Dresser and Wortis, 1967) can be specifically separated on antigen-coated columns, provided that the fractionation is carried out at $4°$C (Wigzell and Andersson, 1969a; Wigzell and Mäkelä, 1970). Separation is less efficient at a higher temperature. We consider this is due to a more rapid turnover and shedding of the membrane-attached antibodies at higher temperatures when the receptors on this type of cell might merely be antibody molecules on their way to becoming humoral antibodies. An example of such an experiment is shown in Table I. Selective retention could be blocked by the presence of free antigen in the medium during filtration. This is in agreement with the concept that free antigen would compete with bead-attached antigen for the surface receptor.

Table I

Separation of PFC by antigen-coated columns in the presence or absence of free antigen

Column	Free antigen	Anti-HSA PFC/ 10^6 cells	Anti-BSA PFC/ 10^6 cells	$\dfrac{\text{Anti-BSA PFC}[a]}{\text{Anti-HSA PFC}}$
−	−	57.2	12.0	0.21 (1.00)
HSA	−	2.1	2.1	1.00 (4.79)
HSA	HSA	25.4	7.2	0.28 (1.33)
BSA	−	20.6	0	0 (<0.05)
BSA	BSA	25.1	10.6	0.42 (2.00)
−	−	18.3	11.2	0.61 (1.00)
HSA	−	1.0	16.5	16.50 (27.05)
HSA	HSA	10.5	7.2	0.69 (1.13)

Only passed and control cells analysed. Concentration of free antigen in columnar fluid in the indicated groups 2 mg/ml. All cells were washed twice after columnar passage to remove free antigen from the cells before testing for PFC. A mixture of anti-HSA and anti-BSA cells was used.

[a] Figures in parentheses = ratio as compared to control (1.00).

Immunological Memory Cells

Immune lymph node or spleen cells were passed through antigen-coated columns, and subsequently tested in a transfer system for immunological memory against the column antigen, and against a non-cross reactive antigen. Specific retention of memory cells could be demonstrated (Wigzell and Andersson, 1969a; Wigzell and Mäkelä, 1970). The results of one experiment are shown in Table II. Free antigen could block this selective retention of the

TABLE II

Separation of immunological memory cells in the presence or absence of free antigen

Cells[a]	Free antigen	Anti-HSA[b]	Anti-OA[b]	Relative anti-HSA[c]
C	–	0.802 ± 0.060	0.277 ± 0.119	100%
C	HSA 1 mg/ml	0.963 ± 0.110	0.494 ± 0.299	100%
HSA-P	–	-1.237 ± 0.480	0.148 ± 0.143	1.2%
HSA-P	HSA 1 mg/ml	0.273 ± 0.079	0.101 ± 0.166	51%

[a] Mixture of anti-HSA and anti-OA spleen cells. C = control cells, HSA-P = cells passed through HSA-coated column. Transfer to 450 r-irradiated mice together with boosting dose of HSA and OA.

[b] Titres at day 10 after transfer. Antigen-binding capacity assessed by ammonium sulphate precipitation. Mean $\pm$ standard error of the mean, all in $\log_{10}$ units.

[c] Mean relative ratio anti-HSA/anti-OA in each group compared to the respective control as 100%.

memory cells with reactivity for the column antigen as it did with high rate antibody-forming cells.

Using cells from animals immunized with either (a) a mixture of two antigens or (b) one of the two antigens followed by *in vitro* pooling of immune cells against the two antigens, similar specific retention of memory cells with the relevant specificity only was observed (Wigzell and Andersson, 1969a) (Table III). These results strongly indicate that the memory cells produce their

TABLE III

Elimination of "singly" or "doubly" immunized immunological memory cells by antigen-coated columns

Cells[a]	Columns[b]	Anti-HSA titre[c]	Anti-OA titre[c]
HSA-OA	–	0.581 ± 0.124	0.569 ± 0.093
HSA-OA	HSA	all negative	0.725 ± 0.231
HSA-OA	OA	0.540 ± 0.176	all negative
HSA + OA	–	0.670 ± 0.239	0.139 ± 0.008
HSA + OA	HSA	all negative	0.098 ± 0.167
HSA + OA	OA	0.236 ± 0.189	all negative

[a] HSA-OA cells from animals immunized with a mixture of HSA and OA. HSA + OA cells from animals immunized with HSA or OA alone, the cells being pooled *in vitro*.

[b] HSA = cells passed through HSA-coated column. OA = cells passed through OA-coated column.

[c] Antigen-binding capacity measured as $\log_{10} \mu g$ of antigen bound/ml of serum. Means $\pm$ standard error of the means. Negative animals = antigen-binding capacity less than $10^{-1} \mu g$ of antigen bound/ml of antiserum.

own receptors, and there was no evidence of passively absorbed antibody playing any role in this system.

NORMAL, POTENTIAL ANTIBODY-FORMING CELLS

Normal spleen cells were sieved through antigen-coated columns, and transferred to 700 r irradiated animals together with antigen(s) in Freund's complete adjuvant (Wigzell and Mäkelä, 1970). At various times afterwards sera were analysed for antibodies against the different antigens. Passage through antigen-coated columns selectively eliminated the potential capacity of the passed cells to form antibodies against the column antigen, whilst leaving the reactivity against other antigens intact (Wigzell and Mäkelä, 1970) (Table IV). The specific unresponsiveness induced by column passage lasted for 4-6 weeks *in vivo* after transfer. As recipient animals had not been thymectomized we consider it quite possible that the loss of immunological paralysis was due to recovery of the host lymphoid system, allowing antibody formation against the column antigen to take place.

TABLE IV

Separation of normal lymphoid cells on antigen-coated columns. Effect on anti-protein antibody formation

Cells[a]	Anti-OA[b]	Anti-BSA[b]	Anti-OA/anti-BSA[c]
C	−0.478 ± 0.083	−1.075 ± 0.039	0.596 (1.00)
OP	all negative	−1.029 ± 0.136	<0.038 (<0.27)
NBP	−0.832 ± 0.037	all negative	>1.168 (>3.73)
BM	all negative	all negative	− −

[a] C = control cells, OP = cells passed OA-coated column, NBP = cells passed $NIP_{10}BSA$-coated column, BM = normal bone marrow cells. C, OP, and NBP = normal spleen cells. All recipients 700 r-irradiated, receiving 10^7 C, OP or NBP cells + 2×10^6 normal marrow cells. Immunization with $NIP_{10}BSA$ + OA in complete Freund's adjuvant.

[b] Titres assessed by ammonium sulphate precipitation at day 13 after transfer. Mean ± standard error of the mean in $\log_{10}$ units.

[c] Anti-OA/anti-BSA = mean of that ratio for the individual serum samples within the group. Ratio in absolute figures in $\log_{10}$ units and (arithmetic ratio compared to control = 1.00).

It was not clear whether this separation of normal lymphoid cells according to potential antibody-forming capacity was acting primarily by removing the thymic or the bursa-type of cell or both (Miller and Mitchell, 1968). To study this we used a thymus-independent antigen, polyvinylpyrrolidone (PVP) (Andersson, 1969). With PVP-coated columns we could demonstrate a selective removal of potential PVP-reactive cells from our passed cells (Table V). We could

TABLE V

Separation of potential PVP antibody-producing cells by filtration through PVP-coated columns

Cells[a]	Anti-PVP[b]	Anti-S.a.[b]	Anti-PVP/anti-S.a.[c]
C	7/7 ($3^{6.0}$)	7/7 ($3^{6.0}$)	1.00
NS-P	6/7 ($3^{5.5}$)	7/7 ($3^{5.4}$)	1.02
PVP-P	1/7 ($3^{3.0}$)	7/7 ($3^{7.0}$)	0.002
–	0/7	0/7	–

[a] C = control cells, NS-P = cells passed through a normal mouse serum-coated column, PVP-P = cells passed through a PVP-coated column. 2×10^7 normal spleen cells into each 700 r-irradiated recipient, receiving simultaneous immunization with PVP and *Salmonella adelaide* (S.a.) heat-killed bacteria.

[b] Passive haemolysis titres determined day 10 after transfer. Number of responders/total number of mice and (mean response of responders).

[c] Mean ratio of anti-PVP/anti-S.a. titres of individual sera within each group. Non-responder sera included, arbitrarily put at titre 1.

thus conclude that antigen-coated columns can separate the cells potentially capable of humoral antibody production according to their immunological potentiality. As will be reported in greater detail later in this paper we still lack evidence that thymus-dependent cells can be separated according to their immunological reactivity.

ANTIGEN-BINDING CHARACTERISTICS OF THE MEMBRANE-ATTACHED RECEPTORS

1. Correlation between avidity of receptor and humoral antibody produced by the same cell. Indirect evidence that the avidity of the receptor for antigen parallels that of the humoral antibody produced by the same cell has accumulated from experiments with antigen-coated columns (Wigzell, 1969). Thus normal cells potentially capable of antibody synthesis have to pass through taller columns to obtain significant specific separation than have immune cells (Wigzell and Mäkelä, 1970; Wigzell, 1969). Also "early" immune cells are more difficult to separate on columns than are "late" cells (Wigzell, 1970). These findings suggest that the binding efficiency of the membrane receptors increases after immunization and with time after immunization. This could be due to an increase in the number of receptors present on immune *versus* normal cells, although present limited evidence on this issue argues against this hypothesis (Ada *et al.*, 1970). Alternatively the avidity of the receptor is similar to that of the humoral antibodies produced by that cell. As a system is now available for testing the avidity of antibodies produced by isolated cells (Andersson, 1970), we now allowed immune cells to pass through antigen-coated columns, and assayed the average avidity of the antibody produced by passed, retained and control cells. When run through a column coated with normal serum only (Fig. 1) the passed and retained cells had the same avidities as the control cells. On the other hand, sieving through a column coated with the relevant antigen (Fig. 2) allowed a selective sneaking through of low avidity antibody-producing

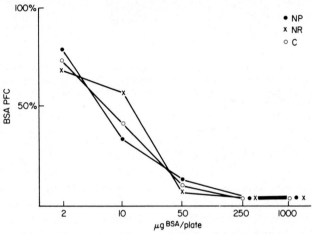

Fig. 1. Anti-BSA cells were passed through a column coated with normal mouse serum. C = control cells, NP = passed cells, NR = retained cells. Same number of cells of each suspension were tested for anti-BSA PFC either without BSA in the agar (100%) or with various concentrations of BSA. Inhibition of plaques by low BSA concentrations indicates high affinity. Each value based on three plates.

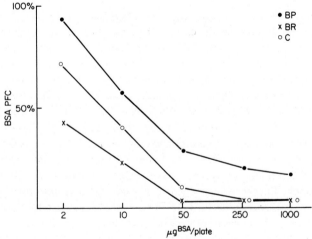

Fig. 2. As in Fig. 1, but passed through a BSA-coated column. BP = passed cells, BR = retained cells.

cells, whereas the high avidity fraction of cells was preferentially retained. Thus it is possible to separate antibody-forming cells according to the avidity of the surface receptor. It remains to be established that normal and memory cells have a receptor with avidity corresponding to the eventual product of the cell, the humoral antibody. Tests to analyse this with antigen-coated columns are in progress.

2. *Size and specificity of antigen-combining site of surface antibody.* Using various hapten-carrier conjugate combinations it is possible to obtain information about the size and specificity of the antigen combining site of the surface receptor as compared to that of the humoral antibody (Wigzell and Mäkelä, 1970). The hapten used was NIP (4-hydroxy-3-iodo-5-nitrophenylacetic acid, Brownstone *et al.*, 1966a) coupled to various proteins.

It was possible to retain selectively anti-NIP cells from normal or immune donors by filtration through NIP-coated columns (Wigzell and Mäkelä, 1970). An example using immune cells is shown in Table VI, the immunogen was $NIP_{11}OA$ (OA = ovalbumin) and the cells were filtered through columns coated with $NIP_{10}BSA$ in the presence or absence of free NIP-epsilon-amino-caproic acid (NIP-cap) (NIP when bound to protein via the azide is primarily attached to lysine, Brownstone *et al.*, 1966a). The anti-NIP cells were specifically retained in the absence of free NIP-cap, but in the presence of free hapten the anti-NIP cells passed through undisturbed. This suggests that anti-NIP cells recognize NIP as the immunodominant group for the membrane-bound antibody. Further support for this is obtained from experiments where anti-NIP-BSA cells were mixed with anti-OA cells and passed through a column coated with BSA. The number of high rate antibody-forming cells against BSA, OA and NIP in control and passed cells was subsequently assessed. If anti-NIP cells from animals immunized with NIP-BSA have any detectable affinity for the bead-attached BSA this should cause a decrease in the NIP/OA ratio of the passed cells. This was never found (Table VII). At the same time a highly significant reduction in anti-BSA cells was seen in the passed cell population. Thus the surface receptor for antigen has a discriminating capacity for haptenic groups similar to that of humoral antibodies in the present hapten system (Brownstone *et al.*, 1966b), with no detectable reactivity for carrier sites being demonstrable within the NIP-combining receptor areas.

CHEMICAL BUILD-UP OF SURFACE RECEPTORS FOR ANTIGEN ON CELLS POTENTIALLY CAPABLE OF HUMORAL ANTIBODY SYNTHESIS

The uptake of radio-isotope-labelled antigen by lymphoid cells (Byrt and Ada, 1969) and "rosette" formation (McConnell *et al.*, 1969) can be blocked by preincubating the cells with anti-immunoglobulin sera. Also induction of the secondary immune response by incubating cells *in vitro* with antigen can be

TABLE VI

Separation of immune anti-hapten cells by passage through hapten-protein-coated columns in the presence or absence of free hapten

Cells[a]	Free NIP-cap[b]	NIP PFC[c]	OA PFC[c]	NIP/OA PFC	Anti-NIP[d]	Anti-OA[d]	Anti-NIP/anti-OA
C	–	426	247	1.72	-0.310 ± 0.146	-0.356 ± 0.089	0.046 ± 0.129
C	30 µg/ml	396	307	1.29	-0.476 ± 0.328	-0.431 ± 0.074	< -1.526
NBP	–	67	167	0.40	all negative	-0.526 ± 0.248	-0.045 ± 0.382
NBP	30 µg/ml	208	102	2.04	-1.102 ± 0.254	-0.894 ± 0.175	-0.208 ± 0.359

[a] Immune anti-NIP-OA cells, C = control cells, NBP = cells passed $NIP_{10}BSA$-coated column.

[b] NIP-cap = NIP-epsilon-aminocaproic acid.

[c] Arithmetic mean of $PFC/10^6$ cells, 2 plates per figure.

[d] Antigen-binding capacity of sera against OA and NIP. Mean ± standard error of the mean in $\log_{10}$ units. Negative animals less than -2.00 in anti-NIP.

TABLE VII

Attempts to demonstrate carrier specificity within the NIP-reactive site on NIP-PFC as assessed by filtration through carrier-coated columns

Cells[a]	Immunogen[b]	Column	NIP-PFC[c]	OA-PFC[c]	BSA-PFC[c]	BSA-PFC/OA-PFC[d]	NIP-PFC/OA-PFC[d]
C	NIP-BSA + OA	–	103	115	378	3.29 (1.00)	0.90 (1.00)
P	NIP-BSA + OA	BSA	57	87	37	0.43 (0.13)	0.66 (0.73)
C	NIP-BSA + OA	–	211	97	518	5.34 (1.00)	2.18 (1.00)
P	NIP-BSA + OA	BSA	69	32	59	1.84 (0.34)	2.16 (0.99)
C	NIP-BSA + OA	–	132	146	195	1.34 (1.00)	0.91 (1.00)
P	NIP-BSA + OA	BSA	95	75	6	0.08 (0.06)	1.27 (1.39)

[a] C = control cells. P = passed cells. Three separate experiments were performed.

[b] NIP-BSA = NIP_5-BSA. A mixture of NIP_5-BSA + OA, 2 mg/ml of each in Freund's complete adjuvant.

[c] PFC expressed as $PFC/10^6$ cells plated. Each figure = mean of two plates.

inhibited by such treatment (Mitchison, 1967). Altogether this suggests that the surface receptor for antigen might be immunoglobulin in nature. We have attempted to study the chemical nature of the receptor using antigen-coated columns, and cells immersed in different anti-immunoglobulin antisera, trying to obtain a selective sneaking through of "blocked" cells (Walters and Wigzell, 1970). Preliminary tests showed that antigen-binding receptors could not be blocked by just any antibody reacting with cell surface antigens. Blocking required anti-immunoglobulin antibodies in experiments with antigen-coated columns (Walters and Wigzell, 1970). Anti-light chain sera blocked the retention of the relevant immune cells by the column. We then tested anti-heavy chain class-specific sera to see whether the receptor and the potential product of the cell, the humoral antibody, had the same heavy chain structure. Mere incubation with these sera did not affect the subsequent ratios of gamma-1/gamma-2a high rate antibody synthesizing cells as tested in a memory test (the antisera used were anti-mouse gamma-1 or gamma-2a). The following design of experiments was used: Immune cells were incubated with either anti-gamma-1 or anti-gamma-2a serum and passed through antigen-coated columns. Control, passed and retained cells (retained cells being mechanically eluted) were subsequently tested for gamma-1 or gamma-2a antibody-producing cells by the indirect plaque assay (Dresser and Wortis, 1967), either immediately after column filtration or in a test for immunological memory (Walters and Wigzell, 1970). Any shift in gamma-1/gamma-2a ratios of antibody-forming cells with specificity for the

TABLE VIII

Effect of rabbit anti-mouse immunoglobulin class-specific antisera on the specific elimination of immunological memory cells by antigen-coated columns.

Cells[a]	NIP-gamma-1 PFC[b]	NIP-gamma-2a PFC[b]	NIP-gamma-1/ NIP-gamma-2a PFC[c]
C	3108	896	3.36 (1.00)
C-gamma-1	1432	579	2.95 (0.88)
P-gamma-1	2772	188	14.63 (4.35)
R-gamma-1	327	238	1.32 (0.39)
C-gamma-2a	1087	536	2.15 (0.64)
P-gamma-2a	290	242	1.34 (0.40)
R-gamma-2a	1219	202	5.41 (1.61)

[a] Immune spleen cells (anti-NIP-BSA). C = control cells. P = passed through a NIP_{10}-OA coated column. Gamma-1 or gamma-2a denote *in vitro* incubation with rabbit anti-mouse gamma-1 or gamma-2a specific antisera prior to column passage. Cells tested for memory by *in vivo* transfer.

[b] Sum of gamma-1 or gamma-2a anti-NIP specific PFC in two spleens in each group.

[c] Absolute gamma-1/gamma-2a PFC ratio in the group and (relative ratio as compared to control = 1.00).

column antigen would suggest a preferential sneaking through of "blocked" cells with the appropriate heavy chain in their humoral antibodies. An example of these experiments (Walters and Wigzell, 1970) is shown in Table VIII. Incubation of memory cells with an anti-gamma-2a antiserum before column filtration selectively let through memory cells of relevant immunocompetence *and* of gamma-2a type, whereas gamma-1 cells reactive towards the column antigen were retained in the usual fashion. Exactly opposite results were obtained when the cells were preincubated with anti-gamma-1 sera. Thus the two types of cells analysed, high rate antibody-forming cells and memory cells, have surface receptors with the same antigen-binding specificity and heavy chains as the eventual products of these cells, the humoral antibody. (As blocking could also be obtained with anti-light chain antisera, although this was not investigated with a criss-cross design, it seems reasonable to assume that light chains are also present in the surface receptor molecule, in accordance with results using other test systems (Greaves *et al.*, 1969; Warner *et al.*, 1970).)

Evidence of Different Antigen-binding Capacity of Receptors on Immune Cells of Thymic and Bursa Type

Our data so far give no information about the possible separation of cells of thymic type by antigen-coated columns. The majority of the experiments described so far involve a mixed thymic-bursa response (Miller and Mitchell, 1968), or have directly shown that cells synthesizing humoral antibodies (Wigzell and Andersson, 1969a; Wigzell and Mäkelä, 1970; Walters and Wigzell, 1970) or potentially capable of doing so (experiments with PVP) can be separated on antigen-coated columns. We chose to study the problem of surface-attached antibodies possibly present on thymic cells using a hapten-carrier system. In these systems there is good evidence in the mouse and guinea pig that immune cells reactive against carrier-specific sites will somehow co-operate with anti-hapten cells, and help the latter to synthesize anti-hapten antibodies (Mitchison, 1969; Rajewsky *et al.*, 1969). The helper cells reacting against the carrier can be shown to be different from antibody-forming cells against the same antigen, and the results to date suggest that the helper cells are of thymic origin (see Mitchison and Raff in this symposium).

We attempted to separate potential humoral antibody-forming cells from helper cells directed against the same antigen by allowing anti-carrier cells to filter through antigen-coated columns. Subsequently control and passed cells were tested for helping activity and for humoral antibody-producing potential against carrier sites in a transfer system. Anti-BSA spleen cells were sieved through a BSA-coated column and then transferred, together with anti-NIP-OA cells to sublethally irradiated animals. Recipients were boosted with soluble $NIP_{10}BSA$, and the subsequent anti-BSA response (dependent on the presence of humoral antibody-producing and helper cells against BSA) was compared to that against NIP (dependent on the presence of humoral antibody-producing cells against NIP and helper cells against BSA). The relative anti-BSA/anti-NIP antibody concentrations would thus indicate whether the BSA column had affected humoral and helper cells against BSA to the same or different extents. In fact there was a highly significant reduction in the anti-BSA antibody concentrations in animals receiving cells that had passed through BSA columns, but no reduction in the anti-NIP titres (Table IX). Thus anti-BSA helper cells

TABLE IX

Separation of anti-carrier immune cells by filtration through carrier-coated columns. Differential impact on humoral antibody-forming cells and helper cells

Exp.	Cells[a]	Anti-BSA[b]	Anti-NIP[b]	Anti-NIP/anti-BSA[c]
1	C	1.890 ± 0.104	2.598 ± 0.289	0.708 (1.00)
	P	0.086 ± 0.171	3.030 ± 0.085	2.944 (0.008)
2	C	2.390 ± 0.141	2.327 ± 0.076	-0.063 (1.00)
	P	0.724 ± 0.270	2.725 ± 0.138	2.001 (0.011)

[a] C = control cells. P = cells passed a BSA-coated column. Cells in exp. 1 immune anti-BSA + NIP-OA cells, only anti-BSA cells passed through column, admixture of anti-NIP-OA to C or P cells thereafter. 10^7 cells of either type into each recipient with boosting dose of NIP_{10}-BSA. Cells in exp. 2 anti-NIP-BSA + OA cells, here only NIP-BSA cells through column, boosting with NIP_{11}-OA + BSA.

[b] Anti-BSA and anti-NIP assessed by ammonium sulphate precipitation. Mean $\pm$ standard error of the mean in $\log_{10}$ units. Sera obtained at day 10 after transfer.

[c] Absolute ratio in $\log_{10}$ units and (relative arithmetic ratios as compared to control = 1.00).

could not be shown to be separable according to immune reactivity by filtration through BSA columns, whereas cells capable of producing humoral antibodies against BSA, present in the same cell populations, were selectively retained. Control experiments using graded cell numbers yielded similar results, thereby excluding the possibility that supraoptimal numbers of helper cells, but not of humoral antibody-forming cells, had been used. In conclusion, in this system we could not selectively retain immune cells of thymic type by column fractionation in the same way that cells of bursa type are retained. Whether this difference in antigen-binding efficiency of the outer surface of the two types of cell is due to qualitative or quantitative factors remains to be seen.

ACKNOWLEDGEMENTS

This work was supported by the Swedish Cancer Society, the Karolinska Institute, the Anders Otto Swärds Stiftelse, and by the Elsa and Sigurd Goljes

Minne (H.W. and B.A.), U.S. Public Health Service Grant GM 12046 (O.M.) and a fellowship from the American Association of University Women (C.S.W.).

REFERENCES

Ada, G. L., Byrt, P., Mandel, T. and Warner, N. L. (1970). *In* "Developmental Aspects of Antibody Formation and Structure" (J. Šterzl and I. Říha, eds), pp. 503-516. Czech. Acad. Sci., Prague.
Andersson, B. (1969). *J. Immun. 102,* 1309-1313.
Andersson, B. (1970). *J. exp. Med. 132,* 77.
Brownstone, A., Mitchison, N. A. and Pitt-Rivers, R. (1966a). *Immunology 10,* 465-479.
Brownstone, A., Mitchison, N. A. and Pitt-Rivers, R. (1966b). *Immunology 10,* 481-492.
Byrt, P. and Ada, G. L. (1969). *Immunology 17,* 503-516.
Dresser, D. W. and Wortis, H. H. (1967). *In* "Handbook of Experimental Immunology" (D. M. Weir, ed.), pp. 1054-1067. Blackwell Scientific Publications.
Greaves, M. F., Torrigiani, G. and Roitt, I. M. (1969). *Nature, Lond. 222,* 885-886.
McConnell, I., Munro, A., Gurner, B. W. and Coombs, R. R. A. (1969). *Int. Archs Allergy appl. Immun. 35,* 209-227.
Miller, J. F. A. P. and Mitchell, G. F. (1968). *J. exp. Med. 128,* 801-820.
Mitchison, N. A. (1967). *Cold Spring Harb. Symp. quant. Biol. 32,* 431-439.
Mitchison, N. A. (1969). *In* "Immunological Tolerance" (W. Braun and M. Landy, eds), pp. 149-151. Academic Press, New York.
Rajewsky, K., Schirrmacher, V., Nase, S. and Jerne, N. K. (1969). *J. exp. Med. 129,* 1131-1143.
Truffa-Bachi, P. and Wofsy, L. (1970). *Proc. natn. Acad. Sci. U.S.A. 66,* 685-693.
Walters, C. S. and Wigzell, H. (1970). *132,* 1233-1279.
Warner, N. L., Byrt, P. and Ada, G. L. (1970). *Nature, Lond. 226,* 942-943.
Wigzell, H. (1969). *Antibiotica et Chemotherapia 15,* 82.
Wigzell, H. (1970). *Transplant. Rev. 5,* 76-104.
Wigzell, H. and Andersson, B. (1969a). *J. exp. Med. 129,* 23-36.
Wigzell, H. and Andersson, B. (1969b). *In* "Cellular Recognition" (R. T. Smith and R. A. Good, eds), p. 275. Appleton-Century-Crofts.
Wigzell, H. and Mäkelä, O. (1970). *J. exp. Med. 132,* 110.

Relationship Between Lymphocyte Receptors and Humoral Antibodies

O. MÄKELÄ,* V. PASANEN† and H. SARVAS*

*Department of Serology and Bacteriology,
Helsinki University
†State Serum Institute, Helsinki,
Finland

We have compared mono-, oligo- and polyvalent conjugates of a hapten NIP (Brownstone et al., 1966) as serological reactants and as immunogens. In a number of experiments (Mäkelä et al., 1969; Sarvas and Mäkelä, 1970) we found that antibodies generally bound NIP in polyvalent form more firmly than in mainly monovalent form. In accordance with this, NIP in polyvalent form was a better immunogen than NIP in mainly monovalent form (Mäkelä et al., 1969; Mäkelä and Kontiainen, 1970). What interested us more and initiated this series of experiments was the finding that 19S anti-NIP fractions had a stronger preference for polyvalency than 7S fractions. In Table I the mean avidity ratio

TABLE I

Combining constant (K'_o) of NIP in the form of polyvalent conjugate $(NIP_{40}BSA)$: NIP in the form of mainly monovalent conjugate $(NIP_{0.5}BSA)$ was used as the reference ligand

Mean of 7 19S fractions	130
Mean of 7 7S fractions	5.7

(Pauling et al., 1944) of 7S fractions of seven rabbit sera is compared with the ratio of 19S fractions of the same sera. 19S fractions had, on the average, a 130-fold preference for NIP in polyvalent form while 7S fractions had only a 5.7-fold preference.

We do not know why 19S antibodies have a particularly strong preference for

243

repeating epitopes on the antigen. But we do know that the average affinity of an IgG binding site is higher than the average affinity of an IgM binding site in the same antiserum. In our experiment the affinity of IgG was on the average four times higher than the affinity of IgM (Mäkelä *et al.*, 1970). Thus IgG might be expected to score better when binding must be based on one site, but IgM may score better when the antigen gives an opportunity for multiple binding. No matter what the explanation was, we argued that if there are special cell lines for the production of IgM and these lines have IgM receptor antibodies, and if the same is true of the 7S antibody classes, our polyvalent antigen might stimulate anti-NIP with a higher proportion of IgM than the monovalent conjugate. The experiments that we have done tend to support the hypothesis. Polyvalent conjugates induced stronger 19S anti-NIP responses than monovalent conjugates, even when their dose was reduced to induce 7S responses similar to those against monovalent conjugates (Mäkelä, 1970).

We did similar immunization experiments with cells coupled with varying concentrations of NIP-azide or NNP-azide. Some of the results appeared to contradict our previous experience. Lightly coupled cells, while less immunogenic than heavily coupled cells, induced more 19S-rich responses than heavily coupled cells. The situation was the opposite of that found with protein conjugates where lightly coupled (monovalent) proteins induced responses devoid of 19S antibodies. The phenomenon was demonstrable in mice with bacterial conjugates as well as sheep or mouse erythrocyte conjugates (Mäkelä, 1970). It was particularly striking with conjugates of syngeneic erythrocytes— CBA erythrocytes treated with 0.01 mg/ml of NNP azide and injected into CBA mice induced pure 19S anti-NNP responses. In our hands, immunization with lightly conjugated syngeneic erythrocytes has been the best method of producing pure 19S anti-NNP antibody.

We believe that the essence of what we have observed is in Fig. 1. As well as data it gives an interpretation of how hapten density may have affected the 19S/7S ratios in the observed way. It is based on the hypothesis that IgM clones have IgM receptors and IgG clones have IgG receptors. According to it the ratio

---

Fig. 1. A summary of our findings and a tentative interpretation of how hapten density on the immunogenic conjugate may affect the IgM/IgG ratio of the resulting anti-hapten response.

Explanations. In the two first columns there are data of two experiments (Mäkelä *et al.*, 1969; Mäkelä, 1970). Antibody titres have been converted to mg concentrations assuming that an IgM anti-NIP concentration of 1 mg/ml corresponds to a titre of 11, while this figure for IgG is 2.0 (Sarvas and Mäkelä, 1970). The third column shows hypothetical combinations of the immunogens with IgG and IgM anti-NIP antibodies. Structures with forks represent humoral and possibly receptor antibodies combining with molecules of NIP (stubs) on the surface of the immunogen.

$NIP_{0.5}OA$ = Ovalbumin carrying 0.5 mole of NIP per mole of OA. 1.0 NIP-bacteria = *Salmonella* bacteria treated with a final concentration of 1 mg/ml of NIP azide.

| ANTIGEN | Observed weight ratio of "IgG"/"IgM" in the response | Combination with | | combination picture is correct can IgM molecules use more combining sites for catching an antigen molecule than IgG molecules? |
		IgG (hypothetical)	IgM	
A. $NIP_{0.5}OA$	55			No
B. NIP_8OA	17			Yes, but IgG can form double bonds
C. 1.0 NIP-BACTERIA	10			Yes, but IgG can form double bonds
D. 0.09 NIP-BACTERIA	1.1			Yes, and IgG cannot form double bonds

of IgG to IgM in an anti-hapten response depends partly on whether the antigen offers IgM receptors an opportunity to utilize their numerous combining sites and their large span. Whenever this opportunity is excluded, for instance by monovalency of antigen molecules, IgG production is favoured (Line A). The opposite situation is created if the topology of the haptens on the carrier surface excludes binding by two sites of an IgG molecule but allows binding by two or more sites of an IgM molecule (Line D). Oversimplifying we can calculate that if we can adjust haptens to have 150 Å distances on the cell surface an IgG molecule cannot bridge two haptens but an IgM molecule can conceivably bind five—one with every subunit.

In situations B and C IgG can form douple bonds and IgM multiple bonds. However, an increase in multiplicity beyond a certain number, say 3 or 4, may have little practical importance. High binding energy may already have been reached, and therefore these responses may have more IgG than the response in situation D. Another factor preventing effective binding by more than perhaps 3 or 4 may be limited flexibility of an IgM molecule. This may explain why in these responses the preference for IgM was less pronounced than in situation D.

An alternative explanation for the varying 19S/7S ratios is offered by findings that the dose of antigen affects the 19S/7S ratio. However, this cannot explain all the data, since weakly immunogenic monovalent protein conjugates stimulated very little 19S anti-NIP, while weakly immunogenic lightly coupled cells (0.09 NIP-bacteria) stimulated 19S-rich anti-NIP. Another fact discrediting this interpretation is that by dose adjustments we could reduce the 7S responses to polyvalent protein conjugate down to the level of the 7S responses against monovalent conjugate. Still the anti-NIP 19S responses to the polyvalent conjugate were 4.4 times as high as those to the monovalent conjugate.

ACKNOWLEDGEMENTS

Some of the work described here was supported by the United States Public Health Service (Grant GM-12046), by the Sigrid Jusélius Foundation and by the Finnish Research Council.

SUMMARY

Immunization of mice with mainly monovalent conjugates of a hapten (NIP) and proteins led to anti-NIP responses which had less 19S anti-NIP than responses against polyvalent conjugates of NIP and the same proteins. This was true even if the total responses had been equalized with dose adjustments. On the other hand, lightly coupled cellular conjugates of NIP stimulated more 19S-rich responses than heavily coupled conjugates.

REFERENCES

Brownstone, A., Mitchison, N. A. and Pitt-Rivers, R. (1966). *Immunology 10*, 465-479.

Mäkelä, O. (1970). *Transplant. Rev. 5*, 3-18.

Mäkelä, O. and Kontiainen, S. (1970). *In* "Developmental Aspects of Antibody Formation and Structure" (J. Šterzl and J. Říha, eds), 565-570. Czech. Acad. Sci., Prague.

Mäkelä, O., Cross, A. M. and Ruoslahti, E. (1969). *In* "Biological Recognition Processes" (R. T. Smith and R. A. Good, eds), p. 287. Appleton-Century-Crofts, New York.

Mäkelä, O., Ruoslahti, E. and Seppälä, I. J. T. (1970). *Immunochemistry 7*, 917-932.

Pauling, L., Pressman, D. and Grossberg, A. L. (1944). *J. Amer. chem. Soc. 66*, 784.

Sarvas, H. and Mäkelä, O. (1970). *Immunochemistry 7*, 933-943.

The Relative Ability of T and B Lymphocytes to See Protein Antigen

N. A. MITCHISON

National Institute for Medical Research,
Mill Hill, London, England

This paper deals with the relative ability of T and B lymphocytes to detect and react to protein antigens. It is concerned with the reaction of the two types of lymphocytes to BSA that leads to immunological tolerance. It is concerned particularly with the difference in threshold for the two classes of cell in their ability to make this kind of response. There are already excellent grounds for believing that a difference in ability to detect antigens exists between the two classes of lymphocyte, the difference being that T lymphocytes are considerably more sensitive.

In the first place, studies on helper cells in the secondary immune response to hapten-protein conjugates indicate that B lymphocytes deprived of help can react, if at all, only to high concentrations of conjugates (Mitchison *et al.*, 1970). In the presence of helper T lymphocytes the response becomes sensitive to doses of antigen lower by approximately three orders of magnitude. This suggests, although it does not formally prove, that the T lymphocyte in this reaction, which leads to humoral antibody production, has a lower threshold of reaction. The only question here is whether the helper lymphocytes do in fact make a response to the low doses of conjugate, or whether they act only in a passive role. Circumstantial evidence, for instance the ability of low doses of antigen to boost immunological memory in the helper cells, indicates that the conclusion that T lymphocytes have a low threshold of response in this reaction is correct.

In the second place, the recent work of Greaves and the Möllers, as reported, for example, in Möller's paper in this symposium, indicates that low doses of sheep red blood cells elicit a response which is largely confined to the T lymphocyte population. This confirms the belief that in the normal immune response T lymphocytes have a low threshold of response and B lymphocytes a relatively high one.

What then about the response that leads to tolerance? Here the most important piece of information comes from the work of Taylor (1969) on the induction of tolerance in thymocytes. This work has shown that doses of antigen at the lower end of the dose range for the induction of low-zone tolerance can induce unresponsiveness in thymocytes. About T lymphocytes after peripheralization, and about B lymphocytes, we are relatively ignorant. The recent work of Miller and Mitchell (1970) with sheep red blood cells suggests that B lymphocytes are relatively difficult to render tolerant, although here again we remain uncertain about the susceptibility of the two types of cell after they have undergone peripheralization, i.e. after T lymphocytes have left the thymus and B lymphocytes have left the marrow.

Klaus Rajewsky and I felt that it was important to study the thresholds for tolerance induction with protein antigens in T and B lymphocytes. Our main reason for doing so was that in studying tolerance we felt that we were measuring an average parameter for the entire population of lymphocytes, whereas experiments on immunization probably throw light on the properties of a small class of reactive cells highly selected in terms of affinity. The second reason for wishing to study tolerance was that we hoped in this way to obtain information about the turnover of the two cell populations. There are grounds for believing that B lymphocytes turn over more rapidly than do T lymphocytes. For example, Nossal and Mäkelä (1962) have obtained indications that, in certain immune responses, the precursors of antibody-forming cells turn over rapidly, as judged by their labelling kinetics prior to administration of antigen.

The experiments of Nossal and Mäkelä have been criticized on the grounds that re-incorporation of the label may have taken place. Furthermore, studies of this sort, which may indicate that the antigen sensitive B lymphocyte proliferates, do not imply that the B lymphocyte population is undergoing constant renewal from stem cells. Indeed the slow recovery of adult rabbits from allotype suppression (Dubiski, 1967) suggests that receptorless stem cells generate B lymphocytes pretty slowly. We therefore felt that information on the overall renewal of the B lymphocytes, such as could be obtained from tolerance experiments, was much needed.

The experimental design that we adopted is as follows: groups of mice were rendered tolerant by irradiation followed by prolonged treatment with antigen. The mice were given 600 r Co^{60} radiation and then 30 injections of BSA three times a week. At the end of 10 weeks the mice were rested for 10 days in order to permit elimination of free BSA (the half-life of free BSA in our CBA mice is 19 hr). The groups of mice were then split. Some remained as controls while others received educated thymus cells. By educated thymus cells we mean the following: normal mice were lethally irradiated with 900 r. They were then immediately injected intravenously with approximately 50×10^6 thymus cells and then immediately afterwards immunized with our standard dose of alum precipitated BSA (800 μg + 2×10^9 pertussis organisms intraperitoneally). Seven to eight days later the spleens of these mice were harvested and approximately

10×10^6 nucleated cells recovered per spleen (spleens from mice that received the same dose of radiation and the same immunization, but did not receive the thymocytes, yielded approximately 1×10^6 nucleated cells). This cell suspension contained 10-20% viable cells with the appearance of lymphocytes or lymphoblasts as judged by phase contrast and trypan blue exclusion, and these we term "educated thymus cells". We inject approximately 5×10^6 viable cells into each host. One day later the hosts and their controls were immunized with the standard dose of alum precipitated BSA + pertussis, and were bled 15-30 days later. The results are expressed as geometric means of the 15-30-day titres for groups of approximately six mice each. Antigen binding capacity was measured by our standard abbreviated procedure at a free antigen concentration of $1\ \mu g/ml$ (Mitchison, 1964).

May I add that at this stage the data are incomplete.

The question that an experiment of this design asks is whether the educated thymus cells find partners with which to co-operate in the recipient mice. The assumption is made that the transferred educated thymus cells do not themselves produce antibody, i.e. that they do not contain antibody-forming cell precursors. For them to do so would be contrary to the accepted picture of the immunological properties of T lymphocytes. Furthermore, I shall introduce experimental evidence based on the effects of host irradiation to justify this assumption. If therefore a response is obtained after the transfer, i.e. if the transferred cells terminate tolerance, we conclude that the B lymphocytes of the recipient have not been rendered tolerant. If on the other hand no response is obtained we conclude that the educated thymus cells have not found a reactive partner, and that the B lymphocytes of the host have therefore been rendered tolerant. Thus we believe that this procedure enables us to test in an animal whose T lymphocytes have been rendered unresponsive whether B lymphocytes have also been rendered tolerant or not.

Since the interest of our findings hinges on the difference between high-zone and low-zone tolerance it is important to establish that the animals that have been rendered tolerant by large doses of antigen do not have enough free antigen still present at the time of transfer to render tolerant the transferred cells. We have two arguments against this possibility. The first is this. We can calculate the amount of free antigen remaining in the high-zone tolerant animals at the time of cell transfer. It is of the order of $1\ \mu g$, and this is not enough BSA to elicit a detectable degree of tolerance in a normal animal (Mitchison, 1964). Secondly, T lymphocytes that have begun to respond to antigen, i.e. that have been educated, ought to be refractory to the induction of tolerance; they should be susceptible to tolerance only in the high zone, if we accept the arguments put forward for a high threshold being a property of an immunized cell population (Dresser and Mitchison, 1968; Mäkelä and Mitchison, 1965).

Why, one might ask, use educated thymus cells rather than normal thymocytes to answer this question? We chose to use educated thymus cells for several reasons. We knew that tolerant mice that have not been irradiated can be reconstituted only to a very limited extent by normal non-immunized lymphocytes (Mitchison, 1968b). Unpublished experiments by Mitchison and R. B. Taylor confirm the expectation that attempts to reconstitute mice rendered tolerant by low doses of BSA by means of normal thymocytes would yield equivocal results. On the other hand, we knew that the education procedure that we adopted had already proved successful in helper experiments (Mitchison, 1969a; Mitchison et al., 1970). We feel therefore that our decision to use educated thymus cells is amply justified.

RESULTS OBTAINED WITH EDUCATED THYMUS CELLS

Logically, the first experiment was to check the ability of our educated thymus cells to reconstitute an immune response in T-lymphocyte-deprived mice. Table I gives the results of an experiment performed with thymectomized,

TABLE I

Reconstitution of $T_x 900$ r mice by educated thymus cells

	BSA binding capacity $\log_{10} \mu g/ml$
Response of $T_x 900$ r mice	−0.56
Response of $T_x 900$ r mice after transfer of educated thymus cells	0.22 ± 0.65
Response of normal mice	0.49 ± 0.45

$P < 0.01$

lethally irradiated, marrow-reconstituted mice. In this experiment thymectomy was performed at approximately four weeks of age; one month later the mice were lethally irradiated and then injected with 5×10^6 syngeneic bone marrow cells. The cell transfer was performed approximately 10 weeks after the irradiation and the mice were then immunized by our standard procedure. Educated thymus cells were obtained by the standard procedure described above. It can be seen that the response of the T_x 900 r group is markedly diminished in comparison with their normal controls, and that their response can be restored to a significant extent by the transfer of educated thymus cells. This finding therefore confirms that our system of education is working satisfactorily.

The first experiment on high- and low-zone tolerance was performed by Rajewsky (1971). It confirmed our hypothesis that educated thymus cells would permit low-zone tolerant mice to make a response to BSA but would not permit high-zone tolerant mice to do so. I shall not document this experiment here since the results have already been published.

My next experiment was performed on low-zone tolerant mice that were exposed to a fairly wide range of irradiation dosages one day before transfer. From now on we shall refer to mice treated with the repeated injection of BSA (10 μg/injection) as low-zone tolerant animals, and mice treated with 10 mg/injection as high-zone tolerant animals. The results of this experiment (Table II) are in some respects unsatisfactory in that the educated thymus cells

TABLE II

Host irradiation

Treatment of low-zone tolerant hosts	BSA binding capacity, $\log_{10}$ μg/ml (1)		(2)
0 r	−0.05		−0.35
150 r	−0.23	P < 0.05	+0.01 ± 0.40
450 r	−0.18		−0.54

(1) Without educated thymus cells
(2) With educated thymus cells

here failed to produce an effect in the non-irradiated recipients. Nevertheless, they did so in the recipients that had received 150 r, thus confirming the earlier finding of Rajewsky's (1970) that low doses of irradiation applied to the prospective recipient before transfer enhanced the potency of the transferred cells. The highest dose of radiation in the present experiment, 450 r, depressed the response of the recipients. We conclude therefore that low doses of radiation encourage the activity of transferred cells via an as yet unexplained space-finding mechanism (Mitchison, 1957). A high dose of radiation on the other hand prevents them from acting, according to our hypothesis by inactivating the B lymphocytes of the host with which they would otherwise co-operate.

This experiment then supports our assumption that the educated thymus cells do not themselves contain antibody-forming cell precursors to any appreciable extent.

The next experiment was to study the effect of educated thymus cells on mice that had been rendered tolerant by various different doses of BSA between the high and low zones. The results of this experiment are shown in Fig. 1. From these results it can be seen that partial paralysis is induced as expected by doses of 1 μg BSA and that paralysis is essentially complete with doses of 10 μg or over unless the mice receive transferred cells. Note that the low titres of antibody detected in these tolerant mice are probably not significant (Mitchison, 1968a). It can also be seen that the transferred educated thymus cells enable the mice that had been made tolerant by doses of 10 μg and 100 μg of BSA to make a response, whereas they were unable to do so in mice that had received doses of

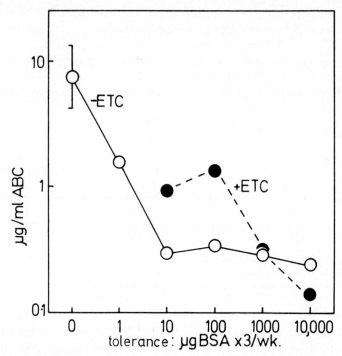

Fig. 1. Ability of educated thymus cells (ETC) to terminate tolerance of BSA as a function of the dose of antigen used to induce tolerance. Tolerance was induced by 600 r followed by 10 weeks of treatment with BSA x3/week, at the doses shown, and the −ETC points indicate the geometric mean of the responses of these mice to immunization. (The bracket illustrates one standard deviation for one of the groups.) The +ETC points indicate the responses of groups of mice that received educated thymus cells.

1000 μg or 10,000 μg. From this finding we conclude that the threshold of tolerance induction for B lymphocytes in these mice is in a dose range between 100 μg and 1000 μg of BSA. We can also conclude that the threshold for inducing tolerance in the intact mouse, which lies in a dose range between 1 μg and 10 μg, represents the threshold for tolerance induction in T lymphocytes. Thus, the thresholds for T and B lymphocytes differ by a factor of 100-1000-fold. This then is our central finding.

We have performed one more experiment so far, a preliminary one on the stability of tolerance in B lymphocytes. In this experiment we asked the following question. How long does it take the B lymphocytes of mice that have been rendered tolerant by high-zone treatment to recover? In order to answer this question we induced high-zone tolerance and then transferred the mice to a low-zone maintenance regime, which we expected would permit B lymphocytes to recover from their state of tolerance. The results (Table III) indicate that

TABLE III

Stability of AFCP-tolerance

	BSA binding capacity $\log_{10}$ μg/ml	
	(1)	(2)
Low-zone tolerance		
(600 r, then 10 weeks BSA 10 μg × 3/week)	−0.53	−0.03 ± 0.75
		P < 0.01
High-zone tolerance		
(600 r, then 10 weeks BSA 10 mg × 3/week)	−0.69	−0.76
High-zone, then 1-3 weeks low-zone	−0.80	−0.84

    (1) Without educated thymus cells
    (2) With educated thymus cells

recovery had not taken place over a period of 1-3 weeks, i.e. that recovery for the B lymphocytes does not proceed rapidly. This finding has very little interest so far; obviously it places us under an obligation to study the recovery of the B lymphocyte population in a more systematic manner. (Later note: further experiments of the type shown in Table III indicate that B lymphocytes in 14-week-old mice tolerant of BSA take 10 weeks to regain appreciable reactivity.)

COMPARABLE RESULTS OBTAINED BY CROSS-REACTIVE TERMINATION OF TOLERANCE

The role of co-operation in the termination of tolerance by immunization with cross-reactive antigens has been much discussed. In his original papers on this phenomenon Weigle (1961, 1962) provides a prescient discussion in which he postulates a two-step recognition process, a matter to which he returns in the present symposium. Later, when the relationship between helpers and antibody-forming cell precursors became clearer this hypothesis took on a more detailed character. It was postulated that immunization with the cross-reactive antigens raised a crop of helper cells with receptors for the new determinants present on the cross-reactive antigens. These helper cells then presented the antigen, via these determinants, to antibody-forming cell precursors, which were thus presented with the original tolerated determinants at a concentration higher than they had hitherto seen. A response to these hitherto tolerated determinants therefore ensued (Boak *et al.*, 1969; Mitchison *et al.*, 1970).

According to this hypothesis the ability of cross-reactive antigens to terminate tolerance would depend on whether a reactive B lymphocyte population is still available. It predicts therefore that treatment with antigen in doses high enough to render B lymphocytes tolerant should also prevent cross-reactive termination. Therefore it predicts a quantitative agreement between experiments with educated thymus cells and experiments on cross-reactive termination. This prediction was tested in an experiment shown in Table IV. Low-zone and high-zone tolerance was induced in these experiments according to the procedures described above, and immunization was carried out by the same procedures except that in some groups of mice dinitrophenylated BSA was used. It can be seen that our hypothesis was verified in this limited comparison. The dinitrophenylated antigen did indeed terminate tolerance of the protein determinants in the low-zone tolerance group, but not in the high-zone tolerance group.

TABLE IV

Cross-reactive termination of low-zone tolerance

	BSA binding capacity $\log_{10} \mu g/ml$	
Low-zone tolerance, immunized with BSA	-0.83	
Low-zone tolerance, immunized with $DNP_{11}BSA$	0.20 ± 0.25	$P < 0.05$
High-zone tolerance, immunized with BSA	-0.55	
High-zone tolerance, immunized with $DNP_{11}BSA$	-0.60	

THE DENSITY OF RECEPTORS ON B AND T LYMPHOCYTES: A PARADOX

The foregoing experiments have verified the proposition that T lymphocytes can see protein antigen better than can B lymphocytes. What is perhaps most remarkable about this finding is that precisely the opposite would have been expected, had we known only about the density of immunoglobulin receptors on the two types of cell. The fluorescence experiments of Martin Raff (in this symposium) indicate that B lymphocytes have more immunoglobulin on their surface than have T lymphocytes.

I can present a little more evidence in favour of Raff's conclusion drawn from a very different kind of procedure. In collaboration with Drs J. H. Humphrey and N. Willcox, experiments were performed on the suicide of BSA-sensitive cells by uptake of highly radioactive BSA. The outcome of three such experiments is shown in Table V. In general these experiments followed the protocols used in our previous adoptive transfers of the secondary immune response (Brownstone *et al.,* 1966; Mitchison, 1957, 1969b). BSA-primed cells were collected from the spleens of mice that had been immunized with BSA alum pertussis 3-10 weeks previously. The cells were incubated with either

TABLE V

Hot suicide of helpers and AFCP

	Exp. 1		Exp. 2		Exp. 3	
	aBSA[a]	aNIP[b]	aBSA	aNIP	aBSA	aNIP
No helpers	—	−0.64	—	−0.31	—	0.55 ± 0.31
Suicided helpers[c]	−0.10 ± 0.41	0.42 ± 0.57	0.10 ± 0.20	0.66 ± 0.83	0.29 ± 1.0	1.15 ± 0.84
Helpers[d]	0.11 ± 0.45	0.47 ± 0.25	0.24 ± 0.59	0.80 ± 0.20	0.78 ± 0.74	1.45 ± 0.57

[a] $\log_{10}$ μg BSA binding capacity/ml.
[b] $\log_{10}$ molar NIP binding capacity/ml $\times 10^{-8}$.
[c] BSA-primed cells incubated overnight at 4°C after BSA-I^{125} 1 μg/ml:
  (1) 300 μc/mg.
  (2) 240 μc/mg.
  (3) 200 μc/mg.
[d] BSA-primed cells incubated after non-radioactive BSA.

radioactive or non-radioactive BSA in medium 199—5% normal mouse serum, at 37° C for 30 min. The cells were then washed once, resuspended in the same medium and incubated overnight at 4° C. They were then again washed once and transferred to 600 r irradiated recipients (approximately 40×10^6 nucleated cells per recipient). The recipients received at the same time non-incubated cells in the same number from donors immunized in the same manner with NIP_3 ovalbumin. Each experiment also included a group of mice that received the NIP-primed cells without the BSA-primed helpers. One day after transfer the recipients were boosted with 100 μg of $NIP_{11}BSA$ and the titres of anti-NIP and anti-BSA antibodies were read by our standard binding test 10 days later.

A comparison of line *1* with line *3* in Table V shows that in each experiment BSA-primed cells enhanced the anti-NIP response on the part of the NIP-sensitive cells, and also generated anti-BSA antibody. These anti-NIP titres can be regarded as a measure of the helper activity of the BSA-primed cells, while the anti-BSA titre can equally be regarded as a measure of their activity as antibody-forming cell precursors, although the anti-BSA titres no doubt also reflect their activity as helpers (Mitchison, 1969b; Mitchison *et al.,* 1970).

The effect of incubation with hot antigen has been shown by Ada and his colleagues, and also by Humphrey and his colleagues to abolish the transfer of the immune response (see, e.g. J. H. Humphrey's contribution to the present symposium). Ada's and Humphrey's experiments involved the polymeric antigens flagellin and TIGAL, and autoradiographic evidence indicates that the uptake of antigen on to lymphocytes is much more effective in their systems than is the case with the non-polymeric protein BSA (unpublished data of J. H. Humphrey, N. Willcox and N. A. Mitchison).

Therefore, efficient suicide by hot BSA was not anticipated in the present experiments. Nevertheless, some effect can be detected. Treatment of the cells with hot BSA results in decline of the anti-BSA titres by 30-70%, and this can be taken, as has been mentioned, as an indication of an effect on antibody-forming cell precursor or B lymphocyte activity.

In comparison the effect on helper activity, as judged by the anti-NIP titre, is smaller, lying in the range 10-50%. These figures are admittedly inconclusive, but a consistent trend can be seen in each of the three replicas. The conclusion may therefore be drawn that suicide affects preferentially antibody-forming cell precursor activity rather than helper activity. This finding, uncertain as it is, is itself open to at least three interpretations.

One interpretation of the difference is that helper activity is normally present in excess, and this has indeed been found to be the case (Mitchison *et al.,* 1970). The second possibility is that cells lethally irradiated by the isotope can still perform as helpers, and this also is in line with previous findings (Mitchison, unpublished data; W. E. Paul, personal communication). The third possibility is that helper cells are indeed less susceptible to hot suicide presumably because, as

Raff's findings would predict, the helper cells have fewer receptors.

Perhaps the only conclusion which can be drawn with any certainty from this experiment is that the results of the hot suicide experiments at least fall in line with Raff's findings. They would hardly be worth quoting did they not extend and confirm the parallel experiments of G. Roelants reported in J. H. Humphrey's contribution to the present symposium.

Conclusion

The work outlined above indicates that T lymphocytes, which appear to carry fewer receptors than B lymphocytes, are nevertheless better able to see antigen. This conclusion is in agreement with several other studies as outlined above.

How then is this paradox to be resolved? I suggest that we are left with two alternative choices. According to one, the differential triggering hypothesis, T lymphocytes are more easily triggered by a combination of antigen with a given number of receptors than are B lymphocytes. Thus, although T lymphocytes, because of their small number of receptors, may bind less antigen at a given external concentration of antigen than do B lymphocytes, they nevertheless take more notice of the small amount that they do bind. This first choice is outlined in G. Möller's contribution to the present symposium.

The second choice lays the emphasis on the binding properties of the receptor itself, rather than on the subsequent reading mechanism of the cell. It postulates a higher affinity for antigen on the part of the T lymphocyte receptor. This we may term the differential affinity hypothesis, and it is interesting to find that it also has already been outlined in the present symposium, this time by R. B. Taylor. According to this hypothesis, at a given concentration of external antigen T lymphocytes bind more antigen than do B lymphocytes, even though they have fewer receptors.

May I finally add two opinions of my own, which I shall not attempt to justify in detail? One is that I prefer the first or differential triggering hypothesis, mainly because of its power to explain the Simonsen phenomenon of a high frequency of T lymphocytes reacting to transplantation antigens (Mitchison, 1968c). The second concerns the relationship of the present findings to the difference between high- and low-zone paralysis. It might be thought that we have, in the differences of threshold that have been encountered between T and B lymphocytes, an adequate explanation of high- and low-zone paralysis. My opinion is that this is more or less a coincidence, and that a better explanation can be found for high-zone tolerance in the difficulty of rendering tolerant a previously immunized cell population rather than in the special needs of B lymphocytes for large amounts of antigen. I have to admit that at this point I part company with Dr Rajewsky. Since it would take another entire paper to argue the point in full perhaps I had better leave our difference at that.

ACKNOWLEDGEMENT

I have to thank Miss Helen Tate for her able technical assistance.

REFERENCES

Boak, J. L., Kölsch, E. and Mitchison, N. A. (1969). *Antibiotics Chemother. 15*, 98.
Brownstone, A., Mitchison, N. A. and Pitt-Rivers, R. (1966). *Immunology 10*, 481-492.
Dresser, D. W. and Mitchison, N. A. (1968). *Adv. Immun. 8*, 129.
Dubiski, S. (1967). *Cold Spring Harb. Symp. quant. Biol. 32*, 311-316.
Mäkelä, O. and Mitchison, N. A. (1965). *Immunology 8*, 549-556.
Miller, J. F. A. P. and Mitchell, G. F. (1970). *J. exp. Med. 131*, 675-699.
Mitchison, N. A. (1957). *J. cell. comp. Physiol. 50*, Suppl. 1, 247.
Mitchison, N. A. (1964). *Proc. R. Soc. B 161*, 275-292.
Mitchison, N. A. (1968a). *Immunology 15*, 509-530.
Mitchison, N. A. (1968b). *Immunology 15*, 531-547.
Mitchison, N. A. (1968c). *In* "Differentiation and Immunology" (K. B. Warren, ed.), Vol. 7, p. 29. Academic Press, New York.
Mitchison, N. A. (1969a). *In* "Immunological Tolerance" (M. Landy and W. Braun, eds), pp. 115-125. Academic Press, New York.
Mitchison, N. A. (1969b). *In* "Organ Transplantation Today" (N. A. Mitchison, J. M. Greep and J. C. M. Hattinga Verschure, eds), pp. 13-23. Excerpta Medica Foundation, Amsterdam.
Mitchison, N. A., Rajewsky, K. and Taylor, R. B. (1970). *In* "Developmental Aspects of Antibody Formation and Structure". (J. Šterzl and I. Říha, eds), pp. 547-561. Academia Publishing House, Praha.
Nossal, G. J. V. and Mäkelä, O. (1962). *J. exp. Med. 115*, 209-230.
Rajewsky, K. (1971). *Proc. R. Soc. B.* (In press.)
Taylor, R. B. (1969). *Transplantation Rev. 1*, 114-149.
Weigle, W. O. (1961). *J. exp. Med. 114*, 111-125.
Weigle, W. O. (1962). *J. exp. Med. 116*, 913-928.

Idiotypes, Autoimmunity and Cell Co-operation

G. MICHAEL IVERSON

*National Institute for Medical Research,
Mill Hill, London, England*

Immunity against both hapten and carrier is required to elicit a maximum secondary anti-hapten response (carrier effect) (Ovary and Benacerraf, 1963). In the mouse carrier effects have been shown to be the result of an act of co-operation, in which a thymus-dependent lymphocyte (the helper) picks up antigen by means of a carrier determinant and presents it to a bone marrow-derived lymphocyte, which is then triggered to produce antibody (Mitchison, 1967; Rajewsky and Rottländer, 1967). For a protein to be immunogenic it must have two determinants (Rajewsky *et al.,* 1969), one acting as the carrier determinant and the other as the "hapten equivalent".

Mouse immunoglobulins have been shown to have two physically separable antigenic areas. One area containing the allotypic determinants is located on the Fc fragment (Mishell and Fahey, 1964). In any given strain of inbred mice all of the immunoglobulin molecules of a given class express, on the Fc region of the immunoglobulin molecule, the allotypic determinants for that class in that strain. On the other hand the Fab fragment has idiotypic determinants (Kunkel *et al.,* 1964), which are located on the variable portion of the Fab fragment. As we shall see these determinants are not unique to certain strains or even to individuals of inbred mice (like allotypes). But they are unique to the immunoglobulin molecules produced by a given myeloma (or at least the idiotypes of different myeloma proteins seldom overlap).

Mice can produce isoantibodies against immunoglobulins that carry allotypic determinants not present on their own immunoglobulins. For instance, the plasmacytoma 5563 (γG_{2a}) arose spontaneously in C3H mice, and expressed the Ig-1^a locus like normal immunoglobulins of the class in this strain. C57BL/6 mice, whose γG_{2a} molecules express the Ig-1^b locus, can produce antibodies reacting with the Ig-1^a determinants if they are injected with 5563 myeloma protein, but C3H mice cannot. Mice producing antibodies against these determinants also produce antibodies against idiotypic determinants (Potter *et al.*, 1966). These mice produce either antibodies against both allotypic and idiotypic specificities, or they do not produce anything detectable against either determinant. Thus mice cannot normally detect idiotypic determinants on native myeloma proteins with allotypic determinants that their own immunoglobulins possess.

Potter *et al.* (1966) interpret these observations as follows: mice of the inbred strain in which the myeloma arose have the genetic capacity to produce the myeloma idiotype, do in fact produce it, become tolerant of it, and therefore are unable to synthesize antibodies against it. Mice of other strains do not produce the idiotype and are therefore able to synthesize antibodies against it. In order to test this hypothesis the serum of normal mice belonging to the strain of origin of a myeloma was accordingly tested for the presence of the myeloma idiotype. This test was performed with the aid of a sensitive procedure based on a solid-phase radioimmuno assay introduced originally for the assay of protein hormones (Catt and Tregear, 1967). With the solid-phase radioimmuno assay, normal mice were shown to possess low levels of an inhibitor of the reaction between 5563 idiotype and antibodies against it. Figure 1 shows the concentration of inhibitor in sera (ng/ml) of individuals of four different inbred strains of mice. This suggests that these mice must possess the structural genes coding for this idiotypic determinant. These findings exclude the hypothesis of Potter *et al.* (1966).

An alternative hypothesis has been advanced by Cohn *et al.* (1969). They suggest that the spectrum of determinants generated as a result of the variable sequence is identical in all strains of mice. This seems to be substantiated by the observations shown in Fig. 1. Cohn *et al.* (1969) also suggested that the Fc portion of the molecule acts as a "carrier" for the Fab portion, in the sense that an otherwise non-immunogenic determinant (the "hapten") becomes immunogenic if attached to a structure which is itself immunogenic (the "carrier").

To test the hypothesis that an antigenic determinant on the Fc portion acts as a carrier for determinants of the Fab portion, 2,4-dinitrophenol (DNP) was conjugated to 5563 myeloma protein and this alum-precipitated conjugate, with *Bordetella pertussis* as adjuvant, was injected into C3H and CBA mice. The unaltered Fc portion of the molecule could not act as the carrier for the reasons stated. But in these experimental conditions the 5563 protein has a DNP determinant to act as the "carrier".

The immunological response to an antigenic determinant is increased if the host is presensitized to the carrier (Ovary and Benacerraf, 1963). This presensitization gives rise to a population of helper cells, specific for the carrier, which has been shown to be thymus-dependent (Raff, 1970). Painting the skin with 1-fluoro-2-4-dinitrobenzene (FDNB) leads to cell-mediated immunity specific for DNP, which is almost certainly mediated by thymus-dependent

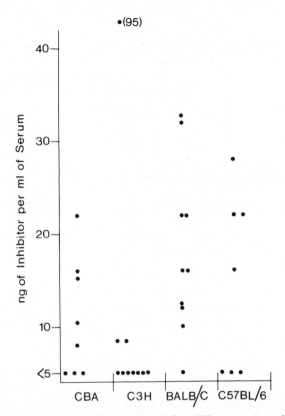

Fig. 1. Levels of inhibitor in individual mice of four different strains, of the binding of [125]I-5563-Fab by CBA anti-idiotype. Plastic test tubes were coated with small amounts of anti-idiotype. Various amounts of 5563-Fab or serum from individual mice were added to the coated tubes and incubated overnight at 37°C. Tubes were then emptied, washed three times with saline and a constant amount of [125]I-5563-Fab was added to each. Tubes were again incubated overnight at 37°C, emptied, washed three times with saline and the bound radioactivity measured. Inhibition of the binding of radioactive 5563-Fab by unlabelled Fab was compared with inhibition of binding by sera from individual mice. Values are given as ng of inhibitor per ml of serum. The lower limit of detectability of the assay was 5 ng/ml. CBA mice were three-month-old males, C57B1/6 mice were six-month-old males, Balb/c and C3H mice were 12-month-old females.

cells. Accordingly the mice were skin painted with FDNB to provide them with a population of thymus-dependent, DNP-specific, helper cells.

Table I compares the anti-idiotype response in FDNB painted mice and in unpainted mice. The observation that the mice have to be presensitized to this new determinant argues against the possibility that the conjugation has merely increased, non-specifically, the immunogenicity of the myeloma protein.

The specificity of these anti-idiotype sera was tested by inhibition of binding

TABLE I

Titres of anti-idiotype sera as a function of presensitization by skin painting

| Strain | Titre[a] ± standard error | |
	Painted with FDNB	Not skin painted
C3H	10.03 ± 0.34	Not detectable
CBA	7.82 ± 0.67	1.20 ± 0.47

Each serum was tested at four dilutions. 10 μl of ^{125}I-5563-Fab, 6×10^{-8} M, was added to 50 μl of antiserum diluted in 1/24 normal mouse serum in borate buffer. The mixture was incubated for 30 min at room temperature, then 50 μl of rabbit anti-mouse immunoglobulin, which had been absorbed with insoluble mouse immunoglobulin Fab, was added to each tube. The contents were thoroughly mixed, incubated for 30 min at 4°C, then centrifuged at 2000 g for 30 min. The supernatant was removed and the radioactivity of the precipitate measured. [a] $\log_2$ of the dilution of antiserum that will bind 50% of the radioactive 5563-Fab at a final concentration of 10^{-8} M.

in the co-precipitation assay. Only 5563 Fab inhibits the binding of radioactive 5563 Fab by C3H anti-5563 (Fig. 2). The immunoglobulin and immunoglobulin fragments that did not inhibit specific binding included immunoglobulins: (1) with the same class and allotypic determinants; (2) with different class and therefore different allotypic determinants; (3) with the same L-chain type (κ); and (4) with a different L chain type (λ). Therefore the antisera must detect those determinants that are unique to 5563 myeloma protein. These unique determinants must be on the variable region of the molecule.

Mice can produce antibodies against these "self" idiotypic determinants. Since these determinants were found circulating in serum why have the antibody-forming cell precursors (AFCP) not been rendered tolerant? Either these AFCP cannot be made tolerant or there is insufficient antigen present to make them tolerant. In order to discriminate between these alternatives CBA mice were injected with 100 μg of particle-free 5563 myeloma protein (unconjugated). About one month later the mice were skin painted with FDNB and primed with various amounts of alum-precipitated DNP-5563 myeloma protein, with *B. pertussis* as an adjuvant. Figure 3 shows that the mice given the particle-free 5563 myeloma protein were tolerant to subsequent injections of an otherwise immunogenic form of the myeloma protein. Thus the AFCP could be made tolerant to the idiotypic determinant. It seems likely that in normal mice there is insufficient antigen circulating to make the AFCP tolerant. But if during immunization the level of an idiotype were to rise to high levels the AFCP could then become tolerant (Iverson and Dresser, 1970).

It is clear from these experiments that strains of mice that do not in normal circumstances respond to an injection of alum-precipitated myeloma protein

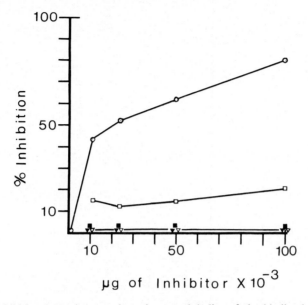

Fig. 2. Inhibition by various myeloma immunoglobulins of the binding between C3H anti-5563 and radioactive 5563-Fab. 10 μl of various concentrations of inhibitor was added to 50 μl of a dilution of antiserum that would bind 80-90% of the radioactive 5563-Fab. After thoroughly mixing 10 μl of ^{125}I-5563-Fab, 6 x 10^{-8} M, was added. Contents of the tube were mixed and incubated for 30 min at 20°C, after which 50 μl of rabbit anti-mouse immunoglobulin Fc was added. To ensure that the rabbit anti-mouse immunoglobulin Fc serum had no specificity whatsoever for 5563-Fab it was absorbed with normal mouse immunoglobulin Fab made insoluble by conjugation to Sepharose, the absorption being repeated until no precipitation of ^{125}I-5563-Fab was obtainable. Mixing was followed by incubation for 30 min at 4°C, centrifugation at 2000 g for 30 min, removal of supernatant, and measurement of the radioactivity in the precipitate. The degree of inhibition was calculated as a percentage of the amount of label precipitated in controls without inhibitor. The various purified myeloma immunoglobulins tested were: 5563-Fab = o——o; 5563-Fc = □——□; LPC-1 (γG_{2a}, Ig-1a) = ■——■; MPC-25 (γG_1, Ig-4a) = ▽——▽; MOPC-104 (γM) = ▼——▼.

plus *B. pertussis*, could be made to respond simply by adding a new carrier determinant (DNP) to the myeloma protein and presensitizing mice to this new determinant. The ability to respond and to be made tolerant indicates that these mice must have possessed antibody forming cells and their precursors for this auto-antigenic idiotypic determinant. These results are compatible with the hypothesis that C3H and CBA mice do not normally respond to the idiotypic determinants of myeloma proteins, not because these determinants are "self" and therefore the mice are incapable of producing specific antibody, but because they do not possess thymus-dependent helper cells specific for a carrier determinant on the molecule. Therefore they cannot present the idiotypic determinant in the appropriate way to the antibody forming cells and their

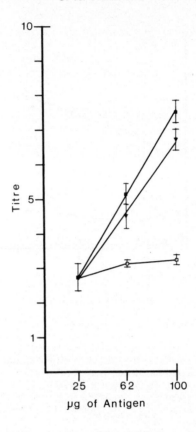

Fig. 3. Anti-5563 idiotype response in CBA mice challenged with different amounts of antigen. Each point indicates the mean titre in six mice ± standard error. ■ = controls; ▼ = mice injected with 0.5 µg particle-free 5563 myeloma protein 4 hr before challenge; □ = mice injected with 100 µg of 5563 myeloma protein 28 days before challenge.

precursors. Each immunoglobulin molecule has two Fab fragments. Each Fab contains at least one idiotypic determinant, therefore the whole molecule must contain at least two antigenic determinants. This raises the question as to why this molecule with two antigenic determinants is not immunogenic. There are at least two possible explanations. The first is that there are no thymus-dependent helper cells with receptors for the idiotypic determinants. This hypothesis has not been definitely excluded but seems unattractive, mainly because it would require a different mechanism for the generator of diversity or at least a different mechanism of selection among helper cells as distinct from AFCP. The second possibility is that two different antigenic determinants are required for immunogenicity (Iverson and Taylor, 1970).

ACKNOWLEDGEMENT

I thank the Welcome Trust for the Welcome Trust Research Travel Grant allowing me to attend the meeting.

REFERENCES

Catt, K. and Tregear, G. W. (1967). *Science, N.Y. 158,* 1570-1572.
Cohn, M., Notani, G. and Rice, S. A. (1969). *Immunochemistry 6,* 111-123.
Iverson, G. M. and Dresser, D. W. (1970). *Nature, Lond. 227,* 274-276.
Iverson, G. M. and Taylor, R. B. (1970). *Proc. R. Soc. B.* (In press.)
Kunkel, H. G., Allen, J. C. and Grey, H. M. (1964). *Cold Spring Harb. Symp. quant. Biol. 29,* 443-447.
Mishell, R. I. and Fahey, J. L. (1964). *Science, N.Y. 143,* 1440-1442.
Mitchison, N. A. (1967). *Cold Spring Harb Symp. quant. Biol. 32,* 431-439.
Ovary, Z. and Benacerraf, B. (1963). *Proc. Soc. exp. Biol. Med. 114,* 72-76.
Potter, M., Lieberman, R. and Dray, S. (1966). *J. molec. Biol. 16,* 334-346.
Raff, M. C. (1970). *Nature, Lond. 226,* 1257-1258.
Rajewsky, K. and Rottländer, E. (1967). *Cold Spring Harb. Symp. quant. Biol. 32,* 547-554.
Rajewsky, K., Schirrmacher, V., Nase, S. and Jerne, N. K. (1969). *J. exp. Med. 129,* 1131-1143.

Cellular Events in the Induction and Termination of Immunological Unresponsiveness

W. O. WEIGLE, J. M. CHILLER and D. C. BENJAMIN

Department of Experimental Pathology,
Scripps Clinic and Research Foundation, La Jolla, California, U.S.A.

The termination of immunological unresponsiveness to heterologous serum proteins in adult rabbits by injections of either cross-reacting antigens or altered tolerated antigen is well established (Weigle, 1967). In this situation, simultaneous injection of the tolerated antigen and the cross-reacting antigens inhibits the termination. In addition, following the termination of the unresponsive state, the rabbits will respond for a limited period of time to subsequent injections of the tolerated antigen, but after several injections, they return to the unresponsive state (Weigle, 1964). None of the antibody produced following the subsequent injections is specific for the tolerated antigen, and can be completely removed or inhibited with the cross-reacting antigen used to terminate the unresponsive state. These phenomena could most readily be explained by the existence of two steps in the immune response, possibly involving two different cell types with only one being unresponsive (Weigle, 1962). The possibility that the termination of immunological unresponsiveness is the result of the presence of unresponsiveness in cell types other than the precursor cell, became more plausible with the demonstration of synergism between thymus-dependent and bone marrow-derived cells in the immune response (Claman *et al.,* 1966; Miller and Mitchell, 1967). The present communication will be concerned with the involvement of thymus-dependent and bone marrow-derived cells in both the termination of unresponsiveness and autoimmunity.

BONE MARROW AND THYMUS CELLS

It was recently shown that both thymus and bone marrow cells of mice become unresponsive during the induction of immunological unresponsiveness to deaggregated human gamma globulin (HGG) (Chiller *et al.,* 1970). In these experiments, thymus and bone marrow cells from both unresponsive and normal mice were used in different combinations in attempts to reconstitute irradiated

C.I.–10

A/J mice. A/J mice that received 850 r whole body irradiation were injected intravenously with bone marrow and thymus cells from normal mice or mice rendered unresponsive by an intraperitoneal injection of 2.5 mg deaggregated (centrifuged at 1×10^5 g for 120 min) HGG (tolerogen). The mice were challenged with aqueous preparations containing 0.4 mg aggregated HGG (immunogen) both immediately after transfer of cells and 10 days later. The spleens were examined five days after the last injection using a modification (Golub *et al.*, 1968) of the hemolytic plaque assay (Ingraham and Bussard, 1964) for indirect plaque forming cells (PFC) to HGG (Table I). PFC to HGG

TABLE I

Specificity of tolerant thymus and bone marrow cells in the immune response to HGG

Treatment	No. of mice	Indirect PFC/spleen to HGG	TGG
NT + NBM + HGG + TGG	7	2,508	9,824
TT + NBM + HGG + TGG	7	0	23,296
NT + TBM + HGG + TGG	7	0	22,425
TT + TBM + HGG + TGG	7	0	9,797
N Donors + HGG + TGG	6	51,684	161,680
T Donors + HGG + TGG	6	0	178,446

NT = Normal thymus—90 x 10⁶ cells
TT = Tolerant thymus—90 x 10⁶ cells
NBM = Normal bone marrow—30 x 10⁶ cells
TBM = Tolerant bone marrow—30 x 10⁶ cells
HGG = Human gamma globulin
TGG = Turkey gamma globulin

were detected only in the spleens of mice that received both normal thymus cells and normal bone marrow cells. If either bone marrow or thymus cells came from an unresponsive donor, synergism was not observed. These results are specific for HGG since all combinations of thymus and bone marrow cells showed a response to turkey gamma globulin (TGG), which does not cross react with HGG. The results demonstrated that both thymus and bone marrow cells probably became unresponsive, and that both cell types had specific receptor sites for HGG. Of equal importance is that only one cell type has to be unresponsive in order to maintain an unresponsive state in the intact mouse. The failure of others (Taylor, 1968; Playfair, 1969) to observe unresponsiveness in both the bone marrow and thymus is probably the result of the particular unresponsive state studied. In the case of SRBC injected into mice treated with cyclophosphamide, one might not expect all of the determinants of this complex and particulate antigen to reach both bone marrow and thymus cells in effective concentrations.

KINETICS OF THE ESTABLISHMENT OF IMMUNOLOGICAL UNRESPONSIVENESS

The establishment of immunological unresponsiveness has been shown to require an induction period which may vary from a few hours to a few days depending upon the antigen and means of inducing unresponsiveness (Golub and Weigle, 1967; Matangkasombut and Seastone, 1968; Mitchison, 1968; Britton, 1969; Diener and Armstrong, 1969). In the mice, four days were required for the establishment of complete unresponsiveness to deaggregated HGG. More recently, the cellular kinetics of the establishment of the unresponsive state to HGG was determined using the modification of the hemolytic plaque assay and a cell transfer system. Adult A/J mice were injected with 2.5 mg of deaggregated HGG and the spleens removed from groups of the mice at various times thereafter. The washed spleen cell suspensions were transferred to irradiated (900 r) recipients. The recipients were challenged with 0.4 mg of aggregated HGG at the time of transfer and nine days later, and their spleens analyzed for indirect plaque forming cells (PFC) on day 14. Similarly, at each time interval, spleens from saline injected donors were transferred in order to obtain the normal response of such a cell transfer system. The response (PFC) obtained in recipients of tolerogen treated spleen cells was expressed as a percentage of that response obtained in the recipients injected with saline, and this percentage plotted as a function of time following injection of tolerogen. Although all of the cells did not become unresponsive until four days in the donor, 75% of the cells became unresponsive after a six-hour *in vivo* exposure to the tolerogen (Fig. 1). Neither direct nor indirect PFC to HGG could be detected in the spleen cells of the *donor* mice at any time during a 20-day period of time following injection of deaggregated HGG, suggesting that the induction of immunological unresponsiveness in this situation is not accompanied by a transient production of detectable antibody and most likely represents the inactivation of specific cells following contact with the tolerogen.

TERMINATION OF IMMUNOLOGICAL UNRESPONSIVENESS

Recent data indicate that adult rabbits rendered unresponsive by neonatal injections of bovine serum albumin (BSA) contain a normal complement of precursor cells with specific receptors for BSA, suggesting that the unresponsive rabbit has a cellular potential to produce antibody to BSA. Normal rabbits and rabbits made unresponsive by neonatal injections of BSA made similar amounts of both binding (Table II) and precipitating antibody to BSA following immunization with aqueous preparations of any one of four cross-reacting albumins (Benjamin and Weigle, 1970a). The avidity of the antibody produced was dependent on the antigen used to terminate the unresponsive state, and not

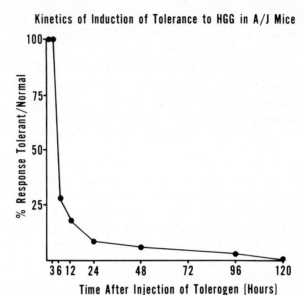

Fig. 1. Kinetics of the induction of immunological tolerance in A/J mice.

TABLE II

Mean antigen-binding capacities[a] of sera from normal and BSA-unresponsive rabbits after two courses of various soluble albumins

Terminating antigen	Status	No. of rabbits	Antigen tested BSA	terminating
PSA	Unresponsive	19	11.6	188.8
	Normal	12	11.7	249.1
	P		0.9	0.2
HSA	Unresponsive	26	9.1	200.8
	Normal	26	7.8	142.7
	P		0.5	0.1
GPSA	Unresponsive	9	7.1	324.4
	Normal	14	8.7	289.3
	P		0.5	0.6
ESA	Unresponsive	15	6.6	170.2
	Normal	26	5.6	139.4
	P		0.5	0.2

[a] μg antigen N bound to globulin (precipitated with 50% saturated ammonium sulphate) present in 1.0 ml of serum.

PSA: pig serum albumin
HSA: human serum albumin

GPSA: guinea pig serum albumin
ESA: equine serum albumin

on whether the antibody came from normal or unresponsive rabbits. Thus, following injection of normal and unresponsive rabbits with these heterologous albumins, no quantitative or qualitative differences in the anti-BSA produced could be detected, suggesting that the unresponsive rabbit had the potential to produce a normal response to the BSA-related determinants on cross-reacting albumins. Once the unresponsive state to BSA is terminated, the rabbits do respond for a limited period of time to subsequent injections of BSA. However, recent data suggest that BSA cannot stimulate differentiation and proliferation of precursor cells in either "virgin" unresponsive rabbits (Benjamin and Weigle, 1970a) or unresponsive rabbits previously immunized with a cross-reacting albumin (Benjamin and Weigle, 1970b), whereas the albumin used to terminate the unresponsive state can.

Rabbits whose unresponsive states to BSA were terminated following the injections of cross-reacting albumins made an immune response to a subsequent injection of BSA. However, none of the antibody produced was specific for BSA, in that it was all removed by absorption with the albumin used to terminate the unresponsive state. These serological studies are supported by the cellular events occurring in the spleens of the rabbits (Benjamin and Weigle, 1970b). Following the injection of BSA, the spleens of the rabbits contained indirect PFC to rabbit red cells conjugated with either BSA or the terminating antigen. Studies on inhibition of PFC by soluble albumins demonstrated that the spleens of normal rabbits, but not the previously unresponsive rabbits, contained cells making antibody specific for BSA. These data suggest that injection of BSA after termination of unresponsiveness does not result in recruitment of precursor cells, but once memory cells are established by the cross-reacting antigens, BSA can cause their conversion to antibody producing cells. This results in antibody only to determinants shared with the cross-reacting albumin.

POSSIBLE CELLULAR EVENTS IN UNRESPONSIVENESS AND AUTOIMMUNITY

The presence of a normal potential of precursor cells to produce antibodies to BSA in rabbits rendered unresponsive to BSA by neonatal injections can most readily be explained if the immune response to BSA in rabbits involves the interaction of two cell types, both containing receptor sites for antigen, and that unresponsiveness lies in a cell other than the precursor cell. It would also be required in normal rabbits that the receptors on the precursor cell (possibly bone marrow-derived) react with a different determinant than does the receptor on the second cell (Mitchison, 1969; Rajewsky et al., 1969). There is reason to suppose that thymus-dependent and bone marrow-derived cells have receptor sites for BSA on their surface, and that these cells react synergistically to produce an antibody response to BSA, since a comparable situation apparently occurs in the mouse with HGG. It has also been shown in the mouse that unresponsiveness induced in adults to HGG requires only unresponsive thymus or unresponsive bone marrow-derived cells, but not both. If the site of unresponsiveness to BSA in the rabbit was at the thymus-dependent cell and the bone marrow was unaffected and contained a normal complement of precursor cells with receptor sites, all of the events observed with the termination of immunological unresponsiveness could be readily explained. If an interaction between thymus-dependent and bone marrow-derived cells via different determinants on BSA and the respective receptors on the two cell types was required before an immune response could occur, unresponsiveness could be the result of the absence of thymus cells with specific receptors for BSA. Since thymus cells with receptors for cross-reacting albumins would be present in rabbits unresponsive to BSA, interaction could take place between the thymus-dependent cells (via determinants specific for HSA) and the bone marrow-derived cells (via determinants related to BSA) resulting in a normal response to the BSA-related determinants. If the bone marrow-derived cell was the precursor cell, it would be stimulated to proliferate, differentiate and synthesize antibody of the specificity dictated by its receptor sites. Subsequent injection of the previously tolerated BSA would be expected to result in an immune response if memory cells could be stimulated directly without a synergistic interaction with thymus-dependent cells, and an exhaustion of these cells in the absence of recruitment of new memory cells would result in a return to the unresponsive state.

This hypothesis is not compatible with the finding in mice of both unresponsive thymus cells and unresponsive bone marrow cells. However, unresponsiveness at the cellular level in animals made unresponsive as adults and those rendered unresponsive during neonatal life may differ. Unresponsiveness in both bone marrow and thymus cells in mice was observed shortly after (2-3 weeks) the induction of unresponsiveness, whereas the termination of unresponsiveness with cross-reacting antigens occurred at least three months after the induction of unresponsiveness. It is possible that the spontaneous loss of unresponsiveness occurs more rapidly in the precursor cells than in thymus-dependent cells (or their equivalent in the rabbit). In experiments showing collaboration between bone marrow and thymus cells, a low level synergism is sometimes seen between normal thymus and tolerant bone marrow cells, but never between tolerant thymus and normal bone marrow, suggesting that the unresponsive state is less stable in the bone marrow cells. The ability of simultaneous injections of the tolerated antigen and the cross-reacting antigens to inhibit the termination of immunological unresponsiveness may be the result of a reinduction of unresponsiveness in the bone marrow cells. In preliminary experiments, it has not been possible to terminate the unresponsiveness to HGG induced in adult mice (where both the thymus and bone marrow cells are unresponsive) by injecting a cross-reacting antigen (bovine gamma globulin).

A series of events very similar to those involved in the termination of immunological unresponsiveness to BSA takes place following the injection of aqueous preparations of either certain thyroglobulins that cross react with rabbit thyroglobulin (Weigle and Nakamura, 1967) or altered thyroglobulin (Weigle, 1965). The natural unresponsive state to the rabbit's own thyroglobulin is apparently terminated; i.e. both thyroiditis and circulating antibody to rabbit thyroglobulin are produced. The production of experimental thyroiditis may be the result of an incomplete unresponsiveness to the antigen at the level of the precursor cell. The new determinants on altered thyroglobulin and the unrelated determinants on the cross-reacting thyroglobulins could thus react with their receptors on thymus-dependent cells. Thymus-dependent cells could then react synergistically with competent precursor cells that have receptors to native thyroglobulin. The reason that thyroiditis, aspermatogenesis, uveitis, encephalomyelitis and possibly glomerulonephritis are so readily induced experimentally may be that the body components involved are not present in the body fluids in sufficient concentration to maintain an unresponsive state at the level of the precursor cells in the bone marrow. Studies of the kinetics of induction, duration and dose response with both bone marrow and thymus-derived cells, will undoubtedly give additional insight into the function of the thymus and bone marrow cells in both autoimmunity and the termination of acquired immunological unresponsiveness.

ACKNOWLEDGEMENTS

Publication No. 420 from the Department of Experimental Pathology of Scripps Clinic and Research Foundation, supported by United States Public Health Service Grant AI 07007, American Cancer Society Grant T-519 and Atomic Energy Commission Contract AT(04-3)-410. W. O. W. is a recipient of United States Public Health Career Award 5-K6-GM-6936. J. M. C. is supported by United States Public Health Service Training Grant GM-00683. D. C. B. is supported by United States Public Health Service predoctoral fellowship 5-F01-GM-29-481.

SUMMARY

Following injection of deaggregated HGG into adult A/J mice both thymus and bone marrow cells were unresponsive, as determined by their inability to restore immunocompetence to aggregated HGG in irradiated recipients. Furthermore it was shown that if *either* the thymus or marrow cells came from unresponsive donors immunocompetence was not restored.

Rabbits made unresponsive at birth to BSA made a normal response to injections of cross-reacting albumin given three months later. The quantity and quality of the antibody produced to the BSA-related determinants were the same in normal and unresponsive rabbits, indicating that unresponsive rabbits had a normal complement of precursor cells to BSA. The events observed during termination of tolerance with cross-reacting antigens can be readily explained if the site of unresponsiveness in rabbits injected at birth is at the thymus-dependent cell and the precursors of antibody-forming cells are not affected. Rabbits injected at birth with BSA may, during maturation, lose unresponsiveness to BSA in the precursor cells, but not in the thymus-dependent cells.

The cellular kinetics of the induction of immunological unresponsiveness in adult A/J mice showed that an unresponsive state was achieved in 75% of the spleen cells within 6 hr after injection of deaggregated HGG. However four days were required before all the cells were affected.

REFERENCES

Benjamin, D. C. and Weigle, W. O. (1970a). *J. exp. Med. 132*, 66-76.

Benjamin, D. C. and Weigle, W. O. (1970b). *J. Immun.* (In press.)

Britton, S. (1969). *J. exp. Med. 129*, 469-482.

Chiller, J. M., Habicht, G. S. and Weigle, W. O. (1970). *Proc. natn. Acad. Sci. U.S.A. 65*, 551-556.

Claman, H. N., Chaperon, E. A. and Triplett, R. F. (1966). *Proc. Soc. exp. Biol. Med. 122*, 1167-1171.

Diener, E. and Armstrong, W. D. (1969). *J. exp. Med. 129*, 591-603.

Golub, E. S. and Weigle, W. O. (1967). *J. Immun. 99*, 624-628.

Golub, E. S., Mishell, R. I., Weigle, W. O. and Dutton, R. W. (1968). *J. Immun. 100*, 133-137.

Ingraham, J. S. and Bussard, A. (1964). *J. exp. Med. 119*, 667-684.

Matangkasombut, P. and Seastone, C. V. (1968). *J. Immun. 100*, 845-850.

Miller, J. F. A. P. and Mitchell, G. F. (1967). *Nature, Lond. 216*, 659-663.

Mitchison, N. A. (1968). *Immunology 15*, 531-547.

Mitchison, N. A. (1969). *In* "Immunological Tolerance" (M. Landy and W. Braun, eds) pp. 115-125. Academic Press, New York.

Playfair, J. H. L. (1969). *Nature, Lond. 222*, 882-883.

Rajewsky, K., Schirrmacher, V., Nase, S. and Jerne, N. K. (1969). *J. exp. Med. 129*, 1131-1143.

Taylor, R. B. (1968). *Nature, Lond. 220*, 611.

Weigle, W. O. (1962). *J. exp. Med. 116*, 913-928.

Weigle, W. O. (1964). *J. Immun. 92*, 791-797.

Weigle, W. O. (1965). *J. exp. Med. 121*, 289-308.

Weigle, W. O. (1967). *In* "Natural and Acquired Immunologic Unresponsiveness". Monographs in Microbiology, World Publishing Company, Cleveland.

Weigle, W. O. and Nakamura, R. M. (1967). *J. Immun. 99*, 223-231.

"Primary" and "Secondary" Reactivity of Lymphocytes to Major Histocompatibility Antigens: A Consideration of Immunologic Memory

DARCY B. WILSON and PETER C. NOWELL

Immunobiology Research Unit
Departments of Pathology and Medical Genetics
University of Pennsylvania School of Medicine
Philadelphia, Pennsylvania, U.S.A. 19104

Responses to Non-histocompatibility Antigens

The distinctive differences between primary and secondary immunologic reactivity—the response of lymphocytes from putatively normal or previously sensitized animals to antigen—are most obvious and pronounced in the humoral immunities. In the primary response to SRBC, a small proportion of cells (10^{-5}-10^{-6}) proliferate clonally, attain a frequency of 10^{-3}, and produce IgM and IgG. In the secondary response, higher titers of IgG are produced, of higher affinity for the immunizing antigens, more promptly (immunologic "memory"), over a more sustained period of time. Numerous recent studies have indicated rather clearly that antibody is made by a lineage of lymphocytes and plasma cells that stem from the bone marrow, and do not depend on the thymus in their maturation to immunologic competence. Even in those instances where an antibody response depends on the presence of a functional thymus, the circulating immunoglobulins are made by B cells; the thymus-dependent lineage of lymphocytes serves a not yet well defined helper function which facilitates the stimulation of proliferation and production of immunoglobulin by B cells (Mitchell and Miller, 1968; Mitchison, 1969).

While the effector mechanisms of delayed type hypersensitivities are less well understood, there are also clear differences in behavior when lymphocytes from normal animals or those with delayed hypersensitivities are confronted with antigen. For example, lymphocytes from sensitized donors cultured with the immunizing antigen release cytoactive substances, which have a variety of effects on other cell populations and may represent the effector substances of cell mediated immunities. Their effects include chemotaxis and inhibition of macrophage migration (chemotactic factor, MIF), stimulation and inhibition of DNA synthesis in lymphoid cells, destruction of fibroblasts (lymphotoxin), and the capacity to promote inflammation and accumulation of polymorphs (skin reactive factor) (Bloom, 1969). The cells involved require the presence of a thymus for their immunologic maturation and, once stimulated by antigens to release their cytoactive factors, these factors are non-specific in their effects and do not require the further presence of antigen. These factors can also be released in cultures of normal lymphocytes, but only by stimulation with (1) the non-specific mitogens PHA, PWM, or other agents with specificity directed against lymphocyte membranes such as ALS, anti-Ig or anti-allotype sera, and (2) other cells bearing major histocompatibility (H) antigenic differences in the mixed lymphocyte interaction (MLI) (Wilson and Billingham, 1967).

With the exception of the last point, all the phenomena discussed above are easily interpretable according to the Clonal Selection Hypothesis. Antigen introduced into a normal animal selects a small population of specifically reactive lymphocytes which may or may not already be making antibody; these proliferate in a clonal manner to increase their frequency by perhaps a thousand-fold so that their immunoglobulin products make a substantial contribution to the circulating immunoglobulin pool (Siskind and Benacerraf, 1969). Similar arguments can be applied to cell mediated immunities. The selective pressure of antigen brings about the proliferation of specifically reactive T cells from an initially small number to the point where they represent a significant frequency in the general lymphocyte population. Inhibition of DNA synthesis and of lymphocyte proliferation with anti-mitotic agents such as colchicine or vinblastine, with purine or pyrimidine analogs, and the destruction of specifically responsive cells with hot suicidal pulses of tritiated thymidine, thereby inhibiting the development of immunity, provide strong evidence on this point (Syeklocha et al., 1966; Dutton, 1966; Dutton and Mishell, 1967).

In both humoral and cell mediated immunities the cellular basis of immunologic memory, the distinction between primary and secondary immune responsiveness, seems to be largely quantitative: the frequency of antigen reactive cells before and after immunization.

RESPONSES TO HISTOCOMPATIBILITY ANTIGENS

Immunologic memory in immunities directed to major H antigen differences seems to present a different problem, and may operate by different mechanisms. The studies of Simonsen (Simonsen, 1967; Nisbet et al., 1969) and others (Szenberg et al., 1962) have shown that inocula consisting of very few immunologically competent lymphocytes are sufficient to produce the symptoms of a GVH reaction, and that prior immunization of the donor does not markedly increase the efficacy of such inocula in producing a GVH reaction involving major histocompatibility differences. These observations are best interpreted by a model in which the number of lymphocytes reactive to transplantation isoantigens of other members of the same species is already large and is not increased significantly by prior immunization. An implication of such a model is that memory in this circumstance does not primarily involve increases in the number of antigen reactive cells, but rather some qualitative alteration in their responsiveness.

The mixed lymphocyte interaction (MLI) has proven to be a useful experimental system for examining the reaction of lymphocytes to the presence of major H antigens. The following discussion summarizes the pertinent information about the immunologic basis of the MLI and includes preliminary results of our attempts to explore the nature of "memory" in the immune response to strong H antigens.

1. Immunologic significance of the mixed lymphocyte interaction (MLI). It is apparent for several reasons that the reactive cells in the MLI are immunologically competent lymphocytes which recognize and are triggered to respond to specific histocompatibility antigens: (1) the immunoproliferative behavior of lymphocytes is similar in the MLI and in the GVH reaction (Wilson and Elkins, 1969); (2) the stimulatory factors are antigens of the major histocompatibility loci in man (Albertini and Bach, 1968), the mouse (Dutton, 1966), and the rat (Wilson and Billingham, 1967; Wilson and Elkins, 1969; Wilson, 1967; Palm, 1964; Schwarz, 1968); (3) the cells that respond are thymus-dependent lymphocytes of the circulating lymphocyte pool (Johnston and Wilson, 1970); (4) the proliferative reactivity of the cells depends on the genetic and immunologic status of the donors—cells from immunologically tolerant animals do not respond to the H antigens to which the donors have been made tolerant, but are responsive to third party H antigen systems; cells from F_1 hybrid donors are nonresponsive to parental strain lymphocytes; and cells from neonatally thymectomized animals display only a limited proliferative response (Wilson *et al.*, 1967).

2. Number of cells reactive to H antigens. Studies with the GVH reaction and with the MLI have clearly indicated that the proportion of lymphocytes in the circulating lymphocyte pool of an immunologically competent individual is a substantially large one—of the order of 2-4% (Simonsen, 1967; Szenberg *et al.,* 1962; Wilson *et al.,* 1968; Nisbet *et al.,* 1969). This figure is at least three orders of magnitude in excess of the frequency of antigen sensitive cells reactive to heterologous erythrocyte antigens and it is therefore difficult to reconcile with the clonal selection hypothesis.

Results of recent studies (Wilson and Nowell, 1970) favor the interpretation that the responding cells in the MLI are unipotential, and that antigen-induced proliferative reactions of this magnitude are restricted to a limited number of special antigen systems such as the major H isoantigens of the species. The reactive cells do not seem to be recruited or activated in some non-specific manner, nor do they appear to be multipotential, reactive to a multiplicity of different antigens.

The possibility of recruitment or non-specific activation was rejected on the basis of results with a three-way MLI involving lymphocytes from normal donors, tolerant isologous donors and F_1 animals. By using tolerant and normal donors of different sexes, the responsive cells were all identified as being derived from the normal parental strain donors. This observation does not rule out the possibility that a few cells might be stimulated in a non-specific manner to divide, but it does indicate that a general phenomenon of recruitment cannot account for the large numbers of reactive cells observed.

Evidence against the possible multipotentiality of reactive cells was obtained with mixed cultures showing that: (1) the magnitude of the proliferative response is increased additively when potentially reactive cells are exposed to two antigen systems simultaneously; (2) a state of induced immunologic tolerance to one H isoantigen system does not alter the response capacity of cells from such a donor to an alternative antigen system; and (3) as discussed below, use of lymphocytes from previously immunized donors results in a more prompt proliferative response against F_1 cells bearing the immunizing antigens, than against others. These data are best interpreted with a model in which separate subpopulations of lymphocytes are responsive to different H isoantigens.

Accepting then, that the cells that react to a given isoantigen system are antigen sensitive cells, unipotential in their response capabilities, and that there is little if any non-specific activation, it follows that the number of antigen systems capable of provoking such a large response, involving 2% or more of the lymphocyte population, must be less than 50. The exact nature of these antigens is not clear, however it appears they represent the strong H antigens of the species. A possible biological basis for the distinction between "strong" and "weak" transplantation isoantigens may be the number of lymphocytes that can recognize and respond to them. Weak antigens may constitute a negligible proliferative stimulus simply because there are smaller numbers of cells reactive to them. Furthermore, lymphocytes from donors previously immunized with a weak antigen, and therefore possessing an expanded clone of reactive cells, might be expected to demonstrate significant proliferation in cultures stimulated by such an antigen.

From these arguments it can be predicted that: (1) the strongly mitogenic, special antigenic systems of one species which can stimulate proliferative activity involving large numbers of cells in homologous cultures would be less effective in stimulating heterologous lymphocytes, and (2) these heterologous antigens should become effective if the lymphocyte donors have been immunized against them. An individual of a given species may have enough circulating lymphocytes to commit significant numbers of them to reactivity to the various H isoantigen systems of his species, but he does not for other species. The results of an experiment to examine this point are presented in Table I. Rat lymphocytes were stimulated with heterologous human leucocytes or with rat leucocytes from homologous parental strain donors. The stimulating cells were pretreated with mitomycin C to inhibit their proliferation. On a cell for cell basis, rat lymphocytes were noticeably less reactive to the surface antigens of human leucocytes than to comparable surface configurations on homologous cells (compare groups A-1 and B-1). Prior immunization against human cells increased the response capacity of rat lymphocytes markedly (compare groups B1 and B2)—an indication that these clones of reactive cells had been amplified significantly by proliferation during immunization.

TABLE I

Proliferative reactivity of lymphocytes stimulated in a homologous and a heterologous mixed lymphocyte interaction (MLI)

Group	Cultures[a]	^{3}H-TdR incorporation (CPM/culture)[b]		
		Day 4	Day 5	Day 6
A	*Homologous MLI:*			
1	rat: BH + DA(m)	2573	3760	1500
2	man: CN + DW(m)	1852	5738	5127
B	*Heterologous MLI:*			
1	BH (normal) + DW(m)	101	980	616
2	BH (immune) + DW(m)	2713	3484	1704
C	*Controls (unmixed):*			
1	rat: BH (normal) ⎫			
	BH (immune) ⎬	50	100	75
	DA(m) ⎭			
2	human: CN	296	715	1649
	DW(m)	<20	<20	<20

[a] (m) denotes that in order to inhibit their proliferation, cells were pretreated with mitomycin C (25 μg/ml/10^7 cells; 37°C; 30 min) then washed three times with medium.
[b] ^{3}H-TdR: incorporation of tritiated thymidine; mean of triplicate cultures; range within 10% of mean values.

The fact that some proliferative activity occurs in the heterologous MLI, even though it is less than in the homologous MLI, indicates that a substantial number of lymphocytes are reacting to the heterologous antigens and this still presents a conflict with the strict interpretation of the clonal selection hypothesis. Two explanations, yet to be tested, might account for the proliferative activity of "normal" lymphocytes against heterologous antigens: (1) that a great many antigens confront as many small subpopulations of reactive lymphocytes; and (2) that the "normal" rat is not truly immunologically virgin, but rather, as a result of contact with multitudinous environmental antigens, many amplified clones of antigen-reactive cells exist in the circulation, and some of these also react to heterologous species antigens.

3. Influence of immunization against major transplantation isoantigens. The effects of prior immunization against isoantigens of the major histocompatibility locus on the proliferative behavior of lymphocytes exposed to these antigens in the MLI is shown in Fig. 1. Rats of the DA strain were inoculated sub-cutaneously or intravenously with dissociated cell suspensions consisting of

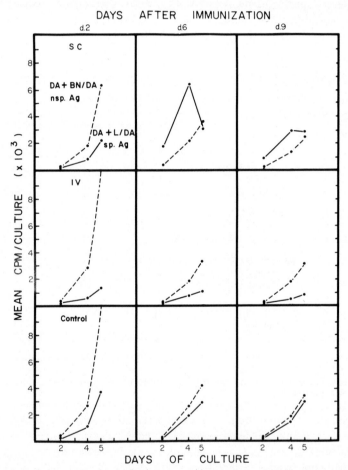

Fig. 1. Influence of prior immunization against a major H isoantigen system on the reactivity of lymphocytes in the MLI. Panels consisting of 3 DA rats were injected subcutaneously or intravenously with Lewis splenic cells, or with isologous DA splenic cells (controls). Cultures with "specific" L/DA F_1 cells or with "non-specific" BN/DA F_1 lymphocytes were set up two, six and nine days after inoculation. Accelerated proliferative responses to the specific antigen system were observed with DA lymphocytes from animals immunized s.c. six days previously.

40 million splenic cells from Lewis strain donors, which are incompatible at the major AgB H locus. As controls, a group of DA animals were inoculated with isologous splenic cells. At various times thereafter, lymphocytes from the sensitized animals were cultured with L/DA F_1 cells bearing the specific immunizing antigens and with non-specific BN/DA F_1 cells bearing an alternative set of strong H isoantigens.

When cultured six days after immunization, lymphocytes from animals inoculated subcutaneously showed an accelerated proliferative response to F_1 cells bearing these antigens. The response was more prompt and could be easily detected during the second day of culture; these same cells cultured with an indifferent antigen or cells cultured from non-immunized animals showed no response at this time. The increased proliferative activity of cells from rats immunized subcutaneously was relatively short-lived; by the ninth day after immunization, it was markedly diminished, and by 15 days it was difficult to detect.

Lymphocytes from rats immunized intravenously, on the other hand, displayed a different response capacity. Cultures stimulated with specific F_1 cells were markedly less reactive than cells from the control-inoculated rats or the subcutaneously immunized ones. Apparently, a significant number of antigen reactive cells disappear from the circulation following intravenous inoculation with alien cells bearing strong H antigens.

In view of Simonsen's observations concerning the "factor of immunization" where cells from previously immunized donors are not noticeably more effective in inducing a GVH reaction involving major H antigen differences (Simonsen, 1967; Nisbet et al., 1969), it was important to determine the number of cells from immunized animals that do respond in mixed cultures when stimulated by these antigens.

For this purpose, lymphocytes from normal, un-immunized DA strain rats and from littermates immunized with BN strain splenic cells were stimulated with DA/BN F_1 cells. Colchicine was included in the medium from the outset of the culture period to prevent sequential divisions by the responsive cells. At various times, ^{3}H-TdR was added to the cultures for a period of 4 hr so that responsive cells were labeled as they entered the DNA synthesizing S phase of the proliferative cycle and could be detected by autoradiography. These procedures permit the direct enumeration, on an hourly basis, of the number of lymphocytes reactive to a particular antigen system as they enter the proliferative cycle for the first time.

The results of this experiment (Fig. 2) show that lymphocytes from previously immunized rats react more promptly to the presence of specific antigen-bearing F_1 cells than lymphocytes from normal animals, but that the absolute number of reactive cells in these two cultures is markedly similar. This is evident from a plot of numbers of reactive cells per hour *versus* hours of

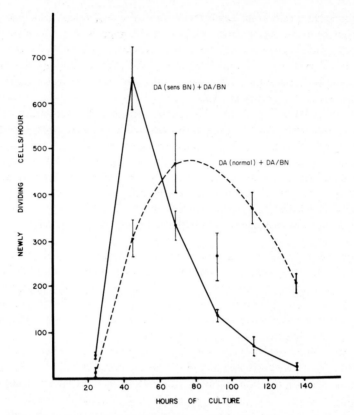

Fig. 2. Number of newly dividing lymphocytes in MLI from normal DA rats, or DA rats immunized seven days previously with BN spleen cells. Based on the area under the curve, which is similar for the two curves, approximately 3-4% of the lymphocyte population of a DA rat is responsive to BN H antigens, regardless of whether the donor has been pre-immunized.

culture, where the area under the two curves is similar and each represents approximately 3-4% of the initial parental lymphocyte population cultured.

4. *Possible explanations for the large number of lymphocytes reactive to transplantation antigens.* Five alternative explanations of a general nature have been offered to account for the substantial number of lymphocytes from an individual that react to H isoantigens of his species. These include: (1) non-specific recruitment; (2) multipotentiality; (3) antigen-presentation; (4) prior immunization with cross-reactive antigens; and (5) the Jerne model. The first two of these have been dealt with above and need not be considered further. The third suggests that the high density of H isoantigen determinants on lymphocyte membranes activates large numbers of cells with low affinity

binding sites, causing them to undergo blastogenic transformation. A weakness of this argument is the comparative inability to induce GVH reactions involving heterologous donor/recipient combinations and the minimal proliferative response of heterologous mixed lymphocyte interactions. The fourth considers that the large number of reactive cells to H antigens represents prior antigenic experience of the animal with cross-reacting environmental or dietary antigens, or tumor-specific antigens on successfully suppressed spontaneously arising neoplasms. The Jerne model proposes that these cells derive from lymphocyte stem-clones bearing antibody-like receptors whose specificity is determined in the germ line, and which is directed against the histocompatibility antigens of a given species (Jerne, 1970).

At present there is no information that supports or denies the last two explanations; however, if the lymphocyte population of germ-free animals were non-reactive to H antigens, this would weigh heavily against the Jerne model and favor the concept of prior immunization. Figure 3 shows the results of

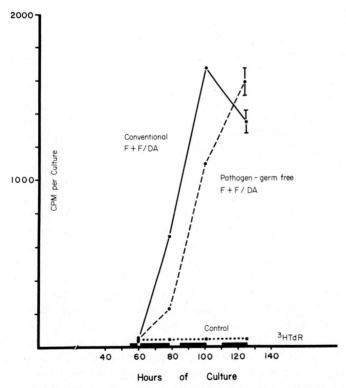

Fig. 3. Proliferative response of peripheral blood lymphocytes from germ-free or conventional Fischer strain rats stimulated with F/DA cells.

stimulating with F_1 cell lymphocytes from pathogen-free Fischer strain rats or littermates raised as weanlings under conventional animal colony conditions. The rats were reared at the Lobund Laboratories, Notre Dame University, South Bend, Indiana by Dr Morris Pollard, and the blood was flown to Philadelphia where the cultures were established. The results of this experiment were very clear; lymphocytes from germ-free rats can respond as well to major histocompatibility isoantigens as those obtained from conventional rats. This information does not, however, rule out the possibility that suppressed tumor-specific antigens or dietary antigens constitute an effective source of cross-reactive antigens which stimulate large clones of cells reactive to H antigens of the species during ontogeny.

CONCLUSIONS

Studies with the GVH reaction and the MLI indicate that the clone of lymphocytes reactive to a given major histocompatibility isoantigen system is already large and that the frequency of its cellular constituents is not significantly increased upon immunization (Simonsen, 1967; Szenberg et al., 1962; Wilson et al., 1968; Wilson and Nowell, 1970; Nisbet et al., 1969). The cellular basis of memory involving immunological reactivity to strong transplantation antigens is therefore operationally different from that for both humoral and cell-mediated immunities to environmental antigens which are characterized by a significant proliferative amplification of clones of antigen-reactive cells as a consequence of immunization.

Two phenomena involving immune reactivity to strong H antigens might be considered under the heading of a memory: (1) the destruction of antigen-bearing target monolayers of fibroblasts by lymphocytes from immunized donors (Wilson and Billingham, 1967); and (2) the well-established second set reaction to skin homografts (Medawar, 1958). The first of these is a short-lived phenomenon. The cytotoxic capacity of sensitized lymphocytes disappears three weeks after immunization (Wilson, 1963) and these kinetics compare favorably with the accelerated proliferative response of sensitized lymphocytes in the MLI discussed above. Furthermore, normal lymphocytes do eventually destroy homologous target fibroblasts after a period of six to seven days, and when subcultured on fresh target monolayers of the same antigenic specificity, do so more promptly—within one to two days (Hirschhorn and Rips, 1965; Ginsburg and Lagenoff, 1968). These considerations prompt the suggestion that memory for immune reactivity to strong H antigens is comparatively ephemeral and reflects an alteration of antigen reactive cells to a temporarily more reactive status. This might involve a temporary period following division when cells already possess the metabolic machinery necessary for subsequent divisions, so that they are primed to respond (G') in the event that they encounter the specific antigen again in the near future. Such a G' cell might also possess a higher density of receptor sites on its surface and therefore be stimulable more promptly and with lower doses of antigen.

Second set reactivity to skin homografts, on the contrary, is not a short-lived phenomenon, but rather has been shown to persist for hundreds of days or for the lifetime of experimental rodents (Medawar, 1958). These circumstances, however, might be quite different. The first application of a skin homograft involves the production of humoral factors, in addition to cell mediated immunities against both strong and weak H antigens and the possibility of immunity against skin-specific antigens not shared by lymphocytes. Consequently, a second skin graft confronts an immune apparatus in which amplified clones of cells reactive against weak H antigens and skin-specific antigens, as well as clones against the major antigens, and also humoral factors may be acting synergistically to produce a second set reaction.

In addition to these quantitative considerations, there is apparently an important qualitative distinction in the behavior of lymphocytes from normal or immunized animals. The studies of Barker and Billingham (1967) have shown that skin homografts survive indefinitely when implanted in raised flaps of skin devoid of lymphatic connections on guinea pig hosts. Apparently the normal lymphocytes circulating in such a host, presumably containing a large proportion that are reactive to the H antigens of the graft, are not capable of fulfilling their immunologic duties. If the flap bearing the homograft is allowed to re-establish lymphatic connections, or if the host is regrafted in the more conventional manner at a site having an intact lymphatic drainage, or if sensitized lymphocytes are adoptively transferred systematically from a previously immunized isologous animal, the test graft is rejected promptly. These observations indicate that the homograft reaction requires that graft antigen material or host lymphocytes exposed to graft antigens must enter the draining lymph node, where newly stimulated lymphocytes emerge possessing the capacity to infiltrate the graft via its vasculature and effect its destruction. Apparently normal lymphocytes do not share this infiltrative-destructive capacity—it may reflect changed surface configurations and/or the altered metabolic status of lymphocytes that have recently divided.

ACKNOWLEDGEMENTS

We are very grateful to Mrs Dianne Fox, Mrs Janet Finnan and Miss Julie Jensen for careful and painstaking assistance in the execution of these experiments, and to Mrs Carol Newlin and Dr Norman Klinman for interesting and provocative discussions dealing with the subject matter of this paper.

This work was supported in part by grants CA-10320, AI-07001 and AI-09275 from the United States Public Health Service. D.B.W. is a Career Development Awardee, USPHS, CA-09873.

SUMMARY

Studies with the graft-*versus*-host reaction (GVH) and the mixed lymphocyte interaction (MLI) indicate that: (1) the behavior of lymphocytes in these systems is a reflection of their immunologic competency to major histocompatibility (H) antigens; (2) that the proportion of lymphocytes reactive to a given H antigen is relatively large (*c.* 2-4%); (3) this proportion is not increased by prior immunization.

Several possibilities have been offered as an explanation of this large number of antigen-reactive cells (ARC). However, these suffer from one or more weaknesses, and these are discussed in turn. Primarily on the basis of exclusion of other possibilities, the most likely interpretation is that the large number of ARC reactive to major H antigens results from either (a) *immunologic* expansion of cells as a result of prior antigenic experiences with cross-reactive dietary antigens or altered somatic determinants on successfully suppressed spontaneous neoplasms, or (b) an innate *non-immunologic* constitutional expansion, perhaps under genetic control, and of significance in ontogeny. An example of this might be Jerne's recently proposed model (Jerne, 1970).

One implication of the fact that the number of ARC is high and not increased by immunization concerns the operation of memory mechanisms in immunities directed to strong H antigens. Unlike immune reactivities to non-H antigens, where the number of ARC before and after immunization is markedly different, memory in respect of H antigens may reflect an altered metabolic status of ARC, or an increased density of receptor sites on their surfaces.

REFERENCES

Albertini, R. J. and Bach, F. H. (1968). *J. exp. Med. 128,* 639-651.
Barker, C. F. and Billingham, R. E. (1967). *Transplantation 5,* 962-966.
Bloom, B. R. (1969). *In* "Mediators of Cellular Immunity" (H. S. Lawrence and M. Landy, eds), pp. 249-262. Academic Press, New York.
Dutton, R. W. (1966). *Bact. Rev. 30,* 397-407.
Dutton, R. W. and Mishell, R. I. (1967). *J. exp. Med. 126,* 443-454.
Ginsburg, H. and Lagenoff, D. (1968). *J. Cell Biol. 39,* 392.
Hirschhorn, K. and Rips, C. S. (1965). *In* "Isoantigens and Cell Interactions" (J. Palm, ed.), pp. 57-63. Wistar Institute Press.
Jerne, N. K. (1970). Brook Lodge Symposium on "Immunologic Surveillance", Academic Press, New York.
Johnston, J. M. and Wilson, D. B. (1970). *Cellular Immunology 1,* 430-444.
Medawar, P. B. (1958). *Proc. R. Soc. B 149,* 145.
Mitchell, G. F. and Miller, J. F. A. P. (1968). *Proc. natn. Acad. Sci. U.S.A. 59,* 296.
Mitchison, N. A. (1969). *In* "Mediators of Cellular Immunity" (H. S. Lawrence and M. Landy, eds), pp. 73-80. Academic Press, New York.
Nisbet, N. W., Simonsen, M. and Zaleski, M. (1969). *J. exp. Med. 129,* 459-467.

Palm, J. (1964). *Transplantation 2*, 603-612.
Schwarz, M. R. (1968). *J. exp. Med. 127*, 879-890.
Simonsen, M. (1967). *Cold Spring Harb. Symp. quant. Biol. 32*, 517-523.
Siskind, G. W. and Benacerraf, B. (1969). *Adv. Immun. 10*, 1-50.
Syeklocha, D., Siminovich, L., Till, J. E. and McCulloch, E. A. (1966). *J. Immun. 96*, 472-477.
Szenberg, A., Warner, N. L., Burnet, F. M. and Lind, P. E. (1962). *Bri. J. exp. Path. 43*, 129-136.
Wilson, D. B. (1963). *J. cell. comp. Physiol. 62*, 273.
Wilson, D. B. (1967). *J. exp. Med. 126*, 625-654.
Wilson, D. B. and Billingham, R. E. (1967). *Adv. Immun. 7*, 189-273.
Wilson, D. B. and Elkins, W. L. (1969). *In* "Proceedings of the Third Annual Leucocyte Culture Conference" (W. O. Rieke, ed.), p. 391. Appleton-Century-Croft, New York.
Wilson, D. B. and Nowell, P. C. (1970). *J. exp. Med. 131*, 391-407.
Wilson, D. B., Silvers, W. K. and Nowell, P. C. (1967). *J. exp. Med. 126*, 655-665.
Wilson, D. B., Blyth, J. L. and Nowell, P. C. (1968). *J. exp. Med. 128*, 1157-1181.

LYMPHOCYTE-LYMPHOCYTE INTERACTIONS
IN IMMUNE RESPONSES

Interaction Between Thymus-dependent (T) Cells and Bone Marrow-derived (B) Cells in Antibody Responses

J. F. A. P. MILLER

Experimental Pathology Unit
Walter and Eliza Hall Institute of Medical Research
Melbourne 3050, Australia

Interaction between antigen and different classes of lymphocytes is a feature of many antibody responses. For instance, in the immune response to hapten-protein conjugates, one class of lymphocytes is involved in responding to the haptenic determinant, whilst other classes are essential for recognition of determinants on the carrier molecule. The two classes must somehow collaborate to allow an antibody response to the hapten (Mitchison *et al.*, 1970). In other interactions, the classes of lymphocytes have been identified as thymus-derived and non-thymus-derived (Claman and Chaperon, 1969; Davies, 1969; Miller and Mitchell, 1969a; Taylor, 1969).

As is now well established, thymectomy at birth, or thymectomy in the adult followed by a heavy dose of total body irradiation and marrow protection, is associated with a diminished antibody response to antigens such as heterologous erythrocytes and serum proteins (Miller and Osoba, 1967). For instance, the antibody response of nnTx* mice and TxXBM mice is low, but can be increased by injecting either thymus cells or TDL (Miller and Mitchell, 1969a). Furthermore in heavily irradiated mice there is no significant response to SRBC within one week post-irradiation unless both thymus and marrow cells are given (Claman and Chaperon, 1969). The question was asked: are the AFC in reconstituted mice derived from the inoculated thymus cells or TDL? It was unequivocally established, by the use of anti-H2 sera and chromosome markers, that the AFC were derived, not from the inoculated lymphocytes, but from the host in the case of nnTx mice, and from the bone marrow in the case of TxXBM

---

* The following abbreviations will be used: AFC = antibody forming cells; BM = bone marrow; BSA = bovine serum albumin; FALG = fowl anti-mouse lymphocyte globulin; FγG = fowl immunoglobulin G; HRBC = horse erythrocytes; LN = lymph node; nnTx = neonatally thymectomized; PFC = plaque-forming cells; SRBC = sheep erythrocytes; TDL = thoracic duct lymphocytes; TL = thymus lymphocytes; TxXBM = adult thymectomized, irradiated, bone marrow protected mice.

mice (Miller and Mitchell, 1969a). Further work to be summarized here, has indicated that the reactivity of thymus-dependent cells (T-cells) with antigen is specific. The nature of the interaction between T-cells and bone marrow-derived* AFC precursors (B-cells) has not yet been established.

INTERACTION BETWEEN T-CELLS AND B-CELLS *IN VIVO*

Interaction between thymus cells and B-cells is essential to enable mice to produce a normal antibody response to heterologous erythrocytes and serum proteins. This is exemplified in Table I for two antigens, SRBC and FγG. The thymus-dependency of the antibody response of mice to FγG has been established in other experiments, and a new plaque technique enabling identification of cells forming antibodies to FγG has been devised (Miller and Warner, 1971). It is evident from Table I that LN (group 5) or spleen (group 10) cells from TxXBM mice lack T-cells, and that LN of TxXBM mice can be as rich a source of B-cells as BM itself (compare groups 4, 5 and 6). Elsewhere it has been established that the B-cells are the AFC precursors (Miller and Mitchell, 1968; Mitchell and Miller, 1968b).

INTERACTION BETWEEN T-CELLS AND B-CELLS *IN VITRO*

Normal mouse spleen cells can respond to SRBC *in vitro* with the production of AFC. So far mixtures of bone marrow and thymus cells have failed to respond to SRBC *in vitro*. If, however, spleen cells from nnTx mice, which give a poor *in vitro* response, are mixed with thymus cells, there is a normal AFC response to SRBC. This is exemplified in Fig. 1, first published in the 1968 Annual Report of the Walter and Eliza Hall Institute of Medical Research. Similar results have since been obtained by Hirst and Dutton (1970) and others. In our system 20 million spleen cells from normal mice gave a peak number, at four days, of 1500-2000 PFC when cultured with one million SRBC. The same number of spleen cells from nnTx mice gave lower responses, from 100 to 800 PFC per culture. Addition of 10-15 million syngeneic or semi-allogeneic thymus cells or TDL increased the *in vitro* response of spleen cells from nnTx mice to levels comparable with those obtained with normal spleen cells. In contrast, addition of BM cells did not materially affect the response or depressed it somewhat.

---

* All haematopoietic cells, including thymus and thymus-dependent cells, are of course, originally derived from stem cells that can be found in the bone marrow. For the sake of simplicity the term "bone marrow-derived cells" has been reserved for those lymphocytes that have NOT differentiated within or migrated through the thymus, or been subjected to a differentiating humoral influence from the thymus epithelial cytoreticulum. To avoid misunderstanding I shall refer to these cells as B-cells and to thymus-dependent lymphocytes as T-cells. It is presumed that the B-cells are analogous to the bursa-derived lymphoid cells of birds.

TABLE I

Interaction between T-cells and B-cells in vivo

Antigenic system	Group	Source of T-cells	Source of B-cells	Cells and antigen inoculated	No. of irradiated recipients	Peak PFC[a] per spleen
	1	none	none	10^8 SRBC	5	10 (15-4)
	2	normal thymus	none	5×10^7 TL + 10^8 SRBC	12	20 (25-15)
	3	none	normal BM	2×10^7 BM + 10^8 SRBC	14	10 (15-5)
SRBC[b]	4	normal thymus	normal BM	5×10^7 TL + 2×10^7 BM + 10^8 SRBC	8	1270 (1600-1010)
	5	none	TxXBM LN	2×10^7 TxXBM LN + 10^8 SRBC	5	4 (7-2)
	6	normal thymus	TxXBM LN	5×10^7 TL + 2×10^7 TxXBM LN + 10^8 SRBC	6	1280 (1660-720)
	7	normal LN	normal LN	2×10^7 LN + 10^8 SRBC	17	1600 (1870-1360)
	8	none	none	FγG	5	8 (15-5)
FγG[c]	9	normal thymus	none	10^8 TL + FγG	5	10 (15-5)
	10	none	TxXBM spleen	5×10^7 TxXBM spleen + FγG	4	40 (400-10)
	11	normal thymus	TxXBM spleen	10^8 TL + 5×10^7 TxXBM spleen + FγG	5	13610 (19600-9460)

[a] Geometric means and upper and lower limits of standard error. Values for direct anti-SRBC PFC and indirect anti-FγG PFC.

[b] Data partly from Miller and Mitchell (1970b).

[c] Unpublished data of Basten and Miller.

For abbreviations see footnote 1.

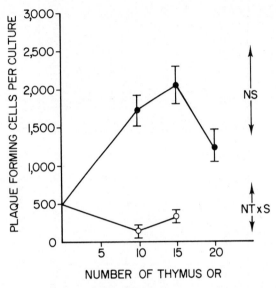

Fig. 1. Interaction between T-cells and B-cells *in vitro*. Peak number of PFC produced in cultures of spleen cells from non-immunized neonatally thymectomized mice after addition of SRBC and various doses of thymus (●——●) or bone marrow cells (○——○). Range of response of spleen cells without added thymus or bone marrow cells is indicated on the right. (NS = normal spleen; NTxS = spleen from neonatally thymectomized mice.) Each point is the average of 4-5 cultures, vertical bars represent the limits of twice the standard error. Data of Miller and Noonan reported in the 1967-68 Annual Report of the Walter and Eliza Hall Institute.

Since semi-allogeneic cells increased the response, anti-H2 isoantisera were used to identify the AFC. The results clearly indicated that all these cells were derived, not from the added thymus cells, but from cells already present in the spleens of nnTx mice, i.e. from B-cells. It is interesting that, in this system, thymus cells heavily irradiated *in vitro*, were as effective as normal thymus cells in restoring the response, in contrast to the situation *in vivo* (Miller and Mitchell, 1968).

"EDUCATION" OF THYMUS CELLS—COLLABORATION WITH B-CELLS *IN VIVO*

It is clear that, in the above experimental situations *in vivo* and *in vitro*, thymus cells are not transformed into AFC, and yet are essential to facilitate or augment the antibody response. Is the reactivity of thymus cells specific? The first experiment suggesting specificity in thymus-dependent cells was reported in 1968 (Mitchell and Miller, 1968a). The ability of T-cells to co-operate with

B-cells could be specifically enhanced by prior exposure to the relevant antigen. The experimental system was as follows: two sets of lethally irradiated hosts were used. The first set received thymus cells intravenously with or without the relevant antigen. After five to seven days the spleens of the first irradiated hosts were removed and a cell suspension of these spleens (which contained a proportion of the inoculated thymus cells or their progeny—Miller and Mitchell, 1969b, 1970a) was injected together with BM cells and the antigen in question into a second set of irradiated mice. Their antibody response was measured at appropriate times. The results of such experiments with two antigens are shown in Table II. In both cases there is a specific enhancement of the response in irradiated recipients of thymus cells previously exposed to the antigen used to challenge the second hosts. This experimental system has been widely used (Claman and Chaperon, 1969; Shearer and Cudkowicz, 1969; Mitchison et al., 1970) and the original observations (Mitchell and Miller, 1968a) confirmed. It was suggested that a similar design would allow thymus cells to become killer cells in an allogeneic system: thus thymus cells from a parent strain A could be injected intravenously into heavily irradiated F1(A x B) hosts and the spleens of these, five days postirradiation, would then kill B target cells *in vitro,* in contrast to normal thymus cells which cannot. This suggestion was tested and shown to be correct (Cerottini et al., 1970; Miller et al., 1970).

"EDUCATION" OF THYMUS CELLS–COLLABORATION WITH B-CELLS *IN VITRO*

Carrier-specificity has recently been demonstrated in secondary antibody responses to the hapten 3,5-dinitro-4-hydroxyphenylacetic acid (NNP) (Breitner and Miller, 1970). As an extension of this work our group, Cheers, Breitner, Little and Miller (to be published), and Mitchison et al. (1970) have now shown that "educated" T-cells can act as carrier-primed cells. The low AFC response to NNP obtained when spleen cells from mice primed to NNP-ovalbumin (NNP-OA) were stimulated *in vitro* with NNP-FγG, could be markedly elevated by adding spleen cells from mice primed with the heterologous carrier, FγG. Spleen cells taken from mice seven days after lethal irradiation and intravenous injection of thymus cells and FγG readily acted as cells primed to the heterologous carrier, and markedly elevated the *in vitro* AFC response (Table III). Clearly an interaction between T-cells primed to the carrier and other cells primed to the hapten has enabled the latter to produce a significant number of AFC to the hapten NNP *in vitro*. Other experiments using semi-allogeneic cell mixtures established that the carrier-primed cells did not produce AFC to NNP.

INDUCTION OF SPECIFIC IMMUNOLOGICAL TOLERANCE IN T-CELLS

Specific immunological tolerance is a property that can be linked to lymphocytes within the thymus (Taylor, 1969) and to thymus-derived recirculating small lymphocytes (T-cells, Miller and Mitchell, 1970b). For example,

TABLE II

"Education" of T-cells: collaboration with B-cells in vivo

Experiment	Cells and antigen given to first irradiated hosts	Cells and antigen given to second irradiated hosts	Peak PFC per spleen in second irradiated hosts[a]
1[b]	1. 10^8 SRBC	One spleen equivalent from first irradiated hosts	1. 310
	2. 10^8 thymus cells		2. 248
	3. 10^8 thymus cells + 10^8 HRBC	+ 10^7 BM cells	3. 133
	4. 10^8 thymus cells + 10^8 SRBC	+ 10^8 SRBC	4. 2103
2[c]	5. 10^8 thymus cells	2×10^7 cells from spleens of first irradiated hosts	5. 0
	6. 10^8 thymus cells + BSA		6. 0
	7. 10^8 thymus cells + FγG	+ 10^7 BM cells + FγG	7. 1130

[a] Average of results from 5 to 20 mice per group. Values for direct anti-SRBC and indirect anti-FγG PFC are given.
[b] Original data of Mitchell and Miller (1968a).
[c] Unpublished data of Basten and Miller.
For abbreviations see footnote 1.

TABLE III

"Education" of T-cells: collaboration with B-cells in vitro

Source of spleen cells	Antigen *in vitro*	4 day NNP PFC per culture[a]
Normal mice primed with NNP-FγG	NNP-FγG	926
Normal mice primed with NNP-OA	NNP-FγG	22
Normal mice primed with NNP-OA and normal mice not primed to any antigen: cells mixed *in vitro*	NNP-FγG	157
Normal mice primed with NNP-OA and normal mice primed with FγG: cells mixed *in vitro*	NNP-FγG	745
Normal mice primed with NNP-OA and heavily irradiated mice given thymus cells and FγG: cells mixed *in vitro*	NNP-FγG	635

Unpublished data of Cheers, Breitner, Little and Miller.
[a] Direct PFC, average of 5-20 cultures.

TDL from mice specifically tolerant to SRBC adoptively transferred reactivity to HRBC but not to SRBC when injected into nnTx mice, which could themselves supply B-cells (AFC precursors) but could not respond normally in the absence of T-cells.*

In further experiments (Tables IV, V—FγG focusing experiment) specifically tolerant populations of T-cells (TDL) were coated with the tolerated antigen (FγG in the form of FALG) and transferred to TxXBM recipients, but the antibody response to the tolerated antigen was *not* enhanced. The antibody response to FγG is thymus-dependent (Miller and Warner, 1971). The anti-FγG response of TxXBM CBA mice could be enhanced significantly by providing 20 million normal (CBA x C57BL)F1 TDL incubated *in vitro* with FALG.†

* It is very likely that there is *not* a complete absence of T-cells in nnTx mice because some may have escaped from the thymus before birth. T-cells are capable of proliferating and differentiating to produce further specific T-cells after antigenic stimulation (Mitchell and Miller, 1968a; Miller and Mitchell, 1969b; Shearer and Cudkowicz, 1969) and this may explain the facts that (1) the antibody response of nnTx mice to some antigens *appears* to be normal (such antigens may be powerful mitotic stimulators for T-cells), and (2) the low antibody response of nnTx mice to some antigens can be augmented simply by increasing the antigen dose (Sinclair and Elliott, 1968) or giving repeated injections of antigens (Takeya and Nomoto, 1967).

† FALG was raised by immunizing fowls with two intravenous injections of 10^9 CBA thymus cells at two-week intervals and bleeding 10 days after the last injection. FγG does not fix mammalian complement and, in mice, intravenously injected FALG-coated mouse TDL were distributed like normal TDL. Furthermore FALG is not immunosuppressive in mice, presumably because it fails to opsonize lymphocytes (Martin, 1969, Ph.D. Thesis). The TDL, in the experiments described here, were incubated with FALG for 30 min at 37°C and extensively washed before injection.

TABLE IV

"FγG-focusing" experiment: effect of incubating TDL with FALG on capacity of TxXBM recipients to produce antibody to FγG

Source of TDL given	Antigens given	Anti-FγG serum antibody: % ^{131}IFγG bound at:			Peak anti-SRBC haemagglutinin titre (reciprocal) at 15 or 22 days
		15 days	22 days	42 days	
No TDL given	FALG and SRBC both *in vivo*	0.5	0.2	1.0	<2 to 16
FγG-tolerant mice	FALG *in vitro* and SRBC *in vivo*	1.2	1.4	12.1	32 to 128
Control (normal) mice	FALG *in vitro* and SRBC *in vivo*	22.6	17.1	42.0	32 to 512

For abbreviations see footnote 1.
Unpublished data of Miller, Sprent, Warner, Basten, Breitner and Martin.
Number of mice per group = 6-19.

TABLE V

"FγG-focusing" experiment: identity of AFC arising in TxXBM CBA recipients of FALG-treated (CBA × C57BL)F₁ TDL.

Source of TDL[a]	Antigen given	Peak anti-FγG PFC response per spleen		% reduction of PFC with:	
		Direct PFC	Indirect PFC	C57BL anti-CBA serum	CBA anti-C57BL serum
None	alum precipitated FγG *in vivo*	510	2		
Normal F₁ mice	FALG *in vitro*	7,690	14,730	99% direct PFC 99% indirect PFC	20% direct PFC 0% indirect PFC
FγG tolerant F₁ mice	FALG *in vitro*	140	2		

For abbreviations and FALG treatment see footnotes 1 and 4.
Unpublished data of Miller, Sprent, Warner, Basten, Breitner and Martin.
[a] 2×10^7 TDL given to each recipient.
Number of mice per group = 4-8.

Anti-H2 serum treatment of the PFC indicated that they were derived, not from the inoculated TDL, but from the BM cells used to protect the irradiated thymectomized CBA. It is important to note that TDL incubated *in vitro* with FALG could restore AFC production in TxXBM mice if the TDL came from normal mice, but not if they came from FγG-tolerant mice. Nevertheless both populations of TDL were agglutinated by the FALG and must therefore have carried FALG on their surface. The failure of FALG-bearing tolerant TDL to restore reactivity cannot be attributed to an inability to recirculate normally or penetrate the correct sites in the spleen, since both normal and tolerant FALG-coated TDL populations were distributed in an identical way in their hosts, as judged by the localization of ^{51}Cr-labelled cells. It might be suggested that T-cells have to proliferate before being capable of interacting, and that T-cells from tolerant donors could not proliferate in response to the tolerated antigen. If this is so, the purpose of this proliferation cannot be solely to provide a mechanism for increasing the number of T-cells capable of "ferrying" the specific antigen and "focusing" it on to those B-cells having specificities for some determinants on the antigen in question. The FγG antigen in the FALG preparation used must have been carried by a vast majority of the injected TDL, and thus there should have been an ample number of TDL able to "focus" the antigen into the correct sites in the spleen. Our findings must therefore be interpreted to indicate that the role of T-cells is linked to a capacity for "recognizing" antigenic determinants, and is specifically dependent, not on the passive transport of such antigenic determinants to B-cells, but upon some *active* process. This presumably involves interaction with the antigen, and further differentiation of the T-cells to produce some factor—a pharmacological factor or an immunoglobulin molecule of a special class (IgX).

REQUIREMENT FOR T-CELLS IN THE EXPRESSION OF SPECIFIC IMMUNOLOGICAL MEMORY

The ability of antigen to "educate" thymus cells, as reported above, implies that the property of an enhanced immune response can be linked to T-cells. The results shown in Table VI indicate that 19S memory to SRBC can be specifically transferred by TDL to TxXBM mice. However, the AFC are not derived from the inoculated TDL (which "carry" the faculty of immunological memory) but from B-cells present in the recipients, which had not been deliberately exposed to SRBC. The specificity of the 19S memory response to SRBC is shown in the experiments in which the mice were challenged with both SRBC and HRBC.

In further experiments, TDL from FγG-primed animals were used. They could transfer 7S memory responses to irradiated animals, even when the TDL were stimulated *in vitro* with *fluid* FγG and washed extensively before transfer. In contrast, normal TDL exposed to fluid FγG *in vitro* and transferred produced

TABLE VI

Adoptive transfer of 19S memory to SRBC in Tx XBM CBA by TDL from (CBA × C57BL)F_1 mice primed to SRBC

Source of TDL[a]	Days after HRBC or SRBC	19S PFC response per spleen		% inhibition of anti-SRBC PFC with:	
		HRBC	SRBC	C57BL anti-CBA serum	CBA anti-C57BL serum
Normal F_1 mice	5	490	2,900		
	6	5650	7,300		
F_1 mice given 5×10^6 SRBC 2 weeks before	5	450	9,000		
	6	4560	37,500	100%	15%

For abbreviations see footnote 1.
Unpublished data of Miller and Sprent.
[a] 4×10^6 TDL given per recipient.
No. of recipients per group = 6-8.

only a weak response (Table VII). The ability of primed TDL to transfer memory when stimulated *in vitro*, was impaired by prior *in vitro* treatment with an anti-H2 serum directed against the histocompatibility antigens of the TDL donor, in the absence of complement (Table VII). This technique readily enabled us to identify the cell types essential for the expression of memory. NnTx CBA mice failed to respond to FγG (Miller and Warner, 1971) and could not be primed to that antigen unless provided with thymus lymphocytes during infancy. Such mice were therefore given 300 million (CBA x C57BL)F_1 thymus cells in separate injections during the first three weeks of life and then challenged with FγG. Three to four weeks later their TDL were obtained and could transfer good 7S memory responses to FγG to irradiated recipients. The AFC in these mice were shown to be CBA and not (CBA x C57BL)F_1 by treatment with appropriate antisera (Table VIII). However, prior treatment of the TDL population of reconstituted nnTx mice *in vitro* with a CBA-anti-C57BL serum impaired ability to transfer memory (Table VIII). These results indicate that the thoracic duct lymph population of thymus cell-reconstituted nnTx mice contains both donor-type T-cells and host-type B-cells and that both classes of cells are essential for the expression of an adoptive memory response. The AFC arising during such a memory response are derived from the B-cells that must have been either recirculating, extracted from the lymphoid tissues of the nnTx mice by the cannulation procedure or both. In order to determine whether the T-cells carry specificity in this system, TDL from thymus cell-reconstituted nnTx mice were exposed *in vitro* to CBA anti-C57BL serum and injected together with *normal* TDL from non-primed, non-thymectomized (CBA x C57BL)F_1 donors into irradiated recipients which were then challenged *in vivo* with either fluid FγG or alum precipitated FγG. The results are shown in Table IX. The failure of normal TDL to reverse the defective anti-FγG response in such an experimental situation argues against the idea that the role of T-cells in allowing the expression of a memory response is non-specific. It seems, therefore, that specific immunological memory is a property that can be linked to T-cells.

Summary and Conclusions

The data outlined above clearly point to the existence of specificity at the level of antigen recognition by T-cells, and to an active role in the differentiation of these cells in response to antigenic stimulation. Both tolerance and memory-specific immunological phenomena can be linked to T-cells. The question as to whether these properties can also be linked to B-cells has not been dealt with above. It is highly probable that it can, but to-date, there is still a paucity of serious experimental designs to determine specificity in B-cells, prior to antigenic stimulation, with unequivocally marked cells. The preliminary but not too extensive data of Chiller *et al.* (1970) are perhaps the best indication

TABLE VII

Adoptive transfer of memory to FγG in heavily irradiated mice by TDL from primed donors stimulated in vitro with fluid FγG

Source of TDL	In vitro treatment of TDL (37°C)		No. of washed TDL given	No. of irradiated recipients	7-day PFC response per spleen		P values for indirect PFC
	First 30 min	Next 60 min			Direct PFC	Indirect PFC	
Normal CBA	normal mouse serum	fluid FγG	10^7	6	50	5	—
CBA primed to FγG	normal mouse serum	fluid FγG	10^7	5	190	216,970	} <0.005
	C57BL anti-CBA serum	fluid FγG	10^7	8	2	460	
	CBA anti-C57BL serum	fluid FγG	10^7	11	3	91,070	<0.005
(CBA x C57BL) F₁ primed to FγG	normal mouse serum	fluid FγG	10^7	10	16,830	29,210	} <0.005
	C57BL anti-CBA serum	fluid FγG	10^7	10	40	150	
	CBA anti-C57BL serum	fluid FγG	10^7	10	10	40	<0.005

For abbreviations see footnote 1.
Unpublished data of Miller and Sprent.

TABLE VIII

Adoptive transfer of memory to FγG in heavily irradiated mice by TDL from nnTx donors

Treatment of nnTx donors	No. of TDL given	Antigen given	No. of irradiated recipients	Peak FγG PFC per spleen	
				Direct PFC	Indirect PFC
None	5×10^6	alum-precipitated FγG *in vivo*	7	250[a]	320[a]
Primed with FγG 4–5 weeks before	5×10^6	fluid FγG *in vivo*	7	10	30
300×10^6 (CBA x C57BL)F$_1$ thymus cells during first 3 weeks of life; primed with FγG at 4 weeks	5×10^6	fluid FγG *in vivo*	6	50	11,820[b]

For abbreviations see footnote 1.
Unpublished data of Miller and Sprent.
[a] One of the nnTx donors cannulated to provide TDL for this group had a minute thymus remnant.
[b] CBA anti-C57BL serum treatment of these PFC reduced the number of PFC by 15%. Similar treatment with C57BL anti-CBA serum caused a 97% reduction.

TABLE IX

Adoptive transfer of memory to FγG in heavily irradiated mice by TDL from (CBA x C57BL)F$_1$ thymus cell-reconstituted nnTx donors: role of T-cells

In vitro treatment of TDL from reconstituted nnTx mice		No. of washed TDL given	No. of normal TDL given[a]	Antigen given *in vivo*	No. of irradiated recipients	Peak PFC response per spleen	
First 30 min	Next 60 min					Direct PFC[b]	Indirect PFC[b]
Normal mouse serum	fluid FγG	5 x 10^6	nil	nil	12	40 (60-25)	2250 (3440-1470)
CBA anti-C57BL serum	fluid FγG	5 x 10^6	nil	nil	9	30 (50-20)	80 (260-30)
CBA anti-C57BL serum	fluid FγG	5 x 10^6	10^7	fluid FγG	4	10 (40-2)	0
CBA anti-C57BL serum	fluid FγG	5 x 10^6	10^7	alum-precipitated FγG	6	30 (80-10)	25 (110-5)
—	—	—	10^7	alum precipitated FγG	8	125 (280-55)	125 (320-50)

For abbreviations see footnote 1.
Unpublished data of Miller and Sprent.
[a] Normal TDL from (CBA x C57BL)F$_1$ donors.
[b] Geometric means and upper and lower limits of standard error.

that we have at present to suggest that specific immune tolerance can be linked to both T-cells and B-cells. In the absence of further unequivocal characterization of the specificity of B-cells, it seems pointless to argue extensively on the nature of the collaboration that takes place between antigen, T-cells and B-cells. If we accept that both T-cells and B-cells are specific, then interaction between T-cells and B-cells could occur in one of several ways:

1. T-cells might react with some antigenic determinants and focus other determinants on the same antigen molecule on to B-cells, so that these could now be switched on effectively to produce antibody. This antigen-focusing hypothesis, favoured by Mitchison, is not supported by the FγG focusing experiment described above. There are, however, no experimental data that would positively exclude such a hypothesis.

2. T-cells might react with some antigenic determinants to produce IgX, an immunoglobulin molecule or part of one, which is not secreted into the serum but rapidly absorbed on to other cell types—B-cells themselves or third-party cells which are non-specific. The IgX would then concentrate other determinants on the same antigen molecule in strategic sites where B-cells could be switched on to produce antibody to such determinants.

3. T-cells might interact with antigen to release pharmacological factors with a multitude of biological activities. Some, such as the migration inhibitory factor, have been identified and play a role in the pathogenesis of the delayed hypersensitivity lesion (David, 1968). Others may facilitate the differentiation of B-cells in response to antigenic stimulation.

In the present state of knowledge, the above are mere fanciful speculations and the mechanism of cell collaboration will remain obscure until serious efforts are made to obtain extensive data on the specificity of B-cells and to isolate molecules such as IgX or pharmacological factors. The data presented in this paper argue strongly in support of the concept that T-cells have specificity in the immunological sense.

This is publication No. 1452 from the Walter and Eliza Hall Institute.

REFERENCES

Breitner, J. and Miller, J. F. A. P. (1970). *Fedn Proc. 29*, 572.
Cerottini, J.-C., Nordin, A. A. and Brunner, K. T. (1970). *Nature, Lond. 227*, 72-73.
Chiller, J. M., Habicht, G. S. and Weigle, W. O. (1970). *Proc. natn. Acad. Sci. U.S.A. 65*, 551-556.
Claman, H. N. and Chaperon, E. A. (1969). *Transplant. Rev. 1*, 92-113.
David, J. R. (1968). *Fedn Proc. 27*, 6-12.
Davies, A. J. S. (1969). *Transplant. Rev. 1*, 43-91.
Hirst, J. A. and Dutton, R. W. (1970). *Cellular Immunology 1*, 190-195.

Miller, J. F. A. P. and Mitchell, G. F. (1968). *J. exp. Med. 128*, 801-820.

Miller, J. F. A. P. and Mitchell, G. F. (1969a). *Transplant. Rev. 1*, 3-42.

Miller, J. F. A. P. and Mitchell, G. F. (1969b). *In* "Lymphatic Tissues and Germinal Centers in Immune Response" (L. Fiore-Donati and M. G. Hanna, Jr., eds), pp. 455-463. Plenum, New York.

Miller, J. F. A. P. and Mitchell, G. F. (1970a). *In* "Control Processes in Multicellular Organisms", Ciba Fdn Symp., pp. 238-250.

Miller, J. F. A. P. and Mitchell, G. F. (1970b). *J. exp. Med. 131*, 675-699.

Miller, J. F. A. P. and Osoba, D. (1967). *Physiol. Rev. 47*, 437-520.

Miller, J. F. A. P. and Warner, N. L. (1971). *Int. Archs Allergy appl. Immun. 40*, 59-71

Miller, J. F. A. P., Brunner, K. T., Russell, P. J. and Sprent, J. (1970). *Transplant. Proc.* (In press.)

Mitchell, G. F. and Miller, J. F. A. P. (1968a). *Proc. natn. Acad. Sci. U.S.A. 59*, 296-303.

Mitchell, G. F. and Miller, J. F. A. P. (1968b). *J. exp. Med. 128*, 821-837.

Mitchison, N. A., Taylor, R. B. and Rajewsky, K. (1970). *In* "Developmental Aspects of Antibody Formation and Structure" (J. Šterzl and I. Říha, eds), pp. 547-561. Publ. House, Czech. Acad. Sci.

Shearer, G. M. and Cudkowicz, G. (1969). *J. exp. Med. 130*, 1243-1261.

Sinclair, N. R. St C. and Elliott, E. V. (1968). *Immunology 15*, 325-333.

Takeya, K. and Nomoto, K. (1967). *J. Immun. 99*, 831-836.

Taylor, R. B. (1969). *Transplant. Rev. 1*, 114-149.

Notes added in proof. (1) It has now been shown that the adoptive memory response obtained by transferring anti-H2-incubated TDL to irradiated recipients (see Table IX) could be elevated towards control levels by supplementing with normal TDL, provided these were derived from CBA and not from (CBA x C57BL)F$_1$ donors. TDL from CBA mice primed specifically to FγG were at least 10 times as effective as TDL from normal CBA mice (Miller, J. F. A. P. and Sprent, J. Results submitted to *J. exp. Med.*).

(2) Unequivocal evidence for immunological specificity in thymus cells and in B cells (as present in spleens of TxBM mice) has now been obtained by the technique of "hot antigen-induced suicide". Exposure of either cell line to ^{125}I-FγG at a concentration of 70 ng FγG/108 cells resulted in a 10-20 fold reduction in anti-FγG PFC as determined in a collaborative co-transfer experiment in irradiated mice. The response to a control antigen, horse erythrocytes, was not reduced (Basten, A., Miller, J. F. A. P., Warner, N. L., Pye, J. Results submitted to *Nature*).

Thymocytes or Thymus Grafts as Reconstituents of Deprived Mice

A. J. S. DAVIES, E. LEUCHARS, V. WALLIS and R. L. CARTER

Chester Beatty Research Institute, Institute of Cancer Research
Royal Cancer Hospital, London, England

There is little doubt that lymphoid cells in some way mediate many immune responses. Consequently, if the lymphocytes that are present in the body of an animal are reduced in number, immunological responsiveness may be diminished. Such reduction can be effected in many ways: by cytotoxic drugs, by irradiation, by antilymphocytic antisera, by manipulation of the source organs of the lymphocyte populations, by removal of mature lymphocytes, or by any combination of these treatments. Unfortunately nearly all of these methods carry with them the chance of functional inactivation as well as numerical reduction of the lymphocyte populations. As a consequence designing experiments intended to elucidate the relationship between the immune response and a particular population of cells in the living animal is always difficult. For example if the thymus is removed from a newborn mouse not only do fewer thymus-derived cells appear in the lymphoid tissues, but any humoral effect the thymus would have had on this or any other population of lymphocytes must also be limited. Similarly, in attempting to reconstitute neonatally thymectomized mice with thymocyte suspensions it is not possible to be certain either that the injected thymocytes are all equivalent to thymus-derived cells, or that any humoral influence of the thymus can be provided by a thymocyte population.

The series of experiments presented here is intended to attack these problems. Deprived CBA/H mice were thymectomized, irradiated (850 r) adult mice injected with 5×10^6 syngeneic bone marrow cells after irradiation (Davies *et al.*, 1969a). These were reconstituted with either 30×10^6 thymocytes or a subcapsular thymus graft, in either instance from CBA/H.*T6T6* donors. The methods have been published elsewhere (Davies *et al.*, 1969a). The immune responses of these mice were determined 50-70 days later and the mitotic responses to certain antigenic stimuli were also analysed in terms of the ratios of T6T6 thymus-derived cells : other dividing cells. In one instance allogeneic chimaeras were prepared by reconstituting syngeneic CBA/Cbi deprived mice with CBA/H.*T6T6* thymocytes or thymus grafts. These mice were sensitized with heterologous erythrocytes and used as donors in a selective transfer test (Davies *et al.*, 1967), which is designed to detect co-operation between thymus and other cells.

Immune Response to I.P. Injection of Sheep Red Blood Cells (SRBC)

1. In deprived mice. Thymectomized and irradiated mice were given only bone marrow cells. Thirty days after irradiation they were injected i.p. with 5×10^8 SRBC. The animals were titrated regularly for circulating anti-SRBC antibodies, and 21 days later they were re-injected with SRBC. The procedure was repeated until five injections of SRBC had been given. There was a slow build-up in titre with increasing numbers of injections of SRBC (Fig. 1). At all times the mean titres had high variance. This was not due to fluctuation in the performances of individual animals, but to the high variability between animals. Some mice produced no detectable antibody, others had moderate titres that built up slowly.

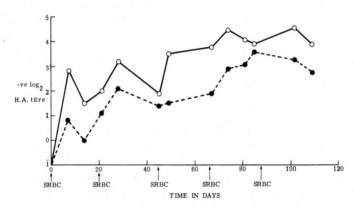

Fig. 1. Mean serum antibody response of deprived mice (see text) following several injections of SRBC. Total haemagglutinating anti-SRBC antibody —O—, mercaptoethanol resistant antibody − −●− −. First injection 30 days after irradiation (n = 8).

2. In mice reconstituted with two thymus lobes. Deprived mice were reconstituted with two thymus lobes immediately after irradiation. Fifty days later they were given a single i.p. injection of 5×10^8 SRBC. Titrations were performed at intervals for 21 days at which time a second injection of SRBC was made. The titration sequence was repeated. The immune response observed (Figs 2 and 3) was slightly less than that of normal mice, but was not otherwise remarkable.

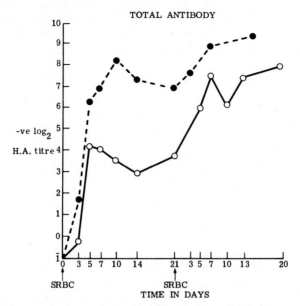

Fig. 2. Mean total serum antibody response of deprived mice reconstituted with either 30×10^6 thymocytes ——o—— (n = 8) or two thymus lobes – –•– – (n = 7) at various times after a single i.p. injection of 5×10^8 SRBC 50 days after irradiation.

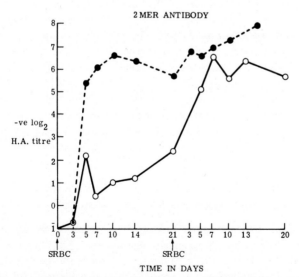

Fig. 3. Mercaptoethanol resistant antibody response of deprived mice reconstituted with either 30×10^6 thymocytes ——o—— (n = 8) or two thymus lobes – –•– – (n = 7) at various times after a single i.p. injection of 5×10^8 SRBC 50 days after irradiation.

3. In mice reconstituted with 30×10^6 *thymocytes.* Deprived mice reconstituted with an i.v. injection of 30×10^6 thymocytes immediately after irradiation were left for 50 days and then subjected to a similar series of injections of SRBC and titrations to those used in the parallel group of animals restored with thymus grafts. The response of the thymocyte-reconstituted animals to a single injection of SRBC was considerably less than that of the thymus-grafted mice (Figs 2 and 3). The difference was largely, but not completely, attributable to a lack of 2-mercaptoethanol resistant (MER) antibodies. There was some evidence that the thymocyte-reconstituted mice were reacting slowly to the antigenic stimulus, since their titres of MER antibody were rising steadily from the 7th to the 21st day. The response to the second injection of SRBC was similar in both groups of mice. The thymus-grafted mice always had slightly higher titres of antibody than did the thymocyte-reconstituted animals.

THE MITOTIC RESPONSE TO AN I.P. INJECTION OF SRBC

1. In mice reconstituted with two thymus lobes. Cytological analyses were made every day for six days of the cells responding by mitosis to a single i.p. injection of SRBC (5×10^8 cells). The proportion of thymus graft-derived cells in the spleen rose from 1-4% to nearly 20% and the response was maximal on days 2 and 3 (Fig. 4). The mitotic response to a second injection given 21 days after the first was brisker than after the first injection and peaked on day 2 (Fig. 4).

2. In thymocyte-reconstituted mice. The same procedure was followed for these animals as had been used for mice reconstituted with thymus grafts. No response to the first injection of SRBC was evident on day 2 (Fig. 4), but by day 3 the response was similar to that of the thymus-grafted mice. The mitotic response to a second injection of SRBC was, however, feeble in the thymocyte-reconstituted mice in comparison to the response in mice with a thymus graft. This is in contrast to the immunological result in which both groups of mice behaved similarly after a second injection of SRBC.

RESPONSE IN DRAINING LYMPH NODES TO A SUBCUTANEOUS INJECTION OF SRBC OR CUTANEOUS PAINTING WITH OXAZOLONE

Quantitation of mitotic responses to antigenic stimuli in the spleen is always rendered more difficult by the relatively high number of dividing non-lymphoid cells in the red pulp. It is possible to dissect out the white pulp follicles and to make analyses on these alone, but apart from the manipulative difficulties of such a procedure it is hard to get the follicles uniformly free of red pulp. Responses in lymph nodes are more readily perceived since there is very little

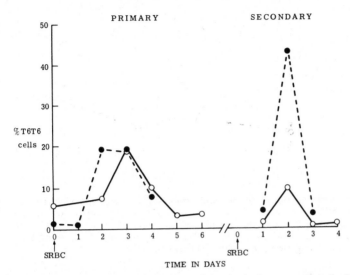

Fig. 4. Average mitotic response in the spleens of deprived mice reconstituted with either 30 x 10⁶ thymocytes ——○—— or two thymus lobes – –●– – at various times after a first or second i.p. injection of 5 x 10⁸ SRBC. First injection 50 days after irradiation, second 21 days after the first. Thymocytes and thymus grafts were from CBA/H.*T6T6* mice, and the results are expressed as the proportion of cells of this type in mitosis (n = 3).

mitotic activity in unstimulated mouse axillary lymph nodes (Davies *et al.*, 1969a). Also the histopathological picture is less complicated. For these reasons the comparison between thymocyte- and thymus graft-reconstituted mice was extended to a study of responses in lymph nodes draining sites of injection of 5 x 10⁸ SRBC or painting of oxazolone. Details of the method have been published elsewhere (Davies *et al.*, 1969a, b).

Both SRBC and oxazolone induced an increase in absolute lymph node weight during the first six days after contact (Fig. 5). The response was less in thymocyte- than in thymus graft-reconstituted mice. Oxazolone was the more powerful of the two stimulants as judged by lymph node weight increase.

The mitotic response to a subcutaneous injection of SRBC in the two groups of mice is indicated in Fig. 6. The proportion of donor cells responding was far higher at day 3 in the animals that were reconstituted with thymocytes, but the duration of the response was similar in both groups. However, very few cells entered mitosis in the thymocyte-injected mice compared with the considerable augmentation of mitotic activity in the thymus grafted animals (Fig. 7).

The response to oxazolone again shows that the proportion of responding cells was higher in the thymocyte-reconstituted mice than in the thymus grafted animals (Fig. 8). No quantitation of the number of cells responding was attempted in this instance.

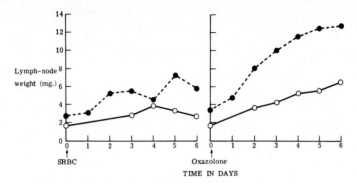

Fig. 5. Mean absolute axillary lymph node weight of deprived mice reconstituted with either 30 x 10^6 thymocytes —o— or two thymus lobes – –●– – at various times after subcutaneous injection into the fore paws or the skin high on the back of 5 x 10^8 SRBC or cutaneous painting in the same sites with 0.1 ml of 3% oxazolone in absolute alcohol. SRBC or oxazolone were given 50 days after irradiation (n = 6).

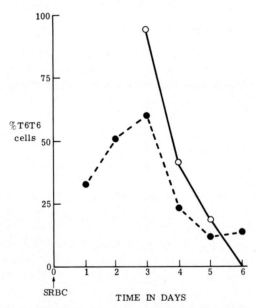

Fig. 6. Mean mitotic response in the draining (axillary) lymph nodes of deprived mice reconstituted with 30 x 10^6 thymocytes —o— or two thymus lobes – –●– – at various times after a subcutaneous injection of 5 x 10^8 SRBC. Injections were 50 days after irradiation. The T6 chromosome marker was present in the thymocytes and thymus grafts and the results are expressed as the proportion of T6T6 donor cells among the cells in mitosis (n = 6).

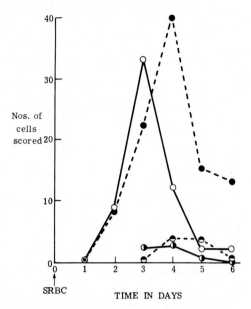

Fig. 7. Mean mitotic response in the draining (axillary) lymph nodes of deprived mice reconstituted with either T6T6 thymocytes or two T6T6 thymus lobes at various times after a subcutaneous injection of 5×10^8 SRBC 50 days after irradiation. The results are expressed as the numbers of T6T6 and unmarked cells responding in the two groups of mice.

—○— T6T6 cells   ⎫ in thymus-grafted mice
– –●– – unmarked cells ⎭

—◑— T6T6 cells   ⎫ in thymocyte-injected mice
– –◕– – unmarked cells ⎭

The histopathological findings in the draining lymph nodes of mice injected subcutaneously with SRBC or painted with oxazolone were as follows.

1. Response to SRBC. In mice reconstituted with thymus grafts, and immunized subcutaneously with SRBC all the regional lymph nodes showed a brisk response, first in the paracortical region, and then in the follicles and medullary cords. The paracortical blast cells began to proliferate after about 24 hr, reaching a peak of activity at four to five days, which then declined. At about the same time (day 4) prominent germinal centres appeared in the follicles, which increased in size and number, and a plasmacytosis developed in the medullary cords. The response in mice reconstituted with thymocytes differed in the following respects. Not all the draining nodes responded, and the reaction, when present, was less pronounced and proceeded at a slower tempo. The sequence of paracortical activity followed by follicular activity was not clearly seen. There was still some paracortical activity as late as six days after immunization. Reactive changes in the follicles and medullary cords were trivial.

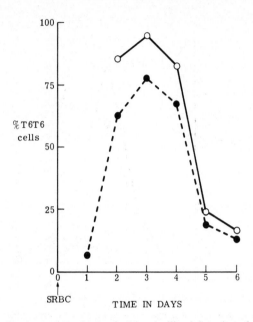

Fig. 8. Mean mitotic response in the draining (axillary) lymph nodes of deprived mice reconstituted with 30 x 10^6 thymocytes —o— or two thymus lobes – –●– – at various times after cutaneous painting with 0.1 ml of a 3% solution of oxazolone in absolute alcohol on the fore paws and high on the back, done 50 days after irradiation. The T6 marker chromosome was present in the thymocytes and thymus grafts and the results are expressed as the proportion of T6T6 donor cells among the cells in mitosis (n = 6).

2. Response to oxazolone. Once again the changes in mice reconstituted with thymocytes differed from those seen in mice reconstituted with a thymus graft. In the latter oxazolone provoked a massive proliferation of paracortical blast cells, which was well under way after 24 hr; the peak was reached by day 4. By day 6 the paracortical proliferation was declining, and obvious reactive changes were seen in the follicles and medullary cords. In contrast, the response of paracortical blast cells in thymocyte-reconstituted mice evolved more slowly, and the biphasic pattern of reaction—affecting paracortex and then follicles and medullary cords—had not become apparent by day 6.

SELECTIVE TRANSFER

The method of selective transfer has been published before (Davies *et al.*, 1967). In this instance CBA/Cbi mice were reconstituted with either a single CBA/H.*T6T6* thymus graft or 15 x 10^6 thymocytes. These primary hosts were injected with SRBC on days 30 and 44 and their spleens were taken for transfer at day 50. The secondary hosts were either CBA/Cbi or CBA/H mice. Some of the CBA/Cbi mice were made isoimmune anti-CBA/H and vice versa. In all four groups of secondary hosts, after irradiation and i.v. injection of 5×10^6 spleen cells from the primary hosts, some mice were injected with 5×10^8 SRBC i.p., others did not receive SRBC. Thus, as before, there were eight groups of secondary recipients. There were no differences between recipients of cells from the two kinds of primary hosts (Table I).

DISCUSSION

The deprived mice in these experiments proved to have immune responses that could be developed slowly by repeated injections of antigen. It is tempting to suppose that the response observed could be regarded as thymus-independent. However, it must be remembered that there are two proven sources of thymus-derived (T) cells (Roitt *et al.,* 1969) in deprived mice (Doenhoff *et al.,* 1970), firstly the bone marrow inoculum, and secondly residual host cells. Neither source is large, but the immunological performance of deprived mice could be related to these T cells, rather than to a direct interaction between antigen and the other population of lymphocytes present (M cells). Similarly, the addition of other lymphoid reconstituents to deprived mice could augment either the numbers or the performance of the residual T cells. The present experiments are not however designed to investigate this point.

In previous publications the immunological reactivity of thymectomized, irradiated, bone marrow-injected mice with thymus grafts has been carefully compared with that of normal mice (Davies *et al.,* 1969a, b). Normal mice gave slightly more vigorous and larger responses than those with thymus grafts. The reason for this difference is not known with certainty. It could be that there is residual immunological damage from the irradiation, that is not apparently improved by injection of bone marrow and implantation of thymus grafts. Alternatively it could be supposed that either the number of bone marrow cells injected is insufficient, or that one thymus lobe is not enough to engender full restoration of immunological activity. Both these latter possibilities are the subject of further study. At the present time it must simply be admitted that radiation might do some permanent damage to the fabric of the lymphoid system, and that any cellular reconstituents may operate in a different manner from their behaviour in the animal from which they were taken. Thus when comparing the behaviour of thymocytes and thymus grafts in deprived mice it must be remembered that the interpretation of the findings may not be simple. Two previous studies have established the feasibility of the reconstitution of deprived mice with thymocytes (Mayhew *et al.,* 1968; Macgillivray *et al.,* 1970).

The most likely reason for the big difference in primary immunological responsiveness between the mice reconstituted with thymus grafts and with thymocytes is that the latter group were deficient in T cells. It is thought that only relatively few of the cells in the thymus are mature and equivalent to the

TABLE I

Haemagglutinin titres of various groups of secondary host mice six days after irradiation and transfer of 5 x 10⁶ spleen cells from primed primary hosts (n = 5). Mean primary host titres at time of transfer were 8.5 (thymus graft-reconstituted mice) and 7.0 (thymocyte-reconstituted mice)

Groups	CBA/Cbi hosts				CBA/H.T6T6 hosts			
	Isoimmune anti-CBA/H.T6T6		Not isoimmune		Isoimmune anti-CBA/Cbi		Not isoimmune	
	+SRBC	–	+SRBC	–	+SRBC	–	+SRBC	–
	1	2	3	4	5	6	7	8
Thymus graft-reconstituted primary hosts	2.2	0.0	6.8	$\bar{1}.8$	$\bar{1}.6$	$\bar{1}.5$	7.6	3.6
Thymocyte-reconstituted primary hosts	2.2	0.4	4.2	0.0	$\bar{1}.8$	$\bar{1}.6$	7.6	3.4

thymus-derived T cell external to the thymus (Raff, 1970). It has been shown that thymocytes are considerably less effective in restoring immunological responsiveness to deprived mice than are peripheral blood lymphocytes (Davies *et al.*, 1971), which are suspected to contain a large proportion of mature T cells (Raff and Wortis, 1970).

The slow maturation of the response, particularly of 2-mercaptoethanol resistant antibody, in thymocyte-reconstituted mice, is reminiscent of the response of normal mice to very small amounts of antigen (Wallis *et al.*, 1970; Wortis *et al.*, 1966), and also to the response of mice long after adult thymectomy (Davies, 1969). These similarities may be fortuitous, but their common feature could be that relatively few T cells are stimulated by antigen and for some reason the response cannot mature until more similar cells have been recruited. Comparison of the histopathological with the cytological responses in the thymocyte-reconstituted mice however does not support such an interpretation. The paracortical response was prolonged, and at the later times it was not only the donor thymocytes that were mitotically active. Indeed in some animals in which paracortical hyperactivity was recorded, only bone marrow-type cells were found in mitosis.

On a more speculative basis it might be that in order for an immune response to proceed at what we regard as a normal pace, T cells must emigrate from the vascular system in increased numbers after antigenic stimulation. If only very small amounts of antigen are given to the animal it may be that such recruitment is not adequate. Alternatively if only very few T cells are present the recruitment may be slow. In adult thymectomized mice, however, some evidence has been presented that numerical insufficiency is not the only cause of the slow immune responses (Davies *et al.*, 1971). Rather more specifically it was proposed that a humoral influence of the thymus might be lacking. Thus in relation to the present experiments the slow immune responses of thymocyte-reconstituted mice could be related not only to the numerical insufficiency of the T cell population, but perhaps also to a lack of a thymic humoral factor. The influence of such a factor could be somehow to affect the ease with which a T cell left the vascular system, and emigrated through the post-capillary venules.

The high proportion of T cells responding in the lymph nodes of thymocyte-reconstituted mice in comparison with the response in mice receiving thymus grafts has been discussed elsewhere (Davies *et al.*, 1969b). It is only necessary to say here that it is thought to be due to the existence, in mice with thymus grafts, of a population of T cells that derive from the bone marrow inoculum via the thymus graft. These cells will "dilute" the ratio of responding T6T6 to normal cells, since their behaviour may well be identical to that of the T6T6 cells known to derive from the thymus graft. In the thymocyte-reconstituted mice no "thymus-processed bone marrow-derived" cells are possible, and the ratio is therefore higher. The 10% of cells that still have the bone marrow marker are thought to be the residual host and bone marrow T cells previously referred to.

The immunological responses to a second injection of SRBC were similar in both groups of reconstituted mice. Also the test of the co-operative function in the selective transfer experiment revealed considerable similarities. It is not possible to be sure whether a T cell can be primed by antigenic stimulus. Certainly the T cell response was quicker and higher after a second injection of SRBC than after a first (at least in thymus-grafted mice). This could have been due to circulating antibody facilitating the contact of a virgin T cell with antigen, or alternatively to the presence of a primed T cell population. The point to be made here is that, whatever the explanation, an initially numerically inadequate population of thymocytes developed the capacity to co-operate in a secondary immune response quite as well as their better endowed experimental (cell) mates from animals with thymus grafts. Thus for immune responses to a second contact with antigen it is not necessary to suppose that any humoral influence of the thymus is operative on T cells, although it will be remembered that some such influence could not be excluded in the primary response.

It might be argued that the finding of antibody in secondary host mice (group 1) in the selective transfer experiment, which should have had no T cells (the isoimmunity being designed to eliminate them), indicates that M cells (Davies et al., 1971) have a memory, and can, independently of T cells, produce some antibody. However, it must sadly be recalled that there are some residual T cells in the primary host (Doenhoff et al., 1970). These might be operative in the secondary host mice. Thus the possibility cannot be excluded that co-operation between T and M cells is a sine qua non for antibody production.

It has been suggested that the group 1 response in mice receiving cells from thymus-grafted animals may be operated by cells derived from the bone marrow graft, that have migrated into the thymus graft in the primary host, and emerged with properties similar to those of cells which, by virtue of their chromosome marker, are known to be T cells (Miller, 1967). This ingenious explanation is somewhat unnecessary, since it seems from the present experiments that the same results can be achieved in group 1 whether a thymus graft is present in the primary hosts or not.

ACKNOWLEDGEMENTS

This work was supported by grants to the Chester Beatty Research Institute (Institute of Cancer Research: Royal Cancer Hospital) from the Medical Research Council and the Cancer Research Campaign.

SUMMARY

Thymectomized, irradiated, bone marrow injected adult mice were given either thymocytes or thymus grafts. The immunological capacities of these reconstituted mice were tested 50-70 days later. The mitotic responses to antigenic stimuli were also analysed with the aid of a chromosome marker present in the reconstituting lymphoid cells. In a separate experiment a selective transfer was made from mice reconstituted with allogeneic thymus grafts or thymocytes. The results are discussed against the background of the possibility that thymocyte-injected mice were not only numerically deficient in T cells, but also lacked a humoral thymic influence that may affect the behaviour of T cells.

REFERENCES

Davies, A. J. S. (1969). *Transplant. Rev. 1*, 43-91.
Davies, A. J. S., Leuchars, E., Wallis, V., Marchant, R. and Elliott, E. V. (1967). *Transplantation 5*, 222-231.
Davies, A. J. S., Carter, R. L., Leuchars, E., Wallis, V. and Koller, P. C. (1969a). *Immunology 16*, 57-69.
Davies, A. J. S., Carter, R. L., Leuchars, E. and Wallis, V. (1969b). *Immunology 17*, 111-126.
Davies, A. J. S., Leuchars, E., Wallis, V. and Doenhoff, M. J. (1971). *Proc. R. Soc. B. 176*, 369-384.
Doenhoff, M. J., Davies, A. J. S., Leuchars, E. and Wallis, V. (1970). *Proc. R. Soc. B. 176*, 69-85.
Macgillivray, M. H., Mayhew, B. and Rose, N. R. (1970). *Proc. Soc. exp. Biol. Med. 133*, 688-692.
Mayhew, B., Macgillivray, M. H. and Rose, N. R. (1968). *Proc. Soc. exp. Biol. Med. 128*, 1217-1221.
Miller, J. F. A. P. (1967). *Lancet ii*, 1299-1302.
Raff, M. C. (1970). *Nature, Lond. 226*, 1257-1258.
Raff, M. C. and Wortis, H. H. (1970). *Immunology*. (In press.)
Roitt, I. M., Greaves, M. F., Torrigiani, G., Brostoff, J. and Playfair, J. H. L. (1969). *Lancet ii*, 367-371.
Wallis, V., Leuchars, E. and Davies, A. J. S. (1970). Submitted for publication.
Wortis, H. H., Taylor, R. B. and Dresser, D. W. (1966). *Immunology 11*, 603-616.

DISCUSSION

FICHTELIUS. Dr Davies, you said that there is no evidence for the existence of a bursal equivalent in mammals, and that there is no use of assuming the existence of such an equivalent.

I think that there is good indirect evidence for the existence of a bursal equivalent in bursaless vertebrates.

Mammals and birds both originate from animals similar to fishes and amphibians. All vertebrates have an immunoglobulin producing apparatus with many common characteristics. Only the birds have a bursa but the bursaless vertebrates have numerous intimate lymphoepithelial relationships with unknown function. Some of them might serve as bursal equivalents.

Antigen Dose and the Avidity of Antibody from Thymectomized Mice

R. B. TAYLOR

M.R.C. Immunobiology Group, Department of Pathology
University of Bristol, Medical School, Bristol BS8 1TD, England

The antibody response of mice to most antigens is depressed by thymectomy. This is because the thymectomized mouse is deficient in thymus-dependent or T-lymphocytes. These cells do not themselves release antibody, but help the thymus-independent or B-lymphocytes to do so (see *Transplantation Rev.,* 1969). The nature of this co-operative function is unknown, but it has been suggested that the T-lymphocytes produce a cell-bound antibody which serves to concentrate antigen at certain critical sites where it can more easily stimulate B-lymphocytes (Mitchison, 1968; Taylor and Iverson, 1970). For certain antigens, such as sheep erythrocytes, the effect of thymectomy can be partly overcome by increasing the dose (Sinclair and Elliott, 1968; Taylor and Wortis, 1968). This suggests that the hypothetical concentrating function of T-lymphocytes may be by-passed by increasing the overall concentration of antigen. If this is true, then we should expect that those B-lymphocytes that can respond to antigen without help would be the ones having relatively high affinity receptors, and therefore making high affinity antibody. Such a conclusion draws support from the sheep erythrocyte experiments, because the effect of thymectomy could be overcome for IgM antibodies much more easily than for IgG. The multivalency of IgM allows it to make a stronger bond with a multivalent antigen such as the erythrocyte (Taylor and Wortis, 1968).

In the experiments to be described an attempt was made to overcome the effect of thymectomy by increasing dosage of a hapten-protein conjugate; and at the same time to study the avidity of the anti-hapten antibody that was produced in thymectomized and control mice.

MATERIALS AND METHODS

Adult CBA mice were used. They were thymectomized, irradiated (900 r) and injected with CBA bone marrow cells as described before (Taylor, 1969). Controls received irradiation and bone marrow alone.

The antigens were NIP-chicken globulin (6 moles/mole) (NIP-CG) and NIP-keyhole limpet haemocyanin (16 moles/100,000 M.W.) (NIP-KLH) (Brownstone et al., 1966a).

Antibody was titrated against a range of concentrations of ^{125}iodine-labelled hapten (N^{125}IP-aminocaproate, NIP-CAP) by a modification of Farr's ammonium sulphate technique (Brownstone et al., 1966b). Although affinity values can be derived graphically from such results, by an extrapolation procedure, the simpler and more reproducible avidity method of Celada et al. (1969) was preferred. These authors titrated anti-human serum albumin antibody by the Farr technique, and observed a linear relationship between log total antigen concentration (bound + free) and the log of the concentration of antiserum required to bind 50% of the antigen. The slope of this line, S, was related to affinity and varied from 0 (zero affinity) to 1 (no dissociation). It can be taken as an index of avidity, although its exact relation with mean affinity is not known at present. This linear relationship was found to hold true for anti-NIP antibody also (Taylor, unpublished work).

EXPERIMENTS AND RESULTS

1. Effect of antigen dose and thymectomy on quantity and avidity of anti-NIP antibody. Thymectomized and control mice were each divided into four groups and immunized with four different doses of NIP-CG alum precipitate i.p. (10, 100, 1000 and 10,000 μg protein) with a constant dose of 2×10^9 killed *B. pertussis* organisms as adjuvant. They were bled five weeks later and the sera of each group pooled, and titrated against 10^{-7} and 10^{-9} M NIP-CAP. The titres rose with increasing antigen dose to a maximum at 1000 μg and then decreased slightly with the 10,000 μg dose (Fig. 1). The response of thymectomized mice was about 10-fold lower than the controls in every group, and there was no evidence within this dose-range that the need for T-lymphocytes was being by-passed.

The avidity values (plotted against antigen dose in Fig. 2) show a tendency for higher avidity with lower doses, and a slightly higher avidity in thymectomized as compared to control sera.

2. Avidity values from individual thymectomized and control mice during the secondary response. Some anti-NIP sera from another experiment were also titrated in a similar way for comparison. In this case the mice had been immunized with 1000 μg NIP-KLH alum-precipitate and pertussis. Three months later they were boosted by i.p. injection of 100 μg NIP-KLH in solution. The sera obtained by bleeding 14 days later were titrated against three concentrations of NIP-CAP. The values of log antiserum concentration required to bind 50% of the hapten (y) were plotted against hapten concentration, and were again found to fall on to straight lines. The slopes of these lines (S) were determined by regression. The value of y corresponding to the highest concentration of

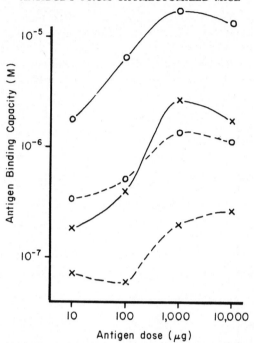

Fig. 1. Effect of increasing antigen dose on the anti-NIP response of thymectomized (x) and control (○) mice. Pooled antibody was titrated against 10^{-7} M NIP-CAP (——) and 10^{-9} M NIP-CAP (— — — —).

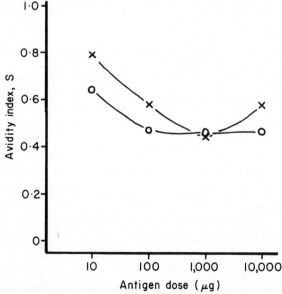

Fig. 2. Effect of varying antigen dose on avidity of pooled antibody produced by thymectomized (x) and control (○) mice.

hapten was then taken as a measure of the total quantity of antibody, and this was plotted against S (Fig. 3). Here it is seen that avidity increased with increase in the quantity of antibody both for thymectomized and control sera. On the average therefore the thymectomized mice made antibody of lower avidity than did the controls.

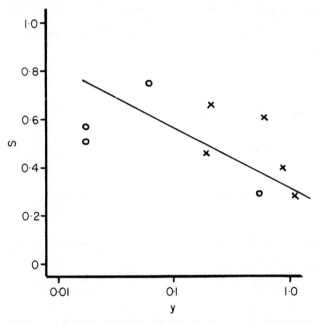

Fig. 3. Avidity and titre of anti-NIP antibodies. Antisera produced in the secondary response of individual thymectomized (x) and control (o) mice were titrated against three concentrations of NIP-CAP (50, 10 and 2×10^{-6} M). The titre is here represented as the log concentration of antiserum required to bind 50% of the hapten at 50×10^{-6} M (y), and is plotted against avidity index (S). The correlation just falls below the 0.05 level of significance.

DISCUSSION

A series of studies of the affinity of anti-hapten antibody in rabbits and guinea pigs (reviewed by Siskind and Benacerraf, 1969) built up a picture of a selective process in the immune response, which favours the division of cells with high affinity for antigen. Thus an initially small population of these increases more rapidly, and finally outnumbers, an initially larger population with low affinity. This selection is strongly influenced by antigen dose, and also by the presence of antibody in the serum. The antibody probably competes with the cells for antigen and thus provides a selective drive, or bias, in favour of high affinity cells. It is generally thought that antigen concentration plays a part in this process, such that high doses of antigen are able to stimulate low affinity cells, which could not be stimulated by lower doses (Siskind and Benacerraf, 1969). However, it is difficult to apply thermodynamic considerations when one considers that the majority of the antigen is in particulate form (in alum precipitate or Freund's adjuvant) and is rapidly taken up by macrophages, and that probably only a very small proportion is ultimately immunogenic. It would be a useful simplification, for the present, if the effects of antigen dose could be explained as being mediated entirely through the rate of increase in the selective bias provided by antibody concentration. Thus, the higher the dose of antigen the more intense the stimulus, and the faster the rate of increase in bias. This will have the effect that the initial rate of selection will be higher and affinity will at first be positively related to titre. However, selection can only continue so long as cell division keeps abreast of the increase in bias. It may be predicted that with intense stimulus, the bias might overtake cell division and essentially shut it off, so that later in the response one might find the condition where affinity would be negatively related to titre. In Fig. 4 the values for antibody

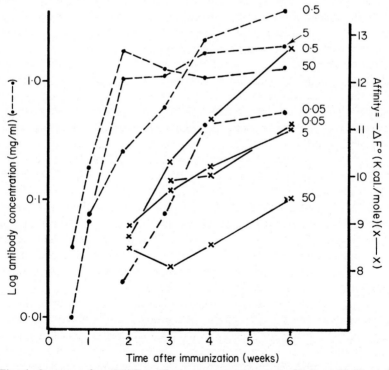

Fig. 4. Increase of anti-DNP antibody concentration and affinity with time after immunization of rabbits with 0.05, 0.5, 5.0 or 50.0 mg doses of DNP-bovine gamma globulin in complete Freund's adjuvant. (Figures taken from Siskind and Benacerraf, 1969).

titre and affinity for various antigen doses given by Siskind and Benacerraf (1969) have been plotted in graphical form. The second condition can be seen at the three-week bleed (compare 0.5, 5 and 50 mg doses). And there is some indication of the first condition at the two-week bleed (compare 0.5 and 5 mg doses). Possibly an earlier bleeding might have shown this more clearly. With suboptimal doses (0.05 mg) the affinity is always positively related to titre, because the bias never overtakes cell division.

Positive relation of affinity to titre has been clearly shown in the secondary response by Celada et al. (1969). Their experiments were done with a cell transfer system, so that the secondary response was initiated in the absence of serum antibody. Under these conditions the antibodies first detected were of low avidity, but avidity rapidly increased in a way suggestive of a brief recapitulation of the primary response. When small numbers of cells (10^6) were transferred, the rate of increase of avidity was much slower—although by three weeks it approached the avidity obtained with larger numbers (10^7).

The predicted effect of thymectomy on affinity will depend very much on whether the antibody is produced as a result of antigen by-passing a thymus-dependent concentrating mechanism, as mentioned in the introduction, or whether it is produced in the regular way, as a result of co-operation between B-lymphocytes and those few T-lymphocytes that remain after thymectomy and irradiation. Certainly some T-lymphocytes do remain, because these mice can still reject homografts—albeit after a prolonged time (Taylor, 1964). They also still contain lymphocytes bearing the thymus antigen theta, although in smaller numbers than control mice (Raff, 1970). If antigen was by-passing the T-lymphocytes, then we should expect thymectomized mice always to produce higher avidity antibody than control mice—perhaps very much higher. But this was clearly not the case, as seen from experiment 2. Therefore, the bulk of the anti-DNP antibody found in these experiments is probably the result of co-operation. This conclusion is reinforced by the failure of increasing dose to narrow the gap between thymectomized and control titres. The small differences in avidity that were found are probably explicable in terms of the considerations on selection that have just been discussed. If T-lymphocytes are a limiting factor in the presentation of antigen, then reduction in their numbers may have the same effect as reduction in the dose of antigen; that is, to reduce the rate of increase in the selective bias produced by antibody.

In experiment 2 we see the simple situation, where avidity is positively related to titre. The thymectomized mice produce less antibody and thus less selective bias—and behave similarly to the mice injected with small numbers of cells in Celada's experiment. In the late primary response on the other hand (experiment 1) the avidity is negatively related to titre, presumably because antigen doses are in the supraoptimal region and have caused bias to overtake cell division. The thymectomized mice act as if they had received a smaller dose of antigen. In this case also their behaviour is explicable in terms of the selective bias provided by antibody, and does not require one to postulate changes in the effective concentration of antigen actually presented to the precursor cells.

Although T-lymphocytes were apparently not by-passed in these experiments, even by large doses of antigen, it should be emphasized again that large doses of sheep erythrocytes did this to a marked extent only for IgM antibodies. It will be interesting to investigate IgM anti-hapten antibodies in thymectomized mice.

In conclusion, we should perhaps not be surprised that thymectomy did not bring any improvement to Nature's selective mechanism!

REFERENCES

Brownstone, A., Mitchison, N. A. and Pitt-Rivers, R. (1966a). *Immunology 10,* 465-479.

Brownstone, A., Mitchison, N. A. and Pitt-Rivers, R. (1966b). *Immunology 10,* 481-492.

Celada, F., Schmidt, D. and Strom, R. (1969). *Immunology 17,* 189-198.

Mitchison, N. A. (1968). *In* "Differentiation and Immunology" (K. B. Warren, ed.), p. 29. Academic Press.

Raff, M. C. (1970). *Immunology 19,* 637.

Sinclair, N. R. St C. and Elliott, E. V. (1968). *Immunology 15,* 325-333.

Siskind, G. W. and Benacerraf, B. (1969). *Adv. Immun. 10,* 1-50.

Taylor, R. B. (1964). Thesis, University of Edinburgh.

Taylor, R. B. (1969). *Transplant. Rev. 1,* 114-149.

Taylor, R. B. and Iverson, G. M. (1970). *Proc. R. Soc. B. 176,* 393-418.

Taylor, R. B. and Wortis, H. H. (1968). *Nature, Lond. 220,* 927-928.

Transplant. Rev. (1969). *1.*

Differential Effects of Corticosteroids on Co-operating Cells in the Immune Response

H. N. CLAMAN, M. A. LEVINE and J. J. COHEN

University of Colorado Medical Center
Denver, Colorado 80220, U.S.A.

INTRODUCTION

The immunosuppressive effects of corticosteroids have been known for many years (reviewed in Dougherty *et al.,* 1964). In general, corticosteroids have been thought to inhibit antibody production because of their lymphocytolytic properties It is not clear, however, *how* corticosteroids affect lymphoid cells, nor even *which* lymphoid cells are affected. In view of recent observations that antibody production may require both thymus-dependent and marrow-derived cells (Claman and Chaperon, 1969), it becomes pertinent to re-evaluate the action of corticosteroids on the immune response. This paper will describe primarily the effects of corticosteroids on the production of circulating antibody, on the initiation of the graft-*versus*-host reaction, and on the response of lymphoid cells to phytohemagglutinin (PHA).

PREVIOUS DATA WITH CORTICOSTEROIDS AND THE IMMUNE RESPONSE

1. Circulating antibody. The inhibition of circulating antibody production by corticosteroids has been documented many times. The depression is related to both the dose of antigen and the dose of steroid (Berglund, 1956). The temporal relation between administration of steroid and antigen is also crucial; steroid given soon before or soon after antigen is strongly immunosuppressive, but once the antibody induction process is under way, it is resistant to steroids. This has been particularly well worked out for the response of rats and mice to sheep erythrocytes (SRBC) (Berglund, 1956; Elliott and Sinclair, 1968). In rats, the inhibition can be overcome by transferring normal spleen or thymus cells to cortisone-treated animals (Berglund and Fagraeus, 1956).

2. Graft-versus-host (GvH) reactions. Relatively little work has been done on the effects of corticosteroids on the initiator cells of the G*v*H reaction. This initiator cell is felt to be a small lymphocyte (Gowans, 1962). Therefore, it was

C.I.–12

somewhat of a surprise when Warner (1964) and Blomgren and Andersson (1969) reported that the cells in the thymus (chicken and mouse respectively) responsible for G*v*H reactions were resistant to corticosteroids. (We have confirmed and extended these experiments—see below.) Because of earlier work of Dougherty *et al.* (1964) and of Ishidate and Metcalf (1963) showing that the thymic cortex is very sensitive to corticosteroids while the medulla is resistant, Warner as well as Blomgren and Andersson felt that the thymic G*v*H cells were medullary.

RECENT EXPERIMENTS IN OUR LABORATORY

1. Cytolytic effects of corticosteroids. Marked effects of corticosteroids were seen after intraperitoneal injection of hydrocortisone acetate suspension. Thymus cell counts made two days later showed a marked decrease as compared with untreated controls. There were also falls of lesser magnitude in the spleen and bone marrow (Fig. 1). In each case, low doses of steroids produced rapid cellular depletion. Increasing the dose of hydrocortisone produced further cellular depletion until a maximum was reached. The curves appear to indicate two populations of cells, one sensitive to hydrocortisone and the other resistant. The relative size of the cortisone-sensitive population in A/J mice varied greatly

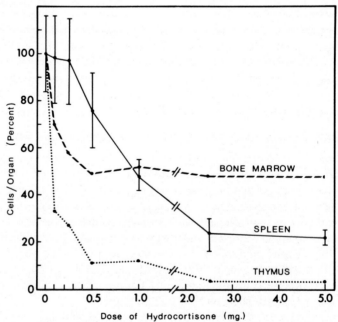

Fig. 1. Fraction of nucleated cells remaining in thymus, spleen and bone marrow of A/J mice after treatment with various doses of hydrocortisone acetate two days before killing. Means and 95% confidence limits given for spleen.

from one organ to another, being about 95% of the thymus, 80% of the spleen and only 50% of the marrow. In addition, the relative size of the cortisone-sensitive fraction of the marrow appeared to vary from one strain of mice to another; 2.5 mg of hydrocortisone depleted A/J femurs to 51% of normal, Balb/c femurs to 79% of normal and CAF_1 femurs to 93% of normal.

The lytic effects of hydrocortisone on mouse thymus cells may also be seen *in vitro* by using cells labeled with ^{51}Cr. When mouse cells are cultured in minimal essential medium with 20% heat-inactivated fetal calf serum, 1 mM glutamine, 100 units penicillin G and 100 μg streptomycin, they are very fragile and there is considerable spontaneous lysis of cells. This spontaneous lysis is temperature dependent; in one experiment, chromium-labeled cells cultured for 17 hr released 48.7% of the label if incubated at 37.5°C, but only 37.5% of the label if cultured at 23°C and only 4.6% of the label if cultured at 5°C. The longer the cells were cultured, the more ^{51}Cr was spontaneously released (Fig. 2). However ^{51}Cr release was increased if the cells were cultured with

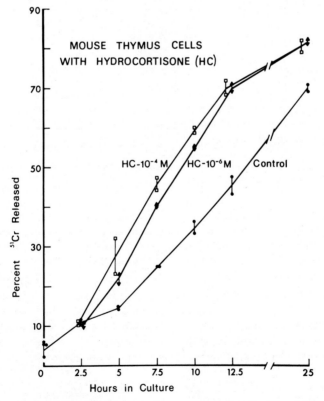

Fig. 2. Release of ^{51}Cr from mouse thymus cells cultured with or without hydro-cortisone sodium succinate (added at time 0).

increasing doses of hydrocortisone sodium succinate. Therefore, the lytic effect of hydrocortisone on mouse thymus cells can be seen *in vitro*, and is time- and dose-related.

The *in vivo* distinction between hydrocortisone-sensitive and hydrocortisone-resistant thymus cells was also confirmed *in vitro*. The release of cellular ^{51}Cr after incubation with hydrocortisone was tested using suspensions of thymus cells from mice pretreated *in vivo* with hydrocortisone 1-2 days before sacrifice. These suspensions were presumably deprived of their corticoid-sensitive cells *in vivo*. The results (Table I) show that normal thymus cells were more sensitive to hydrocortisone than were the thymus cells surviving *in vivo* hydrocortisone treatment.

TABLE I

Effect of hydrocortisone on mouse thymus cells from normal and hydrocortisone-treated mice

| Hydrocortisone | Per cent ^{51}Cr released | |
	Normal mice	Hydrocortisone mice
0	23.2	29.0
10^{-6} M	40.4	39.3
10^{-4} M	51.6	37.6
Change	28.4	8.6

Mice received 2.5 mg hydrocortisone acetate i.p. 48 hr before sacrifice. Cells cultured for 7 hr.
Change = Increased ^{51}Cr release in cells cultured with 10^{-4} M hydrocortisone compared with cells cultured without hydrocortisone.

2. Effects of corticosteroids on antibody formation and PHA response. We have also shown that very large doses of cortisone acetate do not inhibit the marrow precursor of the plaque-forming cell (Levine and Claman, 1970). These results are summarized in Tables II and III. Bone marrow cells from cortisone-treated donors were able to complement normal thymus in the production of cells making antibodies to SRBC. However spleen cells from cortisone-treated donors were unable to transfer plaque-forming ability to irradiated recipients. These results indicate that the marrow precursor of the PFC (Mitchell and Miller, 1968) is resistant to corticoids at least when in the marrow, but that the spleen contains a corticoid-sensitive cell or process that is essential to the production of PFC.

Since the spleen can be thought of as a complete antibody-forming organ containing thymus-dependent and marrow-derived cells, the effects of hydrocortisone on the adoptive response of spleen cells or of thymus-marrow cell

TABLE II

Effect of cortisone on hematopoietic and antibody-forming capacities of spleen and bone marrow

Cells transferred (+SRBC)		Mice/group	^{59}Fe Uptake/spleen	PFC/spleen
A. Cortisone spleen	(11×10^6)	5	1158 (1046-1271)	137 (98-192)
B. Normal spleen	(11×10^6)	6	1524 (1076-1972)	959 (768-1197)
C. Cortisone marrow	(10×10^6)	6	1927 (1509-2345)	3225 (1598-6512)
Normal thymus	(27×10^6)			
D. Normal marrow	(10×10^6)	6	1585 (996-2174)	4004 (3073-5217)
Normal thymus	(27×10^6)			
E. Normal thymus	(27×10^6)	4	139 (18-260)	111 (87-141)

Donor mice were either normal or had received 12.5 mg cortisone acetate i.p. 48 and 12 hr before sacrifice. Lymphoid cells were transferred with 0.1 ml 20% SRBC on day 0. Groups A and B were killed on day 6. Groups C, D and E were boosted with SRBC on day 4 and killed on day 8.
^{59}Fe results are CPM.

TABLE III

Effect of cortisone on hematopoietic and antibody-forming capacities of spleen and bone marrow (derived values)

	^{59}Fe uptake/donor organ	PFC/donor organ
A. Cortisone spleen	758	87
B. Normal spleen	12,705	8029
C. Cortisone marrow + Normal thymus	2,370	3967
D. Normal marrow + Normal thymus	2,425	6126

These values were derived from those in Table II by compensating for loss in cell number in organs of cortisone-treated mice. This Table gives calculated values for ^{59}Fe uptake and PFC produced by transfer of one whole organ, from either a normal or a cortisone-treated donor.

combinations was investigated. Normal LAF_1 cells were transferred to syngeneic lethally irradiated adult mice and stimulated with SRBC. Hydrocortisone was given to the recipients either early or late in the adoptive antibody response. Experiments 1 and 2 in Table IV indicate that the early adoptive response of spleen cells was relatively sensitive to hydrocortisone while the late response was relatively resistant. This finding holds whether one (experiment 1) or two

TABLE IV

Effects of hydrocortisone on plaque-forming cells (PFC) when given "early" or "late" after transfer of spleen or thymus-marrow cell suspensions to irradiated mice

			Days			
Exp. 1	0	1	5	6		
Group	Cells transferred			PFC		
A	30×10^6 spleen +SRBC	OHC	–	176	(75-413)	
B	Same	–	OHC	664	(349-1262)	
C	Same	–	–	753	(421-1277)	
Exp. 2	0	1	4	7	8	
	Cells transferred				PFC	
A	30×10^6 spleen +SRBC	OHC	SRBC	–	1905	(451-8,054)
B	Same	–	SRBC	OHC	6437	(3452-12,000)
C	Same	–	SRBC	–	9170	(5591-15,040)
Exp. 3	0	1	5	8	9	
	Cells transferred				PFC	
A	10×10^6 BM 35×10^6 T +SRBC	OHC	SRBC	–	59	(26-131)
B	Same	–	SRBC	OHC	652	(132-3228)
C	Same	–	SRBC	–	644	(108-3717)

LAF$_1$ male mice. Recipients given 1000 r ^{60}Co irradiation just before transfer. OHC = 10 mg hydrocortisone acetate suspension i.p.

SRBC = 0.05 ml of 40% suspension on day 0; 0.5 ml of 10% suspension on days 4 or 5 when indicated.

PFC = direct PFC/recipient spleen (with 5-95% confidence limits in parentheses) (6 mice/ group).

(experiment 2) doses of antigen were given. The data (experiment 3) also show that combinations of thymus and marrow react to hydrocortisone like spleen cells; that is, they are sensitive to hydrocortisone early but resistant late in the response. In the thymus-marrow system, it appears that the thymus-type cell must be activated with or before the marrow-type cell (Claman and Chaperon, unpublished; Shearer and Cudkowicz, 1969). Taken together, these data indicate that the suppressive effects of corticosteroids occur early in the antibody response to SRBC and might involve antigen handling or thymus-dependent processes.

TABLE V

Phytohemagglutinin response of spleen and bone marrow

		Normal[a]	Cortisone-treated[a]
Spleen	6×10^6	640 ± 66	19 ± 4
Spleen + PHA	6×10^6	2899 ± 374	25 ± 10
Marrow	4×10^6	993 ± 148	981 ± 43
Marrow + PHA	4×10^6	1073 ± 63	2535 ± 501

Donor mice were either normal or had received 12.5 mg cortisone acetate i.p. 48 and 12 hr before sacrifice and culture. PHA-M was added as 0.1 ml of 1-10 for spleen and 1-5 for marrow.

[a] Cells were pulsed with ^{3}H-thymidine for the terminal 5 hr of the three-day culture period. Results are expressed as mean CPM of triplicate cultures $\pm$ S.E.

The effect of corticoid treatment on the PHA response has also been explored. Large doses of corticoids *in vivo* did not inhibit the small response of the bone marrow to PHA but did abolish the spleen's response to PHA (Table V). In fact, the marrow response to PHA was enhanced by hydrocortisone.

3. Effects of hydrocortisone on GvH initiator cells. We also investigated the effects of hydrocortisone on the donors of cells used in the GvH reaction. In these experiments, the ability of thymus cells to initiate GvH was tested by the degree of splenomegaly produced by injecting parental cells into newborn F_1 mice The ability of spleen and marrow cells to initiate GvH was tested by the ability of parental cells to cause splenomegaly in adult F_1 mice. When equal cell numbers were used, suspensions of thymus, spleen or marrow from hydrocortisone-treated donors were more efficient in initiating the GvH reaction than were cells from normal donors (Tables VI and VII, Fig. 3). The results confirm

TABLE VI

Graft-versus-host reactivity of thymus cells from normal and hydrocortisone-treated Balb/c donors in neonatal CAF_1 recipients

Cells transferred	Number of recipients	Recipient spleen index[a]
Normal thymus (6×10^6)	9	$1.28 (1.21-1.35)$[b]
Hydrocortisone-thymus (6×10^6)[c]	8	$2.77 (2.56-2.99)$

[a] Four-day-old CAF_1 mice were injected with thymus cells i.p. on day 0 and killed eight days later. Spleen index is an expression of GvH activity obtained by dividing spleen-to-body weight ratios of the experimental group by mean spleen-to-body weight ratios of littermates receiving F_1 thymus cells.

[b] Mean and 95% confidence limits.

[c] Donors given 2.5 mg hydrocortisone i.p. on day −2.

TABLE VII

Graft-versus-*host reactivity of 30 × 10⁶ bone marrow cells from normal and hydrocortisone-treated Balb/c, A/J and CAF₁ donors transferred to adult CAF₁ recipients[a]*

Group	Cells transferred	Number of recipients	Recipient spleen index
A	Normal CAF_1	14	1.00 (0.88-1.12)[b]
B	Hydrocortisone-CAF_1	7	0.93 (0.82-1.04)
C	Normal Balb/c	16	0.93 (0.82-1.04)
D	Hydrocortisone-Balb/c	18	1.28 (1.07-1.46)
E	Normal A/J	12	1.15 (1.04-1.26)
F	Hydrocortisone-A/J	6	1.47 (1.29-1.65)

[a] Donors of hydrocortisone-cells received 2.5 mg of hydrocortisone acetate on day −2. Marrow cells were transferred on day 0. Recipients killed on day 8.
[b] Mean and 95% confidence limits.

those of Warner (1964) for the chicken thymus and those of Blomgren and Andersson (1969) for the mouse thymus. They also show that hydrocortisone produces similar enrichment of GvH cells in the spleen (*not* found by Blomgren and Andersson) and in the marrow. The simplest interpretation is that the hydrocortisone-sensitive cell is immunologically incompetent with regard to the GvH reaction.

DISCUSSION

This paper reviews some of our knowledge about the activities of corticosteroids upon immunological responses. The material can be looked at from two standpoints: (1) what do the data tell us about the mechanisms of action of the corticosteroids? (2) what do the data tell us about the immune response? The answers to these two questions are closely related, and are quite complex.

The specific biochemical site of action of corticosteroids on lymphoid cells is not known with certainty. There is evidence that corticosteroids *in vivo* and *in vitro* inhibit RNA, DNA and protein synthesis (White and Makman, 1967; Kidson, 1967). Recent data indicate that hydrocortisone interferes with protein synthesis by preventing ATP generation through inhibition of glucose uptake or phosphorylation (Young, 1969). One difficulty with much of the large volume of work concerning the action of corticosteroids on lymphoid cells is that investigators have paid little attention to the types of cells being used. Thymus cells are popular because large numbers of relatively homogeneous cells are easily available, yet the data presented above clearly show the marked functional heterogeneity of thymus cells with regard to their immunological capabilities and their sensitivity to steroids. Lymph node cells have been used, but there has

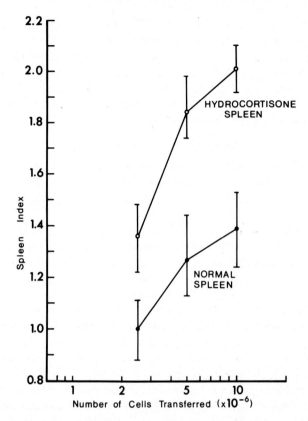

Fig. 3. Graft-*versus*-host reactivity of various doses of spleen cells from normal and hydrocortisone acetate-treated Balb/c mice injected into adult CAF_1 recipients. The spleen index shown is the ratio of relative spleen weights of the experimental group to those of a control group receiving normal CAF_1 spleen cells. Means and 95% confidence limits are given. Each group contained six mice.

been little attempt to characterize the response of these metabolically active cells as compared to more quiescent but still immunocompetent cells, such as those of peripheral blood. Even the problem of corticosteroid receptors is relatively unexplored. Recent evidence indicates that there are two mechanisms for cortisol uptake by thymocytes, a nonspecific low-affinity system and a specific high-affinity system (Schaumburg and Bojesen, 1968). These data, of course, may reflect cell heterogeneity rather than two receptors on one cell.

The results presented above show that corticosteroids have marked cytolytic properties *in vivo* and *in vitro,* but that this varies from one organ to another. The thymus cortex is extremely sensitive to corticosteroids, but the bone marrow is much less so. Previous work has shown that the "lymphocytes" of the guinea-pig bone marrow survive several doses of corticosteroids (Yoffey *et al.*, 1954), but rat marrow lymphocytes are more sensitive (Morrison and Toepfer, 1967). The bone marrow corticoid-resistant cells include the precursors of PFC, the cells that respond to PHA, and the cells that initiate GνH. The thymus and spleen GνH cells are also corticoid-resistant, as are the spleen PFC in the adoptive immune response to SRBC. Nevertheless, the early phase of the mouse's response to SRBC is corticoid-sensitive.

CONCLUSION AND SPECULATIONS

Although various kinds of immune responses are thought to be inhibited by corticosteroids, dissection of these responses has revealed an increasing number of cells that are resistant to corticosteroids. What may we gather from these new facts?

1. The cellular basis of action of corticosteroids upon lymphoid cells is not known. It is quite likely that corticosteroids have different effects upon lymphoid cells depending on (a) their stage in the cell cycle (i.e. G_0, G_1, S, G_2 or M), (b) their location, (c) whether they have met antigen before, (d) their stage of maturation, and (e) unknown variables.

2. In the mouse, cells that initiate the GνH response and cells that are stimulated by PHA are resistant to corticosteroids, at least before the stimulus is applied. Whether they are also resistant after stimulation is not clear. On the basis of these data, we would predict that in the cell-mediated response (tuberculin type), the specific lymphoid cell initiator is corticosteroid resistant. According to this hypothesis, corticosteroids suppress the cell-mediated response by acting on the nonspecific bone marrow-derived mononuclear cell (Lubaroff and Waksman, 1967; Thompson and van Furth, 1970), or (less likely) on the inhibition of some lymphokine.

TABLE VIII

Effects of corticosteroids on the immune mechanism in the mouse

---

RESISTANT

    Marrow precursor of PFC (in the marrow)
    PFC (late in the immune response in the spleen)
    PHA-responsive cell in the marrow
    GνH initiator cell in thymus, marrow and spleen
    ?Specific lymphocyte phase of cell-mediated hypersensitivity

SENSITIVE

    Thymus cortical cells
    Early phase of mouse spleen response to SRBC
    ?Nonspecific (monocyte) phase of cell-mediated hypersensitivity

---

3. The cellular locus of the immunosuppressive effects of corticosteroids on the mouse response to SRBC is not clear.

The marrow precursor of the PFC is corticoid resistant, at least when in the marrow. The PFC itself late in the response is also resistant. The sensitive phase lies somewhere in between and may be: (a) the PFC precursor in the spleen (which may be in a different state compared to the PFC precursor in the marrow); (b) the thymus-derived cell; (c) the *in vivo* counterpart of the "adherent cell" (if there is one). These speculations are summarized in Table VIII.

ACKNOWLEDGEMENTS

Supported in part by USPHS AI-00013, AM-10145, and Canadian Medical Research Council.

We are grateful for the expert assistance of Miss Martha Post, Mr Henry Benner and Miss Lenore Shapiro.

SUMMARY

The cellular bases of the immunosuppressive actions of corticosteroids at the cellular level were reviewed. Experiments in mice show that hydrocortisone depletes various organs of cells to different extents. Two days after 2.5 mg of hydrocortisone, 95% of the thymus cells, 80% of the spleen cells and 50% of the marrow cells were destroyed. Mouse thymus cells labeled with ^{51}Cr can be lysed *in vitro* with hydrocortisone and this lysis is time and dose related.

Corticosteroids *in vivo* inhibit the early but not the late stages of the adoptive antibody response to sheep erythrocytes RBC whether this response is provided by spleen cells or thymus-marrow cell combinations.

The bone marrow precursor of the plaque-forming cell is resistant to corticosteroids and the bone marrow response to phytohemagglutinin *in vitro* is unaffected by prior corticosteroid treatment.

Graft-*versus*-host effector cells in thymus, spleen and marrow are all resistant to corticosteroid treatment of the donors of these cells.

Corticosteroids appear to exert lympholytic activities on immunologically irrelevant cells. The exact locus of their action in immunosuppression is not known.

REFERENCES

Berglund, K. (1956). *Acta path. microbiol. scand. 38*, 311-328.
Berglund, K. and Fagraeus, A. (1956). *Nature, Lond. 177*, 233-234.
Blomgren, H. and Andersson, B. (1969). *Expl. Cell. Res. 57*, 185-192.
Claman, H. N. and Chaperon, E. A. (1969). *Transpl. Rev. 1*, 92-113.

Dougherty, T. F., Berliner, M. L., Schneebeli, G. L. and Berliner, D. L. (1964). *Ann. N.Y. Acad. Sci. 113*, 825-843.

Elliott, E. V. and Sinclair, N. R. (1968). *Immunology 15*, 643-652.

Gowans, J. L. (1962). *Ann. N.Y. Acad. Sci. 99*, 432-455.

Ishidate, M., Jr. and Metcalf, D. (1963). *Aust. J. exp. Biol. med. Sci. 41*, 637-649.

Kidson, C. (1967). *Nature, Lond. 213*, 779-782.

Levine, M. A. and Claman, H. N. (1970). *Science, N.Y. 167*, 1515-1516.

Lubaroff, D. M. and Waksman, B. H. (1967). *Science, N.Y. 157*, 322-323.

Mitchell, G. F. and Miller, J. F. A. P. (1968). *J. exp. Med. 128*, 821-837.

Morrison, J. H. and Toepfer, J. R. (1967). *Acta Haemat. 38*, 250-254.

Schaumburg, B. P. and Bojesen, E. (1968). *Biochim. biophys. Acta 170*, 172-188.

Shearer, G. M. and Cudkowicz, G. (1969). *J. exp. Med. 130*, 1243-1261.

Thompson, J. and van Furth, R. J. (1970). *J. exp. Med. 131*, 429-442.

Warner, N. L. (1964). *J. exp. Biol. med. Sci. 42*, 401-416.

White, A. and Makman, M. H. (1967). *Adv. Enzyme Regul. 5*, 317-322.

Yoffey, J. M., Ancill, R. J., Holt, J. A. G., Owen-Smith, B. and Herdan, G. A. (1954). *J. Anat. 88*, 115-130.

Young, D. A. (1969). *J. biol. Chem. 244*, 2210-2217.

The Role of a Small Pool of Immunocompetent Thymus Cells in the Humoral Antibody Response:
Comparison between Thymus-dependent and Thymus-independent Systems

BIRGER ANDERSSON and HENRIC BLOMGREN

Department of Tumour Biology, Karolinska Institutet
S-104 01 Stockholm 60, Sweden

INTRODUCTION

The thymus plays a central role in the development of immunological reactivity in mammals. Thymectomy during the neonatal period or adult thymectomy together with irradiation primarily affects the animal's capacity to react with cellular immunity, such as graft rejection and delayed hypersensitivity. In some systems humoral antibody formation is also impaired by thymectomy. In these cases the humoral antibody forming cell is derived from the bone marrow, but cannot synthesize antibody unless thymus-dependent cells are also present acting as helper cells (Mitchell and Miller, 1969). Thymectomy deprives the animals of thymus-dependent cells, and therefore the marrow-derived cells cannot form antibody when confronted with the antigen. The mode of action of the thymus during maturation of immunocompetence is not completely clear. There is experimental evidence that soluble extracts from the thymus can induce immunological competence in non-competent cells from thymectomized animals (Trainin *et al.,* 1968). We will present evidence that immunological maturation of immunocompetent cells can take place within the thymus, indicating that the highest concentration of the soluble factor is probably inside the thymus. We shall also summarize our experiments to characterize the immunocompetent cells present in small numbers within the thymus. Finally a series of experiments will be presented that indicate that in some antigen systems the presence of thymus cells is not necessary for the induction of humoral and antibody formation in the marrow-derived population of cells.

IMMUNOCOMPETENT CELLS IN THE THYMUS

It is known that the thymus is heterogeneous with regard to cellular composition. It consists of a cortical region with small, rapidly dividing lymphocytes and a medullary area with larger cells that do not divide within the thymus. The cortical cells form approximately 95% of the lymphoid cells in the thymus and the medullary cells 5%. We used the different sensitivity of these two cell populations to cortisone to discriminate between them. When 125 mg of cortisone acetate per kg body weight was injected into mice we found that within two days the small cortical cells disappeared, whereas the larger cells remained unaffected. The remaining, cortisone-resistant cells were very efficient in a graft-*versus*-host test, which is a measure of cellular immunity (Blomgren and Andersson, 1969). They were approximately 10 times more reactive than cells from normal thymus. Furthermore, they were also very efficient as helper cells for humoral antibody formation in thymus-dependent systems (Andersson and Blomgren, 1970). We tested the humoral antibody response of adult thymectomized, irradiated and bone marrow reconstituted mice against sheep erythrocytes, bovine serum albumin, ovalbumin and the NIP hapten (4-hydroxy-3-iodo-5-nitrophenyl-acetic acid) given in the form of NIP_{10}ovalbumin. In all these cases the cortisone-resistant thymus cells were more efficient in enhancing humoral antibody production as compared to equal numbers of cells taken from untreated thymus (Andersson and Blomgren, 1970). We also found that the cortisone-resistant thymus cells cannot themselves produce humoral antibody when injected into lethally irradiated mice together with antigen. This excludes the possibility that peripheral lymphocytes migrate to the damaged areas of the cortisone-involuted thymus. Another possibility is that non-competent lymphoid cells migrate to the cortisone damaged thymus and there, under the influence of some thymus factor(s), rapidly gain immunological competence of the cellular type. We think however that we have shown that this is not the case. Two days after cortisone treatment the reactivity per cell had increased greatly (Fig. 1, Table I), but the total immunological reactivity of the organ remained unaffected, and thus no new immunocompetent cells entered the thymus during this time period.

The experiments shown in Table II were designed to test whether the demonstration of a distinct immunocompetent cell population within the thymus might be relevant to the origin of reactive cells of this type. Adult mice were lethally irradiated and injected with bone marrow. At various times afterwards different lymphoid organs were tested for the presence of cells that could enhance the humoral antibody response against sheep erythrocytes in thymectomized, irradiated and bone marrow reconstituted recipient mice. We found that reactive cells of this type appeared first in the thymus, and thereafter in the spleen and lymph nodes. Our interpretation is that under these conditions immunocompetent cells of the cellular immunity type seem to be generated within the thymus, and probably later seed out into spleen and lymph nodes. Whether this occurs normally cannot be ascertained from the present data but the possibility exists.

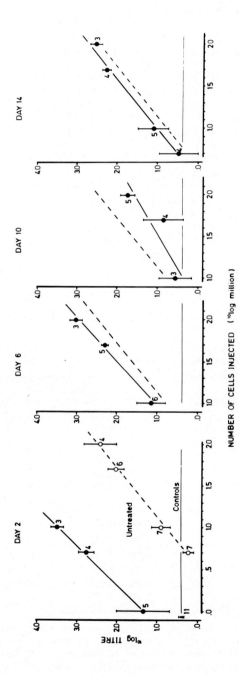

Fig. 1. Effect of cortisone treatment on the capacity of thymus cells to act as helpers for humoral antibody formation against SRBC in thymectomized, irradiated and bone marrow reconstituted mice. (——) Cortisone-treated mice. (— — —) untreated mice. Linear regression lines for the dose-response relationship were determined at different times (day 2, day 6 etc.) after cortisone injection into the thymus cell donor mice.

TABLE I

Effect of cortisone treatment on the mouse thymus

Days after cortisone	Linear regression of dose-response line[a]	Cell no. per thymus $(\log_{10} \times 10^6)$	Total reactivity per thymus $(\%)$[b]	Relative reactivity per cell $(\%)$[c]
0	$y = 1.67x - 0.88$	2.29	100	100
2	$y = 2.12x + 1.32$	0.75	94	1416
6	$y = 1.83x - 0.74$	0.75	0.5	133
10	$y = 1.19x - 0.85$	1.68	2	65
14	$y = 1.64x - 0.62$	2.29	155	142

[a] $y = \log_{10}$ haemolytic titre in thymectomized mice eight days after injection of thymus cells and SRBC. $x = \log_{10}$ cell dose (millions).

[b] y value obtained when x = the total cell no. in the thymus.

[c] x value obtained when y is given a constant value (e.g. 1.00).

TABLE II

Variations in the immunocompetence of lymphoid cells of mice irradiated with 800 r and injected with bone marrow

Days after irradiation	Cell type[a]	Antibody response in recipients $\log_{10}$ haemolytic titre ± S.E.	Responders/total
10	T	0.24 ± 0.11	4/14
	S	0.22 ± 0.13	3/13
	L	0.21 ± 0.14	2/7
14	T	0.95 ± 0.40	5/8
	S	—	0/8
	L	—	0/8
25	T	3.10 ± 0.50	4/4
	S	1.43 ± 0.58	3/4
	L	1.55 ± 0.53	3/4
Unirradiated controls	T	1.47 ± 0.43	6/7
	S	2.92 ± 0.46	7/7
	L	2.20 ± 0.19	7/7

Immunocompetence was assessed by the ability of the lymphocytes to induce antibody production in thymectomized, irradiated mice given marrow cells and 4×10^8 SRBC. Recipients were bled 5, 10 and 15 days after immunization and the peak titres determined.

[a] T = thymus cells, S = spleen cells, L = lymph node cells. 10^7 cells injected i.v.

THYMUS-INDEPENDENT ANTIBODY FORMATION

During our studies of the ability of thymus cells to reconstitute the humoral antibody response of adult thymectomized, irradiated and bone marrow reconstituted mice, we found that humoral antibody formation did not always depend on an injection of thymus cells together with the antigen. This was the case in two antigen systems: polyvinyl pyrrolidone and *E. coli* endotoxin. In order to establish whether there was genuine thymus-independent antibody production against these antigens we tested them over a wide dose range and in the presence or absence of thymus cells. At no dose level did the injected thymus cells act as helpers for humoral antibody formation (Tables III and IV). Furthermore, if the thymectomized mice were reconstituted with fetal liver instead of bone marrow (to avoid the possibility of thymus-dependent cells being transferred with the cell inoculum) and were also treated with repeated ALS injections (to eradicate possible thymus-dependent cells surviving irradiation or being transferred with the fetal liver) they nevertheless still responded to *E. coli* endotoxin. We find it important to describe these systems of apparently completely thymus-independent antibody formation, since earlier reports of thymus independence of antibody formation have either involved neonatally thymectomized animals or a single antigen dose (Humphrey *et al.*, 1964). Neither the possibility of competent cells having left the thymus before

TABLE III

Antibodies against E. coli *endotoxin in thymectomized, irradiated, bone marrow reconstituted mice. Lack of effect of thymus cells*

Antigen dose ($\log_{10} \mu g$/mouse)	Antibody response in:	
	BM[a]	BM + T[b]
−6	0.67 ± 0.42	1.80 ± 1.11
−5	2.60 ± 0.24	2.00 ± 1.10
−4	4.00 ± 0.55	2.00 ± 0.63
−3	4.33 ± 0.66	3.86 ± 0.83
−2	6.40 ± 0.51	6.00 ± 0.32
−1	7.25 ± 0.25	8.25 ± 0.48
0	6.20 ± 0.86	7.17 ± 0.40
No antigen	<0.00	<0.00
4×10^8 SRBC	<0.00	6.50 ± 0.29

Log_3 haemolytic titre against endotoxin-coated SRBC day 8 after immunization. Mice immunized with SRBC were tested against uncoated SRBC. Mean of five mice per group ± S.E.

[a] Thymectomized, irradiated, bone marrow reconstituted mice given antigen at day 0.

[b] Thymectomized, irradiated, bone marrow reconstituted mice given antigen and 4.5×10^7 syngeneic thymus cells at day 0.

TABLE IV

Antibodies against polyvinyl pyrrolidone in thymectomized, irradiated, bone marrow reconstituted mice. Lack of effect of thymus cells

Antigen dose ($\log_{10}$ μg/mouse)	Antibody response in: BM[a]	BM + T[b]
−4	$<$−1.45	$<$−1.45
−3	−1.42 ± 0.16	−1.44 ± 0.03
−2	−0.70 ± 0.16	−1.05 ± 0.15
−1	0.09 ± 0.22	−0.12 ± 0.15
0	0.06 ± 0.21	−0.15 ± 0.47
+1	−1.04 ± 0.07	−0.79 ± 0.16

$\log_{10}$ antigen binding capacity of antisera (μg of antigen bound per ml of undiluted serum) measured by radio-iodinated polyvinyl pyrrolidone day 8 after immunization. Mean of five mice per group ± S.E.
[a] and [b] See footnotes to Table III.

thymectomy nor the possibility that antibody formation is thymus-dependent only at certain antigen dose levels was excluded.

Finally we want to report results that suggest a role of thymus cells in the induction of immunological memory. The primary antibody response to sheep erythrocytes in mice is dependent upon the presence of thymus cells over a wide range of antigen dosage (Table V). In contrast the antibody response to a second injection of antigen was not dependent upon the presence of thymus cells during priming if the first antigen dose was large (Table VI). This indicates that the

TABLE V

Antibodies against sheep erythrocytes in thymectomized, irradiated, bone marrow reconstituted mice. Effect of thymus cells

Antigen dose SRBC/mouse	Antibody response in: BM[a]	BM + T[b]
4×10^5	$<$0.00	$<$0.00
4×10^6	$<$0.00	$<$0.00
4×10^7	$<$0.00	2.80 ± 0.86
4×10^8	$<$0.00	2.80 ± 0.80
4×10^9	$<$0.00	4.33 ± 0.67

$\log_3$ haemolytic titre against SRBC day 8 after immunization. Mean of five mice per group ± S.E.
[a] and [b] See footnotes to Table III.

TABLE VI

Secondary antibody response to sheep erythrocytes in thymectomized, irradiated, bone marrow reconstituted mice. Effect of the antigen dose used for priming on the thymus dependence of the secondary response

Primary antigen dose (SRBC/mouse)	Secondary antibody response BM[a]	BM + T[b]
4×10^5	2.00 ± 2.00	8.33 ± 0.66
4×10^6	2.25 ± 2.25	9.67 ± 0.33
4×10^7	5.20 ± 1.02	9.75 ± 0.62
4×10^8	3.60 ± 1.60	5.40 ± 0.68
4×10^9	7.33 ± 0.33	8.00 ± 0.00

Log_3 haemolytic titre against SRBC day 7 after a second injection of 4×10^7 SRBC given 60 days after the primary injection.

[a] and [b] See footnotes to Table III.

thymus cells might act as antigen concentrating cells. It can be seen that such an effect of thymus cells was not observed in the secondary response to polyvinyl pyrrolidone, further proving the complete thymus independence of the humoral antibody response to this antigen (Table VII). In other experiments we found that the priming for an anamnestic response against *E. coli* endotoxin was not dependent upon the presence of thymus cells, even at very low antigen dosage levels (Blomgren and Andersson, unpublished observations).

TABLE VII

Secondary response to polyvinyl pyrrolidone in thymectomized, irradiated, bone marrow reconstituted mice. Lack of effect of thymus cells injected with antigen of the primary response

Primary antigen dose (log_{10} µg/mouse)	Secondary antibody response BM[a]	BM + T[b]
-4	-1.09 ± 0.02	-0.93 ± 0.13
-3	-0.97 ± 0.13	-1.08 ± 0.05
-2	-0.82 ± 0.08	-1.05 ± 0.04
-1	-0.76 ± 0.23	-0.74 ± 0.14
0	-0.52 ± 0.09	-0.24 ± 0.11
$+1$	-0.29 ± 0.04	-0.47 ± 0.13

Log_{10} antigen binding capacity of antisera (µg of antigen bound per ml of undiluted serum) measured seven days after a secondary injection of 10^{-3} µg of antigen given 60 days after the first antigen injection.

[a] and [b] See footnotes to Table III.

SUMMARY

The mouse thymus consists of 95% small lymphocytes which are sensitive to cortisone and immunologically inert. The remaining 5% are relatively resistant to cortisone and immunologically reactive, as measured by the graft-*versus*-host assay and the capacity to enhance humoral antibody production in thymectomized mice in thymus-dependent antigen systems. Evidence is presented for the formation of these immunocompetent cells in the mouse thymus, and subsequent migration into the peripheral lymphoid organs. In some other antigen systems no evidence for thymus dependence of the humoral antibody production could be found, although a wide range of doses of antigen was tested. It was also shown that in thymus-dependent systems immunological memory was dependent upon thymus cells for induction at low antigen doses, whereas in thymus-independent systems no indication of an impact of thymus cells upon memory could be obtained at any antigen dose.

ACKNOWLEDGEMENT

This work was supported by the Swedish Cancer Society, Anders Otto Swärds Stiftelse and Jane Coffin Childs Memorial Fund for Medical Research.

REFERENCES

Andersson, B. and Blomgren, H. (1970). *Cellular Immunology 1*, 362-371.
Blomgren, H. and Andersson, B. (1969). *Expl Cell Res. 57*, 185-192.
Humphrey, J. H., Parrott, D. M. V. and East, J. (1964). *Immunology 7*, 419-439.
Mitchell, G. F. and Miller, J. F. A. P. (1969). *J. exp. Med. 130*, 765-775.
Trainin, N., Small, M. and Globerson, A. (1968). *J. exp. Med. 128*, 821-837.

Analysis of Hapten- and Carrier-specific Memory in a System of Co-operating Antigenic Determinants

VOLKER SCHIRRMACHER*

*Institut für Genetik der Universität Köln,
Köln 41, Weyertal 121, West Germany*

In terms of the co-operation hypothesis, the carrier specificity of the secondary immune response to haptens reflects co-operation between hapten- and carrier-specific receptors that are located on different cells (Mitchison *et al.*, 1970). Carrier-primed "helper" cells with carrier-specific receptors interact with the hapten-carrier complex. They then present the hapten determinants of the complex to hapten-specific receptors on hapten-primed antibody forming cell precursors. This stimulates them to anti-hapten antibody production. There is increasing evidence (Mitchison *et al.*, 1970; Raff, 1970) that co-operation between hapten-specific and carrier-specific cells is co-operation between thymus-independent bone marrow-derived (B) and thymus-dependent (T) lymphocytes (reviewed in *Transplant Rev.*, 1969).

The experiments reported in this paper are concerned with the function of these two cell types. In particular, the following questions are investigated:

1. How do the quantities of hapten and carrier determinants in the secondary antigenic stimulus affect co-operation?

2. Which cell type determines the class distribution of the secondarily stimulated anti-hapten antibody?

3. Can high affinity antibody forming cell precursors be stimulated in the absence of helper cells?

THE EXPERIMENTAL SYSTEM

The experimental system used is described in detail elsewhere (Rajewsky *et al.*, 1969; Schirrmacher and Rajewsky, 1970). Carrier proteins were bovine serum albumin (BSA) and human gamma globulin (HGG), and the hapten was sulphanilic acid coupled to N-chloracetyl-l-tyrosine (sulf-tyr). The hapten-carrier complexes (sulf-BSA and sulf-HGG) were prepared by coupling diazotized sulphanilic acid to the proteins. Anti-hapten antibodies were measured by

* Present address: Karolinska Institutet, Dept. of Tumor Biology, 5-10401 Stockholm 60, Sweden.

passive haemagglutination with sulphanilic acid coupled to sheep red cells, or by a binding assay based on the Farr technique (Farr, 1958). For this assay radioactively labelled hapten was synthesized by reacting diazotized ^{35}S-labelled sulphanilic acid with N-chloracetyl-l-tyrosine. The compound was purified by column chromatography and had a specific activity of about 1 C/mmole (Schirrmacher, 1970b). The stimulation of antibodies in the secondary response is expressed as the difference in haemagglutination titre ($\log_2$) or hapten binding capacity between sera at the peak of the response and on day 0, the time of secondary stimulation (Table I). The means and standard deviations are calculated on the assumption that the haemagglutinin titres and binding capacities are log-normally distributed within each group of rabbits. Rabbits primed with a sulf-BSA complex in Freund's complete adjuvant produce a good secondary response to the sulf group when stimulated with sulf-BSA (homologous system), but not when stimulated with sulf-HGG (heterologous system). In the latter situation however a good secondary response is obtained if the rabbits are pretreated with free HGG in adjuvant (co-operative system).

TABLE I

Analysis of secondary anti-hapten antibody production by passive haemagglutination and hapten binding in different experimental systems

A. *Experimental scheme*

Group	Primary (day -28)	Secondary (day 0)	System
A	sulf$_{19}$BSA 1000 μg	—	—
B	sulf$_{19}$BSA 1000 μg	sulf$_{19}$BSA 1000 μg	homologous
C	sulf$_{19}$BSA 1000 μg	sulf$_{18}$HGG 1000 μg	heterologous
D	sulf$_{19}$BSA 1000 μg +HGG 10 μg	sulf$_{18}$HGG 1000 μg	co-operative

B. *Results*

Group	No. of rabbits	Day 7 minus day 0 haemagglutination titre $\log_2$ units	Day 7 minus day 0 hapten binding capacity[a] $\log_2$ units
A	10	0.4[b]	0.8
B	6	2.3	2.6
C	7	0.4	1.1
D	10	2.6	3.7

[a] Difference in serum dilutions ($\log_2$) at which 25% of labelled hapten was bound at a concentration of 10^{-8} M (Schirrmacher, 1970b).

[b] Arithmetic mean, standard deviation varied from 0.04 to 0.12.

The Quantity of Hapten and Carrier Determinants Required for Immunogenicity in the Secondary Response

In the co-operative system the extent of the secondary anti-hapten response is constant over a dose range from 1 to 10^3 μg of sulf-HGG (Schirrmacher and Rajewsky, 1970). The stimulation is carrier-dependent even with the highest dose of the heterologous complex.

The number of sulf groups per molecule of secondary carrier strongly influences the intensity of stimulation of secondary anti-hapten antibodies. Table II shows how the number of antigenic determinants on a variety of

TABLE II

Variation of the number of sulf groups per molecule of secondary carrier

Group	Complex	Sulf determinants	Carrier determinants (%)	Day 7 minus day 0 haemagglutination titres	
				anti-sulf[a]	anti-carrier
A	$sulf_3 BSA$	0.05	100	1.1	2.9
B	$sulf_{13} BSA$	1.4	100	2.1	2.4
C	$sulf_{31} BSA$[b]	4.5	33	3.2	2.4
D	$sulf_{53} BSA$	45	3	1.9	0.9
E	$sulf_{18} HGG$	5	50	1.5	0.0
F	$sulf_4 HGG$	1	100	1.5	1.7
G	$sulf_{18} HGG$	5	50	3.1	1.3
H	$sulf_{52} HGG$	15	10	5.1	1.8
I	$sulf_{208} HGG$	46	0.14	2.3	0.0

Primary stimulation 100 μg $sulf_{19} BSA$ (groups A-I) and 10 μg HGG (groups F-I) in adjuvant on day −28. Secondary 10 μg of indicated complexes on day 0. Sulf groups determined by analysis of sulphur content. Sulf and carrier determinants assayed by comparing the molar concentration of the complex required for 50% inhibition of haemagglutination with that of sulf-tyr or HGG (compare Fig. 1). Titres (log_2 units) are arithmetic means of values from five rabbits. Standard deviation varied from 0.1 to 0.4.

[a] Determined under conditions of mild reduction (0.01 M dithiothreitol; for characterization of 7S antibodies, see Schirrmacher and Rajewsky, 1970).

[b] $Sulf_{19} BSA$ induced a somewhat smaller anti-sulf response.

hapten-carrier complexes affects their immunogenic capacity in the secondary response. The hapten and carrier determinants were assayed by comparing the molar concentrations of the complexes required for 50% inhibition of haemagglutination with those of free hapten and carrier (Fig. 1). The following observations can be made:

1. In both the homologous (groups A-D) and co-operative systems (groups

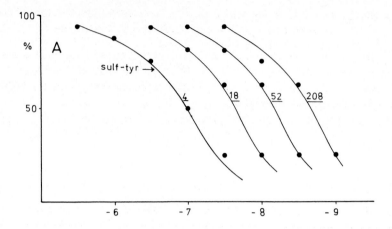

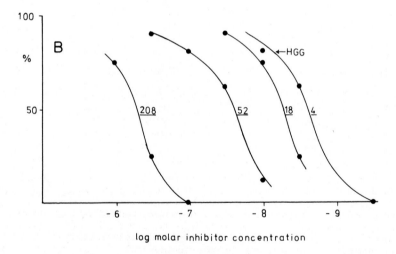

log molar inhibitor concentration

in hemagglutination assay

Fig. 1. Assay of hapten and carrier determinants on hapten-carrier complexes by inhibition of passive haemagglutination activity of anti-sulf antibodies (A) and anti-HGG antibodies (B). Anti-sulf and anti-HGG antisera were day 0 sera of animals primed with 100 μg sulf$_{19}$BSA or 10 μg HGG in adjuvant on day -28. Underlined numbers indicate numbers of sulf groups on the various hapten-carrier complexes. For example the molar concentration of sulf$_4$HGG at 50% inhibition is the same as that of sulf-tyr (A) or HGG (B). Thus the complex has, on average, one hapten determinant and 100% carrier determinants per molecule.

F-I) there was a certain ratio of hapten to carrier that gave the best stimulation of hapten-specific cells. The complexes with one or less hapten determinants per molecule (groups A and F) did not stimulate the anti-hapten response (compare with group E). Up to a certain point stimulation increased with increasing hapten density on the carrier molecules. However the complexes with the highest number of hapten determinants per molecule had only a low immunogenic capacity (groups D and I).

2. The optimal stimulation in the co-operative system was 3-4 times greater than the optimal stimulation in the homologous system. This was paralleled by the three-fold higher number of hapten determinants at the optimum point for sulf-HGG (group H) than at the optimum point for sulf-BSA (group C). Similar numbers of hapten determinants on BSA and HGG gave the same degree of stimulation of anti-hapten antibodies (groups C and G).

3. The complexes with the highest number of hapten determinants per molecule (groups D and I) had lost many carrier determinants.

The latter observation suggests that the cellular receptor on the carrier-specific helper cell has a specificity similar to the humoral anti-carrier antibody. This was supported by the fact that coupling carrier determinants to the $sulf_{208}HGG$ complex considerably increased its ability to stimulate anti-sulf antibodies (Table III). Cross-linking of the complex with itself did not have such an effect (group F).

These experiments can be interpreted as follows:

a. Limiting numbers of helper cells could explain the finding that varying the secondary dose of the complex by a factor of 1000 had little effect on the immune response, but varying the number of hapten determinants per molecule of secondary carrier by a factor of 13 (Table II, groups F-H) considerably altered the titre of anti-hapten antibodies. Alternatively, stimulation of hapten-specific memory cells may require simultaneous triggering of two or more neighbouring receptors.

b. In these experiments neither the environment of the hapten determinants nor the exact distance between them on the primary carrier seemed to play an important role in determining the extent of carrier specificity. The optimal response in the co-operative system is considerably higher than the optimal response in the homologous system. This was reproducible with two differently coupled sulf-BSA complexes used for priming (Tables II and III). It is unlikely that different numbers of helper cells in the co-operative and homologous systems can account for this finding, since the same number of hapten determinants on BSA and HGG induced the same degree of stimulation (groups C and G, Table II). Instead it seems that in these systems the intensity of stimulation of anti-hapten antibodies is correlated with the number of hapten determinants per molecule of secondary carrier (compare groups F, B, C, G and H, Table II). This may be an expression of the probability of interaction of

TABLE III

Effect of the number of carrier determinants per molecule of secondary hapten-carrier complex

Group	Primary day −28		Secondary day 0		Rabbits per group	Day 7 minus day 0 anti-sulf haemagglutination titre ($\log_2$ units)
A	sulf$_{31}$ BSA	1000 µg	sulf$_{31}$ BSA	1000 µg	5	2.8 ± 0.1[a]
B	sulf$_{31}$ BSA	1000 µg	sulf$_{52}$ HGG	1000 µg	5 ⎱	0.5 ± 0.1
C	sulf$_{31}$ BSA	100 µg	sulf$_{52}$ HGG	100 µg	2 ⎰	
D	sulf$_{52}$ BSA +HGG	100 µg 10 µg	sulf$_{52}$ HGG	100 µg	3	4.1 ± 0.5
E	as D		(sulf$_{208}$HGG)—(HGG)[b]	100 µg	4	3.7 ± 0.4
F	as D		(sulf$_{208}$HGG)—(sulf$_{208}$HGG)[b]	100 µg	4	1.7 ± 0.5

[a] Arithmetic mean plus standard deviation.
[b] Bisdiazotized benzidine was used as cross-linking agent for coupling.

hapten determinants presented by the helper cells with hapten-specific receptors.

c. The helper cells seem to recognize the carrier determinants by means of receptors, whose specificity is similar to the specificity of humoral antibodies (compare groups H and I, Table II and E and F, Table III).

ANTIBODY CLASSES AND HELPER CELLS

Hapten-specific 19S and 7S memory was analysed earlier using the present system (Schirrmacher and Rajewsky, 1970). The priming dose of sulf-BSA and the time interval between primary and secondary stimuli seemed to determine the class distribution of the secondary anti-hapten antibody. The specificity and class of the antibody produced by the hapten-specific memory cells (B cells) therefore seems to be predetermined. Secondary stimulation of 7S and 19S anti-hapten antibody with heterologous complexes is strictly dependent on presensitization with free secondary carrier. However neither the dose of HGG nor its timing relative to the time of secondary stimulation influences the class distribution of secondary anti-hapten antibody to any considerable extent. The helper cells seem to be responsible only for the extent of secondary stimulation.

According to the co-operation hypothesis of antibody induction the carrier-specific memory cells can be divided into different cell lines: (a) the helper cells (presumably T cells); and (b) the antibody forming cell precursors (B cells). As in the experiments of Mitchison (1968), helper activity developed very rapidly in the sulf-carrier system after priming with HGG in adjuvant. The helper activity was already maximum on the 10th day after HGG priming, whereas HGG antibody titres increased for at least four weeks. In the co-operative system there was no correlation between the class distribution of secondary anti-carrier and anti-sulf antibodies (Fig. 2). There was also no correlation between the strength of the secondary anti-hapten response and the amount of anti-carrier antibodies present on day 0. Definite co-operation was observed even when anti-carrier antibody was not detectable in the serum at the time of secondary stimulation (Fig. 3). Furthermore there was no correlation between the extent of secondary stimulation of anti-hapten and of anti-carrier antibodies (Fig. 4). This presumably indicates a great fluctuation in the ratio of anti-hapten to anti-carrier antibody forming cell precursors in individual animals and under different experimental conditions. The data are thus entirely compatible with the concept that in this system and in others helper cells and antibody forming cell precursors belong to different cell populations.

ANTIBODY AFFINITIES AND HELPER CELLS

Can heterologous complexes in the sulf-carrier system stimulate cell clones with high affinity to the hapten without helper cells? In a first approach to this

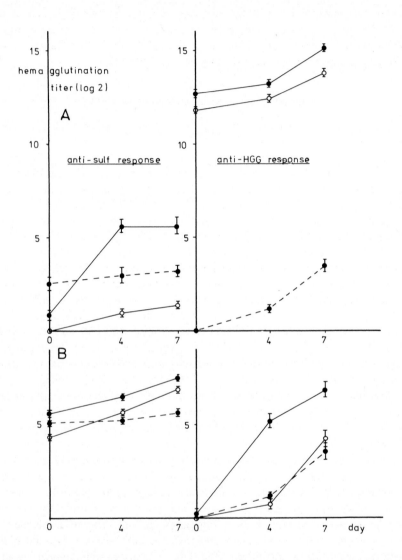

Fig. 2. Non-correlation between the class distributions of secondary anti-sulf and anti-HGG antibodies. Primary, 1000 μg sulf$_{19}$BSA on day −10 (A) or −28 (B) and 10 μg HGG on day −28 (A) or day −5 (B) in adjuvant (in the heterologous system − dashed lines − no HGG was given). Secondary, 1000 μg sulf$_{18}$HGG on day 0. Symbols and vertical bars represent arithmetic means and standard deviations of haemagglutinin titres (log$_2$) before (●) and after (○) reduction (Schirrmacher and Rajewsky, 1970). The small reductions in 7S titre (A right and B left) are unspecific as tested by sucrose gradient analysis.

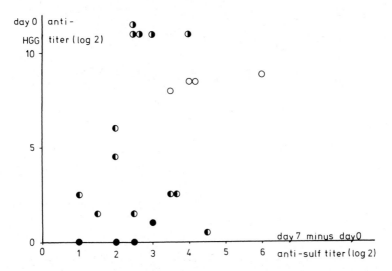

Fig. 3. Non-correlation between the secondary stimulation of anti-hapten antibodies in the co-operative system and the amount of anti-carrier antibody in day 0 sera. Passive haemagglutination titres. Symbols represent individual values from four different experiments:

Symbol	day −28	day −10	day −5	day 0 (secondary)
●	1000 µg sulf$_{19}$BSA		10 µg HGG	1000 µg sulf$_{18}$HGG
○		1000 µg sulf$_{19}$BSA	10 µg HGG	1000 µg sulf$_{18}$HGG
◐	1000 µg sulf$_{19}$BSA + 10 µg HGG			1000 µg sulf$_{18}$HGG
◑	100 µg sulf$_{19}$BSA + 10 µg HGG			10 µg sulf$_{52}$HGG

problem, the haemagglutinating activity of anti-sulf antibodies in sucrose gradient fractions was analysed by hapten inhibition (Fig. 5). Changes in sensitivity to hapten inhibition after secondary stimulation must be due to changes in relative affinity for the hapten. In the heterologous system the values for 50% inhibition before and after secondary stimulation were always identical. Thus relative affinity had not increased. However in the co-operative system affinity increases were detected in both antibody classes. In a second approach the average intrinsic association constants for 7S antibodies were determined by the hapten binding assay. This method does not detect 19S antibodies in the sera

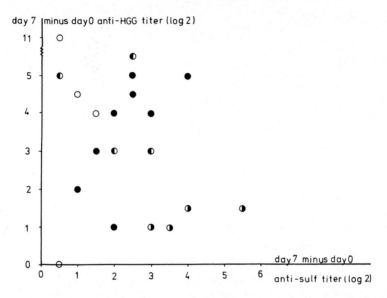

Fig. 4. Non-correlation between secondary stimulation of anti-hapten and of anti-carrier antibodies in the co-operative system. Passive haemagglutination titres of 7S sucrose gradient fractions. Symbols represent individual values from four different experiments:

Symbol	day −42	day −28	day 0 (secondary)
○		200 μg sulf$_{19}$BSA + 0.1 μg HGG	1000 μg sulf$_{18}$HGG
◐		200 μg sulf$_{19}$BSA + 1 μg HGG	1000 μg sulf$_{18}$HGG
◑		100 μg sulf$_{19}$BSA + 10 μg HGG	10 μg sulf$_{52}$HGG
●	30 μg sulf$_{19}$BSA + 0.1 μg HGG		10 μg sulf$_{18}$HGG

or the gradient fractions (Schirrmacher and Rajewsky, 1970). Figure 6 shows the Sips plot (Karush, 1962) of the antibody-hapten interaction for representative sera, Table IV gives the complete results. K_0 values of 10^6-10^7 M^{-1} were obtained. In the heterologous system there was no difference in the K_0 values before and after secondary injection, but the association constants increased

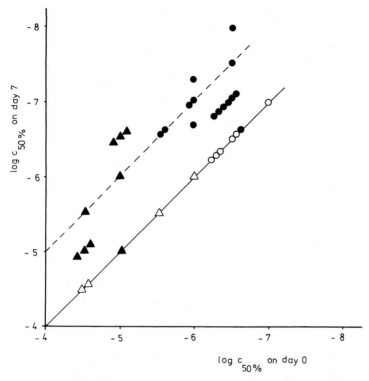

Fig. 5. Hapten inhibition of haemagglutinating activity in sucrose gradient fractions. Molar hapten concentrations at 50% inhibition (log $c_{50\%}$) of day 7 antibodies are plotted against those of day 0 anti-sulf antibodies. ▲ : ● represent individual values for 19S and 7S antibodies in the co-operative system and △ : ○ represent those in the heterologous system. —— indicates no increase and – – – – indicates a 10-fold increase in relative affinity.

significantly in the co-operative system. The results thus provide no evidence for a selective stimulation of high affinity cells in the absence of helper cells.

The differences in the antibody concentration and in the free energy change ($\Delta F°$) of the antibody-hapten interactions between day 7 and day 0 sera in the co-operative system are very similar to those obtained by Paul *et al.* (1967). They used DNP coupled to bovine gamma globulin. Under certain experimental conditions they found an impressive stimulation of high affinity antibodies in the *heterologous* system ($\Delta(\Delta F°)$ from 2 to 5 Kcal/mole). The differences between the sulf and DNP systems can possibly be explained by differences in the affinities of the cellular receptors for the hapten determinant. Low affinities of the cellular receptors for the hydrophilic sulf determinant are perhaps one of the reasons for the very pronounced carrier specificity in this system. This will be discussed more extensively elsewhere (Schirrmacher, 1970a).

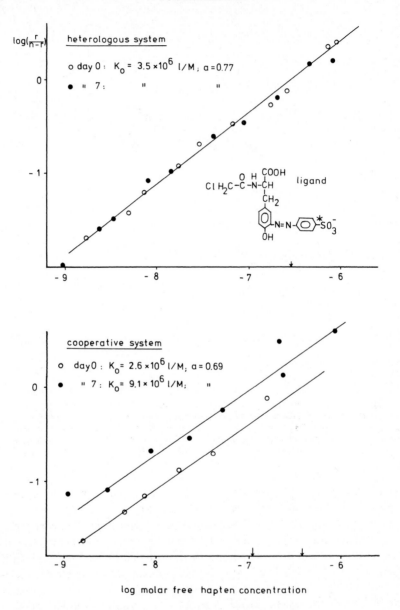

Fig. 6. Binding of anti-sulf antibodies with ^{35}S-labelled hapten at equilibrium, plotted according to the equation: log r/n = a logK$_0$ + a log c. Where r is the number of hapten molecules bound/molecule of antibody, n is the valency (assumed to be 2), a is the Sips heterogeneity index, K$_0$ is the average intrinsic association constant and c is free hapten concentration in moles/litre. Rabbits 348 and 336 (Table IV) are shown.

TABLE IV

Affinity increase of anti-hapten antibodies in the co-operative system

Rabbit no.	Day 0		Day 7			$\Delta(\Delta F^\circ)$[d]
	mg/ml[a]	$K_0 \times 10^6$ M^{-1} [b]	mg/ml[a]	$K_0 \times 10^6$ M^{-1} [b]	f$_{[Ab]}$[c]	
A. Co-operative system						
Primary 1000 μg sulf$_{19}$ BSA + 10 μg HGG day -28						
335	0.59	1.5	1.80	5.5	3.0	-0.79
336	0.56	2.6	2.02	9.1	3.6	-0.74
337	0.46	3.5	1.55	11.5	3.4	-0.72
Secondary 1000 μg sulf$_{18}$ HGG day 0						
338	0.37	3.6	1.24	11.5	3.3	-0.68
339	0.56	2.2	1.39	7.3	2.5	-0.71
S34	0.37	1.9	1.55	2.3	4.2	-0.13
S35	0.046	3.8	0.18	6.9	4.0	-0.34
				mean	3.4	-0.57
B. Heterologous system						
Primary 1000 μg sulf$_{19}$ BSA day -28						
345	0.65	3.8	1.21	5.2	1.9	-0.18
346	0.93	8.4	1.02	10.5	1.1	-0.12
347	0.56	2.9	1.21	2.9	2.2	0.00
Secondary 1000 μg sulf$_{18}$ HGG day 0						
348	0.65	3.5	1.39	3.5	2.1	0.00
349	0.42	2.3	0.93	2.3	2.2	0.00
S20	0.37	1.3	0.37	1.3	1.0	0.00
S21	0.21	6.0	0.18	11.5	1.5	-0.38
				mean	1.7	-0.10

[a] Calculated from hapten binding capacity at an excess of free hapten (extrapolated to $>10^{-6}$ M) (Schirrmacher, 1970b).
[b] Average intrinsic association constant, determined by the Farr technique.
[c] Factor for increase of antibody concentration $^d\Delta F^\circ$ of hapten interaction with day 7 antibody$-\Delta F^\circ$ of hapten interaction with day 0 antibody, in Kcal/mole. $\Delta F^\circ = -RT \ln K_0$.

ACKNOWLEDGEMENTS

I am grateful to Dr K. Rajewsky and Dr H. Seiler for helpful discussion and advice. This work was supported by the Deutsche Forschungsgemeinschaft and the Stiftung Volkswagenwerk. In partial fulfilment of doctoral thesis requirements, University of Cologne.

SUMMARY

Hapten- and carrier-specific memory was analysed in a system of co-operating antigenic determinants. The data confirm the co-operation hypothesis of antibody induction. The experiments demonstrate:

1. That the carrier determinants and the hapten density on the secondary hapten-carrier complex play a dominant role in determining the extent of the secondary anti-hapten response.

2. That the hapten-primed antibody forming cell precursors are committed as far as antibody class (19S/7S) is concerned, whereas the carrier-specific helper cells only influence the extent of the secondary response.

3. That there is no evidence in this system that antibody forming cell precursors with high affinity for the hapten can be stimulated without helper cells. Presensitization of the animals to carrier determinants can entirely account for the carrier specificity of the secondary anti-hapten response. This holds for all the parameters investigated: amounts, classes (19S/7S) and affinities of secondary anti-hapten antibodies.

REFERENCES

Farr, R. S. (1958). *J. infect. Dis. 103*, 239-262.
Karush, F. (1962). *Adv. Immun. 2*, 1-40.
Mitchison, N. A. (1968). *In* "Differentiation and Immunology" (K. B. Warren, ed.), Vol. 7, pp. 29-00. Academic Press, New York.
Mitchison, N. A., Rajewsky, K. and Taylor, R. B. (1969). *In* "Developmental Aspects of Antibody Formation and Structure" (J. Šterzl, ed.), pp. 547-561. Academic Press, New York.
Paul, W. E., Siskind, G. W., Benacerraf, B. and Ovary, Z. (1967). *J. Immun. 99*, 760-770.
Raff, M. C. (1970). *Nature, Lond. 224*, 378-379.
Rajewsky, K., Schirrmacher, V., Nase, S. and Jerne, N. K. (1969). *J. exp. Med. 129*, 1131-1143.
Schirrmacher, V. (1970a). Doctoral thesis, University of Cologne.
Schirrmacher, V. (1970b). Manuscript in preparation.
Schirrmacher, V. and Rajewsky, K. (1970). *J. exp. Med. 132*, 1019-1033.
Transplant. Rev. (1969). *1.*

Absence of Carrier Specificity in an Adoptive Anti-NIP Response in the Rat

SIRKKA KONTIAINEN

Department of Serology and Bacteriology
University of Helsinki, Helsinki, Finland

Hapten-specific plasma cell precursors in mice and rabbits seem to be stimulated when hapten determinants are presented to them by carrier-specific thymus-dependent cells (Mitchison, 1969; Rajewsky *et al.*, 1969). The help of carrier-specific cells is important, and usually poor secondary anti-hapten responses are obtained with hapten coupled to a heterologous carrier (Ovary and Benacerraf, 1963). Sometimes it has been possible to overcome carrier specificity. One way of doing it is to increase the time between the two antigen injections (Paul *et al.*, 1967; Rittenberg and Campbell, 1968), another to increase the antigen concentration (Brownstone *et al.*, 1966b).

I decided to study carrier specificity in an adoptive secondary response for two reasons. First, in adoptive circumstances the amount of preformed specific antibodies is greatly reduced. This should increase the effective antigen concentration available for plasma cell precursors. Also carrier specificity might be less in IgM than in IgG anti-hapten production. This was based on the assumptions that multivalent IgM receptors act more efficiently as their own concentrators than IgG receptors, and that reaction between IgM receptor and antigen leads to secretion of IgM antibodies. These assumptions were supported by the finding that the IgM anti-sheep red blood cell response is less thymus-dependent than the IgG response (Taylor *et al.*, 1967). In adoptive circumstances a secondary IgM response is easily obtained (Kontiainen and Mäkelä, 1968), and carrier specificity of IgM and IgG responses could be studied.

THE EXPERIMENTAL CONDITIONS

Inbred rats originating from the Sprague-Dawley stock were immunized with alum-precipitated NIP-conjugates (Brownstone *et al.*, 1966a) together with *H. pertussis* bacteria (Brownstone *et al.*, 1966b). Three weeks or four months later aliquots of pooled spleen cells were transferred to 760 r irradiated syngeneic recipients. The recipients were stimulated with either homologous or heterologous conjugates of NIP (fluid preparation).

Anti-NIP antibodies were measured by NIP-T2 phage inactivation (Mäkelä, 1966), and the contribution of IgM and IgG to the anti-hapten titre by the combined reduction-hapten-inhibition method (Kontiainen and Mäkelä, 1967).

RESULTS

Stimulation with either homologous or heterologous conjugate resulted in anti-NIP titres (expressed in the Tables as antilogs of $\log_{10}$ means) which were not significantly different. This occurred whether the second antigen injection was given four months (Table I) or three weeks (Table II) after the first antigen injection. The 11 day sera were also titrated using the Farr technique (Brownstone et al., 1966b). The titres ($\log_{10}$ means of ABC ± SE) when 1000 or 100 μg of the conjugate were used are included in Tables I and II.

TABLE I

Adoptive secondary responses by spleen cells primed four months earlier with NIP-HSA

Recipients stimulated with	Anti-NIP titres				Anti-NIP titres at day 11 using Farr assay
	Days after cell transfer and restimulation				
	6	8	11	13	
NIP-HSA, 1000 μg	2900	57,000	683,000	740,000	1.42 ± 0.07[c]
NIP-HSA, 100 μg	1200[b]	22,000	100,000	147,000	0.64 ± 0.17
NIP-HSA, 10 μg	390[b]	1,900[b]	6,000	5,100	ND
NIP-CG, 1000 μg	830[b]	14,000	163,000	330,000	0.85 ± 0.16
NIP-CG, 100 μg	620[b]	5,800	54,000	210,000	0.61 ± 0.12
NIP-CG, 10 μg	330[b]	840[b]	6,500	22,000	ND
—	420	840	1,200	1,700	ND
NIP-HSA[a], 100 μg	170	200	190	200	ND
NIP-CG[a], 100 μg	130	150	180	320	ND

[a] No cells given.

[b] Not significantly greater than corresponding controls.

[c] An average of 1.4×10^{-8} moles of ^{125}NIP-aminocaproic acid was bound by 1 ml of serum at the final free hapten concentration of ca. 10^{-8} M.

ND not done

c.o.v. of individual means varied between 1.08 and 1.96.

The dose of spleen cells was 23 million per recipient.

TABLE II

Adoptive secondary responses by spleen cells primed 21 days earlier with NIP-HSA

Recipients stimulated with	Anti-NIP titres				Anti-NIP titres at day 11 using Farr assay
	Days after cell transfer and restimulation				
	6	8	11	15	
NIP-HSA, 1000 μg	8,800	41,000	109,000	115,000	0.41 ± 0.14
NIP-HSA, 100 μg	11,600	47,000	76,000	75,000	0.35 ± 0.17
NIP-CG, 1000 μg	2,000[a]	13,600	78,000	82,000	0.19 ± 0.14
NIP-CG, 100 μg	800[a]	2,900[a]	27,000	41,000	0.66 ± 0.08
—	580	630	960	990	ND

[a] Not significantly greater than corresponding controls.
ND not done.
c.o.v. of individual means varies between 1.10 and 1.62.
The dose of spleen cells was 100 million per recipient.

When the contribution of IgM and IgG to the anti-NIP titres were compared in groups having equally high total titres, using NIP-T2 phage inactivation, a marked difference was seen (Tables III and IV). Heterologous conjugate induced a response more rich in IgM than homologous conjugate. Thus 100 μg of homologous NIP-HSA and 1000 μg of heterologous NIP-CG induced anti-NIP titres of 76,000 and 78,000 at day 11 (Table II). IgM made up 8% of the total in the NIP-HSA stimulated group and 73% of the total in the NIP-CG stimulated group (Table IV). The lowest dose of either homologous or heterologous conjugate failed to induce an IgM anti-hapten response (Table III).

The factors that control an anti-hapten response might include (1) hapten-specific antibodies, (2) hapten-specific cells, (3) carrier-specific antibodies, and (4) carrier-specific cells. The results in this paper show that when the role of (1), (3) and (4) was reduced a good stimulation of hapten-specific cells could be obtained whether homologous or heterologous conjugate was used. This is in contradiction to the results obtained with these antigens in a normal secondary response (Kontiainen and Mäkelä, 1970). In addition, IgM receptors were less dependent on the concentration help of carrier-specific cells than IgG receptors.

ACKNOWLEDGEMENTS

These experiments were supported by grants from the Yrjö Jahnsson and Paulo Foundations. The skilful technical assistance of Mrs Aila Åkerlund is gratefully acknowledged.

TABLE III

Adoptive secondary IgM and IgG responses of spleen cells primed four months earlier with NIP-HSA

Recipients stimulated with	Anti-NIP titres				IgM at day 11 (% of total titre)
	Days after cell transfer and restimulation				
	11		13		
	IgM	IgG	IgM	IgG	
NIP-HSA, 1000 µg	35,000	643,000	24,000	710,000	5
NIP-HSA, 100 µg	4,600	93,000	4,400	143,000	5
NIP-HSA, 10 µg	1,400[b]	4,100	700[b]	4,100	—
NIP-CG, 1000 µg	34,000	110,000	51,000	260,000	24
NIP-CG, 100 µg	23,000	29,000	35,000	170,000	44
NIP-CG, 10 µg	800[b]	5,500	1,100[b]	20,000	—
—	580	540	700	900	—
NIP-HSA[a] 100 µg	170	20	190	10	—
NIP-CG[a] 100 µg	160	20	300	20	—

[a] No cells given.
[b] Not significantly greater than the corresponding controls, c.o.v. of individual IgM means varied between 1.12 and 1.62 and IgG means between 1.05 and 2.47.
The dose of spleen cells was 23 million per recipient.

TABLE IV

Adoptive secondary IgM and IgG responses of spleen cells primed 21 days earlier with NIP-HSA

Recipients stimulated with	Anti-NIP titres				IgM at day 11 (% of total titre)
	Days after cell transfer and restimulation				
	11		15		
	IgM	IgG	IgM	IgG	
NIP-HSA, 1000 µg	19,000	86,000	6,500	107,000	18
NIP-HSA, 100 µg	6,000	70,000	3,800	72,000	8
NIP-CG, 1000 µg	54,000	20,000	26,000	53,000	73
NIP-CG, 100 µg	13,000	11,000	11,000	28,000	54
—	700	960	250	990	—

c.o.v. of individual IgM means varied between 1.15 and 1.78, and IgG means between 1.13 and 1.71.
The dose of spleen cells was 100 million per recipient.

REFERENCES

Brownstone, A., Mitchison, N. A. and Pitt-Rivers, R. (1966a). *Immunology 10*, 465-479.

Brownstone, A., Mitchison, N. A. and Pitt-Rivers, R. (1966b). *Immunology 10*, 481-492.

Kontiainen, S. and Mäkelä, O. (1967). *Ann. Med. exp. Fenn. 45*, 472-476.

Kontiainen, S. and Mäkelä, O. (1968). *Int. Arch. Allergy 34*, 417-427.

Kontiainen, S. and Mäkelä, O. (1970). *Immunology 20*, 101-108.

Mäkelä, O. (1966). *Immunology 10*, 81-86.

Mitchison, N. A. (1969). *In* "Immunological Tolerance" (M. Landy and W. Brown, eds), pp. 149-151. Academic Press, New York.

Ovary, Z. and Benacerraf, B. (1963). *Proc. Soc. exp. Biol. Med. 114*, 72-76.

Paul, W. E., Siskind, G. W., Benacerraf, B. and Ovary, Z. (1967). *J. Immunol. 99*, 760-770.

Rajewsky, K., Schirrmacher, V., Nase, S. and Jerne, N. K. (1969). *J. exp. Med. 129*, 1131-1143.

Rittenberg, M. B. and Campbell, D. H. (1968). *J. exp. Med. 127*, 717-730.

Taylor, R. B., Wortis, H. H. and Dresser, D. W. (1967). *In* "The Lymphocyte in Immunology and Haemopoiesis" (J. M. Yoffey, ed.), pp. 242-244. Edward Arnold, London.

Interaction of Bone Marrow-derived Cells and Thymus-dependent Cells in the Immune Response Against Erythrocytes *in vitro*

KLAUS-ULRICH HARTMANN

Max-Plank-Institut für Virusforschung
Tübingen, West Germany

I do not intend to discuss all the experimental evidence suggesting that cells of different types co-operate during the induction of a humoral immune response. A number of excellent reviews have been published recently (*Transplant. Rev.,* 1969), and further evidence has appeared at this symposium.

I have tried to learn more about the interaction between thymus-dependent cells and bone marrow-derived cells using the *in vitro* culture system introduced by Mishell and Dutton (1967). In this system suspended spleen cells are incubated for four days together with antigen (erythrocytes). Then the number of 19S haemolysin producing cells—direct plaque-forming cells (PFC)—is counted in the Jerne assay.

Thymus and bone marrow cells could not be kept in culture under these conditions. Therefore these cells were transferred first to irradiated animals and sometime later the spleens were used as a source of thymus-dependent or bone marrow-derived cells (Claman *et al.,* 1966; Mitchell and Miller, 1968a). In detail: mice were irradiated with 700-800 r and injected intravenously with 5×10^7 syngeneic thymus cells; eight days later a suspension was prepared from their spleens. This T cell suspension contains lymphoid cells that are probably descendents of the injected thymus cells, and certainly also some radiation-resistant host spleen cells. In most experiments 10^7 sheep (SRBC) or horse (HRBC) erythrocytes were injected together with the thymus cells. Because the T cells had been in contact with the erythrocytes they are called "educated T cells" in this report, SRBC-educated T cells etc. Little is known about this "education": it might be selection of antigen sensitive cells in the presence of the immunogen. It might lead to differentiation or multiplication of the antigen sensitive cells under the influence of the immunogen.

The B cells were prepared from the spleens of animals that had been thymectomized, lethally irradiated (850 r) and injected with 3×10^7 bone marrow cells 3-4 weeks earlier. Again it should be made clear that these suspensions contain various radiation-resistant host cells as well as descendents of the injected bone marrow cells.

None or only very few PFC could be detected after four days' incubation of either B or T cells alone, or a mixture of B and T cells, together with SRBC (Table I). But if B cells were kept four days in culture together with SRBC-educated T cells and SRBC then a great number of PFC developed (Table I).

TABLE I

Co-operation between thymus-dependent and bone marrow-derived spleen cells in vitro

Cell suspensions		PFC per 10^6 cells
B cells	T cells	
8×10^6	—	0
—	11×10^6 (I)	0
4×10^6	5×10^6 (I)	10
—	10×10^6 (II)	32
4×10^6	5×10^6 (II)	265

B cells: mice thymectomized, irradiated (850 r) and injected with bone marrow cells 24 days earlier.
T cells: mice irradiated (700 r), injected with 5×10^7 thymus cells (group I) or with thymus cells and 10^7 SRBC (group II).
Spleen cell suspensions were cultivated with 3×10^6 SRBC, PFC assayed four days later.

The education of the T cells was quite specific. SRBC-educated T cells together with B cells and SRBC gave rise to anti-SRBC PFC, but SRBC-educated T cells together with B cells and HRBC did not produce anti-HRBC PFC. HRBC-educated T cells gave comparable results (Table II).

The PFC were descendents of the B cell population (Table III): B cells and T cells of different strains were cultivated together and the harvested cells were treated with anti-H-2 serum before the PFC assay (Mitchell and Miller, 1968b; Mishell *et al.*, 1970). Antiserum directed against the B cell population inhibited the PFC. It might be noted in Table III that more PFC could be detected in DBA cultures than in C57Bl cultures. This probably depends on the T cells. DBA T cells with C57Bl B cells produced *more* PFC than C57Bl T cells with C57Bl B cells, whereas C57Bl T cells with DBA B cells produced *less* PFC than DBA T cells with DBA B cells. So far these results agree with results of *in vivo* experiments on thymus/bone marrow interactions.

Continuing these experiments cultures of B cells and SRBC-educated T cells were prepared. Addition of HRBC did not stimulate the formation of anti-HRBC

TABLE II

Co-operation between sensitized thymus cells and bone marrow-derived cells in vitro

| Spleen cell suspension | | RBC in culture | PFC/10^6 cells assayed with: | |
B cells	T cells	(3×10^6)	SRBC	HRBC
1.1×10^7	–	SRBC	7	0
1.1×10^7	–	HRBC	0	0
–	3.2×10^6 (I)	SRBC + HRBC	0	0
6×10^6	1.6×10^6 (I)	SRBC	420	4
6×10^6	1.6×10^6 (I)	HRBC	0	20
–	4×10^6 (II)	SRBC + HRBC	0	0
6×10^6	2×10^6 (II)	SRBC	20	0
6×10^6	2×10^6 (II)	HRBC	5	475

B cells: mice thymectomized, irradiated (850 r) and injected with bone marrow cells 17 days earlier.

T cells: mice irradiated (650 r) and injected with 5×10^7 thymus cells and 10^7 SRBC (group I) or thymus cells and 10^7 HRBC (group II) seven days earlier.

TABLE III

Inhibition of PFC with anti-H-2 sera in cultures of bone marrow-derived and thymus-dependent spleen cells from different strains of mice

| Cell suspension | | PFC per culture on day 5, assayed after treatment with: | | |
B cells	SRBC-educated T cells	NMS	Anti-H-2^b	Anti-H-2^d
C57Bl (15×10^6)	–	45	0	30
–	C57Bl (8×10^6)	10	0	10
C57Bl (8×10^6)	C57Bl (4×10^6)	110	0	145
C57Bl (8×10^6)	DBA (2×10^6)	515	75	380
DBA (9×10^6)	–	40	30	0
–	DBA (4×10^6)	20	30	0
DBA (5×10^6)	DBA (2×10^6)	840	700	10
DBA (5×10^6)	C57Bl (4×10^6)	260	210	20
Normal spleen cells:				
C57Bl spleen (17×10^6)		800	105	900
DBA spleen (15×10^6)		2500	2500	230

C57Bl/6 H-2^b DBA/2 H-2^d cells were cultured with 3×10^6 SRBC for five days then harvested, centrifuged and resuspended in the original volume. 0.2 ml cell suspension was incubated with 25 μl normal mouse serum (NMS) or anti-H-2 serum for 30 min at $0°$C. After washing in BSS (5 ml), 25 μl guinea pig serum was added, and the mixture incubated for 15 min at $35°$C. After washing in 5 ml ice cold BSS PFC were assayed.

B cells: mice thymectomized, irradiated (850 r) and injected with 3×10^7 syngeneic bone marrow cells.

T cells: mice irradiated (750 r) and injected with 5×10^7 thymus cells and SRBC.

PFC. But if HRBC were added together with SRBC then both anti-SRBC and anti-HRBC PFC could be detected (Tables IV and V). (It should be mentioned that there is no cross reaction between PFC against SRBC and HRBC.) Similarly if B cells were cultivated with HRBC-educated T cells, anti-SRBC PFC could be detected after simultaneous stimulation with SRBC and HRBC (Table VI).

The T cells are antigen-specific—they might be selectively enriched or primed—and only in the presence of the proper antigen can clones of PFC develop from the B precursor cells. Still we do not know what the T cells are adding during the triggering and the proliferation of the precursor cells. The most favoured suggestion would be that the antigen sensitive T cells produce a

TABLE IV

Co-operation between sensitized thymus cells and bone marrow-derived cells in vitro

B cells	SRBC-educated T cells	RBC in culture	PFC/10^6 cells at day 4 against: SRBC	HRBC
10×10^6	–	SRBC + HRBC	22	8
–	7×10^6	SRBC + HRBC	49	0
5×10^6	3.5×10^6	SRBC	450	21
5×10^6	3.5×10^6	HRBC	14	29
5×10^6	3.5×10^6	SRBC + HRBC	1010	250

T cells: F_1 mice irradiated (600 r), injected with 5×10^7 thymus cells and 10^7 SRBC nine days earlier.

TABLE V

Co-operation between sensitized thymus cells and bone marrow-derived cells in vitro

Spleen cell suspensions B cells	SRBC-educated T cells	RBC (3×10^6) in culture	PFC/10^6 cells on day 5 assayed against: SRBC	HRBC
15×10^6	–	SRBC + HRBC	31	0
–	1.1×10^6	SRBC + HRBC	0	0
7×10^6	0.6×10^6	SRBC	1820	0
7×10^6	0.6×10^6	HRBC	0	10
7×10^6	0.6×10^6	SRBC + HRBC	980	140

B cells: mice thymectomized, irradiated (800 r) and injected with 3×10^7 bone marrow cells three weeks earlier.
T cells: mice irradiated (850 r) and injected with 5×10^7 thymus cells, 10^7 SRBC and one day later with 10^9 killed *B. pertussis*. Spleen cells taken nine days later.

TABLE VI

Co-operation between sensitized thymus cells and bone marrow-derived cells in vitro

| B cells | Spleen cell suspensions | | Stimulation with SRBC + HRBC PFC/10^6 cells at day 4 against: | |
	SRBC-educated T cells	HRBC-educated T cells	SRBC	HRBC
6×10^6	2.5×10^6	–	136	9
6×10^6	–	3×10^6	528	273
1.2×10^7	–	–	26	22
–	5×10^6	–	0	0
–	–	6×10^6	10	0

B cells: mice thymectomized, irradiated (500 r), injected with 3×10^7 bone marrow cells 21 days earlier.

T cells: mice irradiated (600 r) and injected with 4×10^7 thymus cells and 10^7 SRBC or HRBC eight days earlier.

factor in the presence of antigen. This helps the B precursor cells (in the presence of their antigen) to differentiate and form clones of PFC. Even if this is true we still do not know whether physical contact between T and B cells is necessary in addition to the factor, whether the factor acts on the B cells directly or on other cells (? macrophages), and whether the factor is necessary during triggering of the precursor cells or later to push clone formation.

In addition it should be remembered that all my experiments involve erythrocytes as antigen—therefore special antigen focusing might not play an essential role, that the results concerned *in vitro* cultures—effects of compartmentation, follicle formation and microenvironments might not be visible, and that only 19S PFC were measured. I would not like to generalize from these experiments.

With antigens other than erythrocytes it should be possible to learn more about the role of the thymus-dependent cells and—more interesting—about the activation, stimulation to antibody secretion and to multiplication of the resting precursor cells.

ACKNOWLEDGEMENTS

This work was supported by the Deutsche Forschungsgemeinschaft grant Ha 569/3. I would like to thank Mrs Reeg, Miss Mehner and Mrs Zappe for their skilful technical assistance and co-operation and Dr G. F. Mitchell for demonstrating the thymectomy technique. Part of this work has been reported at the symposium "Genetics of the Antibody Response" (1969) in Brügge and at the symposium (1969) in Kronberg (*Behringwerk-Mitt. 49*, 208).

SUMMARY

Interaction between bone marrow-derived lymphoid cells (B cells) and thymus-dependent lymphoid cells (T cells) during the immune response (formation of haemolysin-producing cells) *in vitro* is shown. The precursor cells of the PFC were present in the B cell suspension. The helper function of the T cells was specific : only SRBC-educated T cells could interact with B cells and SRBC to form anti-SRBC PFC. Anti-HRBC PFC resulted from incubation of B cells, HRBC-educated T cells and HRBC. When SRBC plus HRBC were present both anti-SRBC PFC and anti-HRBC PFC developed in cultures of B cells and SRBC-educated T cells.

REFERENCES

Claman, H. N., Chaperon, E. A. and Triplett, R. F. (1966). *J. Immun. 97*, 828-832.
Mishell, R. I. and Dutton, R. W. (1967). *J. exp. Med. 126*, 423-442.
Mishell, R. I., Dutton, R. W. and Raidt, D. J. (1970). *Cellular Immunology 1*, 175-181.
Mitchell, G. F. and Miller, J. F. A. P. (1968a). *Proc. natn. Acad. Sci. U.S.A. 59*, 296-303.
Mitchell, G. F. and Miller, J. F. A. P. (1968b). *J. exp. Med. 128*, 821-837.
Transplant. Rev. (1969). *1*.

In vitro Response of Partially Purified Spleen Cells

J. S. HASKILL and J. MARBROOK

Department of Pathology, Queen's University
Kingston, Ontario, Canada

The co-operation of two or more cell populations in the *in vitro* response to sheep erythrocytes has been amply confirmed by a number of workers (Haskill *et al.*, 1970; Mishell *et al.*, 1970; Mosier and Coppleson, 1968; Osoba, 1970; Shortman *et al.*, 1970). We have employed partially purified populations of spleen cells derived by equilibrium density gradient centrifugation methods (Haskill *et al.*, 1970). This method has permitted us to characterize the *in vitro* response to sheep cells in terms of three apparently different cell populations. Mishell *et al.* (1970) have obtained similar results using a BSA centrifugation method. On the other hand, Mosier and Coppleson (1968) have suggested the involvement of three cell types in the immune response based upon results with separation of adherent and non-adherent cell populations. More recently, Shortman *et al.* (1970) have used not only the equilibrium density gradient centrifugation method, but also adherence and filtration through glass bead columns to partially purify spleen cell populations. They have compared the *in vitro* response to sheep erythrocytes and to the Salmonella antigens, and have concluded that the *in vitro* response appears to depend upon different cell populations for these two different antigens. In few of these studies, however, has there been a direct opportunity to measure the synergism observed in terms of immunological phenomena. That is, in most of the experiments described it is conceivable that the enhanced immune response observed *in vitro* is, at least in part, a function of improved tissue culture conditions. It is the purpose of this report to describe a new method which permits us to measure the numbers of progenitors of plaque forming cells and of focus inducing cells, and to measure the average immunological burst size or number of antibody forming cells per focus. We have employed this method to extend our studies on the synergistic phenomena observed *in vitro,* and have been able to draw some positive conclusions as to the nature of the interactions.

It would be advantageous to summarize briefly some of the results obtained using the partially purified spleen cell fractions (Haskill *et al.*, 1970). The results of a typical experiment comparing the ability of fractionated mouse spleen cells to respond to sheep erythrocytes when assayed either by the adoptive immunity assay *in vivo* or by the Marbrook liquid culture *in vitro* test (Syeklocha *et al.*, 1966; Marbrook, 1967) are given in Fig. 1. It is clear that there is a region of the spleen cell gradient where the *in vitro* response is quite marked. There is also a denser region of the gradient (density 1.070) where there is little, if any,

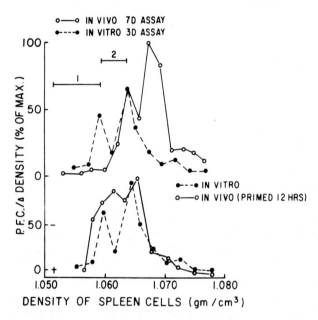

Fig. 1. Density distribution profiles of the immunologically competent cell populations in spleens of C57-CBA mice. Responses were measured either by the adoptive transfer method or by the Marbrook liquid culture system.

immune response to sheep erythrocytes *in vitro,* and yet there is the major portion of the *in vivo* response. The density position of the sheep erythrocyte *in vitro* response is similar to results observed in studying the populations of antigen sensitive cells present in the rat 10 hr after antigenic stimulation *in vivo* (Haskill, 1969). A comparison of the *in vitro* profile from normal animals with the *in vivo* profile 10 hr after an injection of antigen indicated that the two profiles were virtually identical. This suggested that a major portion of the *in vitro* response could be the result of a primed population of cells, or at least a population of cells which had already been put into cell cycle (Haskill, 1969). Furthermore, it was observed that the dose response relationship of this density fraction (fr (2)) was apparently linear (Haskill *et al.*, 1970). This was not characteristic of a response dependent on three cell populations for immune competence.

The region of the spleen gradient which gave such a poor response *in vitro* could, however, respond *in vitro* if it were mixed with the appropriate fraction of spleen cells. The light density end of the gradient (fr (1)) contains cells which permitted the denser region of the spleen gradient to give a good *in vitro* response (Fig. 2) (Haskill *et al.*, 1970).

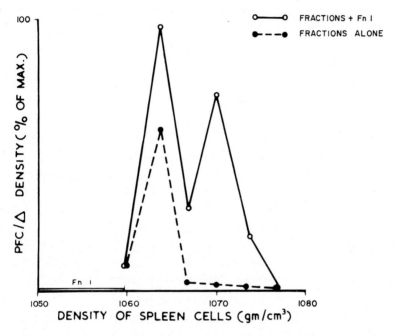

Fig. 2. *In vitro* response of fractionated mouse spleen cells cultured either alone or with added fr (1) cells. For conditions see Haskill *et al.* (1970).

It has been of interest to analyse more closely the *in vitro* response of fr (2), and to attempt to answer the question of whether or not it represents the immune response of a single class of cells, or of two or more cell populations.

Use has been made of the method of Miller and Phillips (1969), which permits the separation of cells on the basis of their sedimentation velocity in a nonlinear gradient of fetal calf serum at unit gravity. By doing this, it has been possible to achieve a two dimensional cell separation, which permits segregation of two populations differing from each other by either their density or their volume. After velocity separation of fr (2) and *in vitro* culture of each fraction it was found that all of the immune response was recovered in a single symmetrical peak of activity which contained less than 2% of the total phagocytes for sheep erythrocytes (Fig. 3). The data still suggested that more than one cell type need not be involved in this response. Also, although there was a significant depletion in the phagocytic population, there still had not been a loss in activity. Utilizing the prior experience that the light density fr (1) cells frequently were capable of

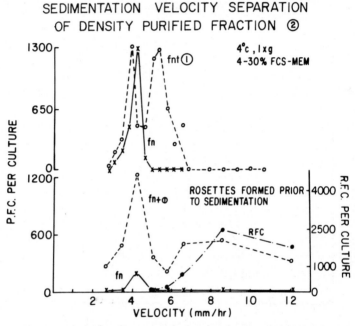

Fig. 3. Sedimentation velocity separation of fr (2). In the top diagram fractions were cultured alone or with 2×10^6 fr (1) cells. 2×10^7 SRBC were added per culture. For sedimentation method see Miller and Phillips (1969). In the bottom diagram the conditions were similar except that 5×10^8 SRBC were mixed and centrifuged to a pellet with the fr (2) cells prior to sedimentation. Approximately 50×10^6 fr (2) cells were used in each experiment.

acting synergistically, the previous experiment was repeated culturing the velocity fractions either in the absence, or the presence, of 2×10^6 cells from fr (1). There were two peaks of activity following culture in the presence of fr (1). This suggested that the *in vitro* response of fr (2) was in fact the result of the overlap of two populations of cells. Osoba (1970) has recently indicated that one of the cells responsible for *in vitro* immunity is capable of directly recognizing antigen; that is, is a rosette forming cell (RFC). The above experiment was repeated following prior rosette formation. Rosettes were formed to sheep erythrocytes with fr (2) and then the sedimentation separation was carried out and the cells cultured either in the presence or absence of fr (1). Two things should be observed. Firstly, the overall recovery of activity of plaque forming cells following rosette formation was very small; it was less than 20%. Secondly, when the fractions were cultured with fr (1) rosette formation did not affect the lower velocity peak, but the second peak at about 5 mm/hr advanced into the high velocity region of the gradient characteristic of rosette forming cells. It was therefore concluded that one of the populations involved in this immune response is a rosette forming cell. We were then interested in what kind of a synergism was occurring between fr (1) and our rosette population. It should be noted that this enriched population of rosette forming cells is not, however, a pure population and it contains a very high concentration of phagocytic cells (98% of the total). So the enriched rosette population contains both phagocytes and rosette forming cells but is incapable of a significant immune response. However, when fr (1), or the residual population of fr (2) cells depleted of RFC (which later is defined as the focus inducing cell (FIC)) was separated from the RFC region by a Nucleopore membrane (0.4 μ) (Table I) it was possible to induce an immune response in the rosette population after three days of culture. Therefore, the rosette population contains the antibody forming cell progenitors, and they can be induced into an immune response as a result of some diffusable factor produced by the FIC fraction. It was therefore concluded that the immune response of fr (2) is apparently a result of the interaction of two cell types, one of which is a rosette forming cell and one of which appears to be capable of secreting some substance that can assist in the immune response of the rosette population.

TABLE I

Effect of mixing RFC fractions with FIC

PFC/culture	(±30%)
FIC	3
RFC	1.3
RFC 220 ‖	FIC 54
RFC + FIC 100	

RFC fractions (velocity 7-14 mm/hr) mixed with the 4 mm/hr region of fr (2) depleted of RFC. This latter population has been defined as the focus inducing cells (FIC). Fractions were separated by a nucleopore membrane (0.4 μ). The RFC population were maintained 2 mm above the FIC population. Cultures were continued for three days.

At this stage it was necessary to develop a means of quantitating both the number of "clones" of antibody forming cells and the average number of antibody forming cells occurring in each "clone". Marbrook (1970) has developed a means of culturing spleen cells which consists of 49 limiting dilution assays combined into one culture under conditions of high cell density. One to 10×10^6 spleen cells with antigen are cultured in a moulded acrylamide dish. The cells settle down into discrete dimples and each of the 49 individual dimples is assayed at the end of the experiment for antibody forming cells. This permits the enumeration of both the number of positive areas or foci of antibody formation and the average number of antibody forming cells per focus (burst size) (Fig. 4). Two additional features of this system should be noted. First, in

0	0	0	4	0	3	4
0	2	5	0	0	5	0
4	0	3	4	0	0	0
0	3	1	0	0	15	14
31	16	0	0	0	0	15
11	10	0	0	0	0	0
0	0	2	5	0	5	

6	0	0	0	0	0	0
0	0	7	0	0	3	0
0	0	0	0	0	0	0
0	4	0	0	0	13	0
0	0	0	0	4	0	0
0	0	0	0	0	0	24
0	14	2	0	0	0	

TOTAL PFC   162

21 FOCI

PFC PER FOCUS 7.7

TOTAL PFC   77

9 FOCI

PFC PER FOCUS 8.5

Fig. 4. Examples of the spatial distribution of antibody forming cells (PFC) in each of 48 dimples in an acrylamide culture of fr (2) cells. The average number of PFC per focus and the number of foci can be calculated. A table of coincidence values for increasing numbers of positive foci is used to correct for overlapping foci.

each dimple the lymphoid cells apparently form themselves into localized aggregates or lymphoid areas, which are quite characteristic of the system and which can easily be quantitated (20 per dimple cultured at a concentration of 4×10^6 cells). Secondly, the plaque morphology in a positive dimple usually is very homogeneous. Perhaps even more important, is the fact that individual dimples, even though they are neighbours, frequently show plaques not only of identical morphology within themselves, but different from each other. This further emphasizes the clonal nature of this response.

The acrylamide method of Marbrook has been used to investigate the immune response of fr (2) cells. 1-4 x 10^6 cells of fr (2) were cultured with either

2×10^8 or 2×10^7 sheep erythrocytes, and the number of foci per culture and the average number of antibody forming cells per culture were measured (Fig. 5). Once again, the number of foci per culture was directly proportional to the number of cells cultured (Haskill *et al.*, 1970). Interestingly enough, the average number of antibody forming cells per focus at day 3 was not significantly changed by altering the number of cells per culture from 1×10^6 to 4×10^6.

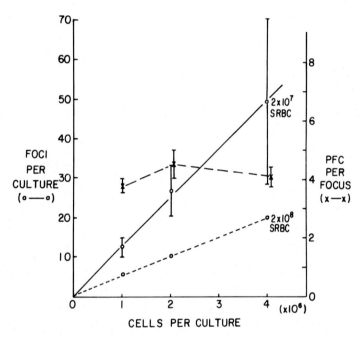

Fig. 5. Three day dose response data for fr (2) cells cultured by the acrylamide method. Standard deviations for the results of 2×10^7 SRBC are given (four cultures). The average of triplicate cultures is given at 2×10^8 SRBC per culture.

The kinetics of the immune response of either 8×10^6 unfractionated spleen cells, or 2×10^6 or 4×10^6 fr (2) cells were investigated (Fig. 6). The concentration of fr (2) cells affected the time at which the peak immune response was observed. With 4×10^6 fr (2) cells the maximum number of foci per culture actually appeared by day 2. The number rose from a very low value of 10 at 24 hr, reached a maximum of about 80 in the next 24 hr, and fell off again within the next 24 hr. This suggested that these antibody forming cell foci, or clones, are exerting their influence on the immune response for only a very short period of time. The average life span of these antibody forming cells may

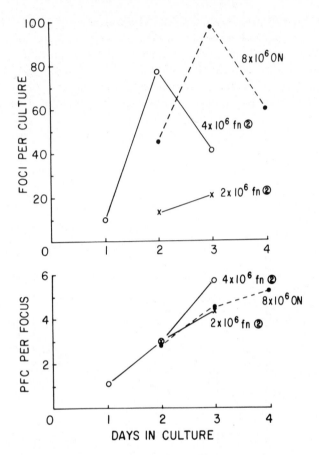

Fig. 6. Kinetics of the immune response of unfractionated spleen cells (ON) and fr (2) cells. Each point is an average of 3-4 cultures.

well be no more than 12 hr. This is in contrast to the generally accepted idea of exponentially expanding numbers of antibody forming cells (Jerne *et al.*, 1963; Dutton and Mishell, 1967). This early response is not solely a function of the number of foci per culture, for with 8×10^6 unfractionated spleen cells, the peak number of foci per culture was delayed till day 3. Throughout this period of time, regardless of the number of cells cultured, the average number of antibody forming cells per culture continually increased. It would therefore appear that the *in vitro* immune response of fr (2) is a function of continually appearing clones of antibody forming cells, rather than the more orthodox view of exponentially expanding clones.

It has been noticed in Fig 5 that the antigen dose apparently had some effect on the number of foci observed. We therefore did a more extensive experiment where we cultured fr (2) cells at 4×10^6 cells per culture with either no sheep cells added to the culture, or with increasing doses of antigen from 2×10^5 to 2×10^8 sheep cells per culture (Fig. 7). The number of foci per culture was directly affected by the antigen concentration, whereas there was no significant difference in the average number of antibody forming cells per focus. The fact

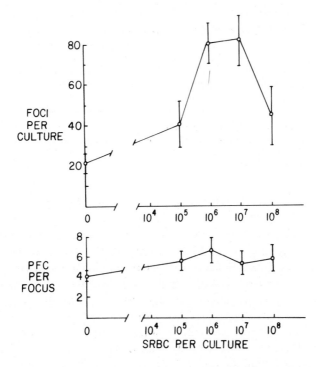

Fig. 7. Effect of SRBC concentration on the number of foci and burst size of fr (2) cells assayed at day 3. Standard deviations of four cultures per point are given.

that, even in the absence of directly added sheep cells, the fetal calf serum, itself, could stimulate clones of normal burst size is rather interesting. This suggests that the focus inducing cell population is sensitive to the level of antigen in the culture. Depletion of the RFC population of fr (2) results in an overall decrease in the immune response (Fig. 3). It was of interest to know if this decreased response in the 4 mm/hr velocity region was a result of simply fewer foci per culture or if, in fact, the average number of antibody forming cells per focus was also decreased, or both. The 4 mm/hr velocity region of fr (2) cells following rosette formation was isolated. The average number of foci per culture per

4×10^6 cells, both before and after rosette formation, was measured (Table II). In contrast to the effect of antigen concentration on the immune response, both a decrease in the number of foci per culture and a decrease in the average number of antibody forming cells per focus were observed. This was in contrast to the results obtained by varying the antigen concentration.

TABLE II

The response of fr (2) depleted of RFC's (2D assay)

	+RFC	−RFC
Foci (per 4×10^6 cells)	20.3 ± 7	6.0 ± 3.6
RFC/focus	3.5 ± 0.1	2.1 ± 0.1

Comparison of the number of foci per 4×10^6 cells and the burst size of normal fr (2) cells (velocity 3-6 mm/hr) and the same population depleted of RFC. Refer to Fig. 3. Four cultures each were assayed at day 2.

The data suggested that the antigen, the time of culture and the concentration of cells could all have an effect on the number of responding units or foci per culture and on the immunological burst size. This therefore seemed to be an excellent opportunity to go back to the results shown in Fig. 3 which demonstrated the interaction of two cell types, and to see if there was any difference in the properties of the immune response, that is the number of foci or the burst size, as a function of the region of overlap of the two cell populations involved. The average number of antibody forming cells per focus and the number of foci per culture were measured as a function of the sedimentation velocity of fr (2) cells (Fig. 8). In the low velocity end of the gradient where the focus inducing cell is in the highest concentration there was a low number of antibody forming cells per focus. At the region of the optimal immune response, however, there was a maximum value of the burst size, and in the region of the higher number of rosette forming cells, we observed a slightly lower value of antibody forming cells per focus.

DISCUSSION

These data suggest that two cell populations are involved in the immune response of fr (2), and that one of these is a rosette forming cell. The other cell type, defined as a focus inducing cell, is capable of initiating the immune response by means of a diffusable substance. To summarize the data presented in this paper, the following simple model for the immune response of the fr (2)

SEDIMENTATION VELOCITY SEPARATION OF
DENSITY PURIFIED FRACTION ②

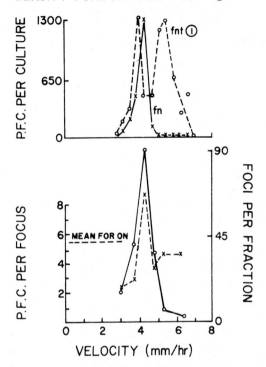

Fig. 8. Influence of relative proportions of the FIC and RFC populations on the immunological burst size (lower diagram). Fr (2) cells were sedimented as usual, and the number of cells per culture was kept within $1\text{-}4 \times 10^6$. The value of PFC per focus for unfractionated fr (2) cells (ON) is also included. ---x--- PFC/focus ○——○ foci/fraction. The upper diagram is the upper part of Fig. 3 included for reference.

cells in the acrylamide system is given in Fig. 9. Each square represents an individual dimple in the acrylamide culture. If, for instance, 2×10^6 fr (2) cells are cultured, each dimple on the average contains eight rosette forming cells. Based upon the sedimentation velocity and density of these cells, it is possible to calculate their effective cellular diameter to be *c.* $8.3\,\mu$ (Miller and Phillips, 1969; Haskill and Moore, 1970). Secondly, we can obtain a first approximation to the number of focus inducing cells present by simply counting the number of foci obtainable from two million cells per culture. There is, on the average, one focus inducing cell for every two dimples in a culture. These cells are considerably smaller than the rosette forming cell population and are about $7.5\,\mu$ in diameter. This difference in size between the rosette forming cell and

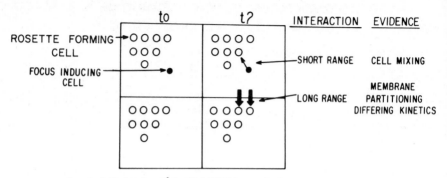

Fig. 9. A simple model based on the preceding data to describe the immune response of fr (2) cells in acrylamide cultures.

the focus inducing cell, incidentally, could well explain the results obtained by Nossal et al. (1967) and Shortman et al. (1970) on the glass bead size separation of the cells involved in immune responses. By selecting out the medium size lymphocytes, as they have done, they would severely deplete the spleen cells or thoracic duct cells of the rosette forming cell population. If, for instance, the number of rosette forming cells in the Salmonella antigen system were to be different from that in the sheep erythrocyte system, it would be possible to obtain differences in the sensitivity to depletion of the larger cell populations based simply on the relative numbers of the rosette forming cell populations to the two antigens. Another point which should be made about this model is based on the experiments where we have separated the focus inducing cell from the rosette forming cell by Nucleopore membranes. It is quite clear that the interaction between these two cell types need not take place by direct cell contact and, in fact, it may be possible to stimulate neighbouring areas, or neighbouring dimples, as a result of diffusion of the products of this focus inducing cell between the acrylamide partitions.

Nothing very positive has been found concerning the necessity for phagocytic populations in these responses. Two points should be noted however. First, the peak immune response of fr (2) cells after velocity separation does come in a region which is depleted by 98% of its original level of phagocytic cells, and this fails to produce a significant loss in activity. Second, in experiments where rosette forming cell regions were isolated, although they contained the bulk of the phagocytic cells, they failed to give any significant immune response. That is, the lymphocyte-type rosette forming cells in the presence of antigen and phagocytes are incompetent to give an immune response.

The final point which should be made about the results presented here is that of the nature of the immune response in fr (2). It appears from the kinetics of the response that the foci can appear and disappear within a very short period of time (probably no more than 12 hr). This data would appear to be explained best by an immune response which consists, not of an exponentially increasing population of dividing antibody forming cells, but rather the response of a large number of rosette forming cells, each of which gives rise to clones of antibody forming cells with as many as 30 antibody forming cells per progenitor.

ACKNOWLEDGEMENTS

This work was supported by the National Cancer Institute of Canada and the Medical Research Council of Canada.

SUMMARY

Investigation of the *in vitro* response of the fr (2) type of cell from non-immunized mice leads to the following conclusions:

1. Cell mixing experiments indicate the involvement of at least two cell populations
   (A) one population directly binds antigen (RFC),
   (B) the other population is defined as a focus inducing cell. This cell apparently produces a diffusable substance capable of stimulating the response.
2. The parameters measured were
   (A) the number of foci
   (B) immunological burst size
   (C) rate of development of (A) and (B)
3. The number of foci *not* the burst size is affected by the antigen dose.
4. Selective removal of RFC reduces both the number of foci and the burst size.
5. The immune response may consist of a series of bursts of developing clones of PFC, each with a lifetime which might be as short as 12 hr.

REFERENCES

Dutton, R. W. and Mishell, R. I. (1967). *Cold Spring Harb. Symp. quant. Biol. 32*, 407-414.
Haskill, J. S. (1969). *J. exp. Med. 130*, 877-893.
Haskill, J. S. and Moore, M. A. S. (1970). *Nature, Lond. 226*, 853-854.
Haskill, J. S., Byrt, P. and Marbrook, J. (1970). *J. exp. Med. 131*, 57-76.

Jerne, N. K., Nordin, A. A. and Henry, C. (1963). *In* "Cell Bound Antibodies" (B. Amos and H. Koprowski, eds), pp. 109-125. Wistar Institute Press.

Marbrook, J. (1967). *Lancet ii*, 1279-1281.

Marbrook, J. (1970). Manuscript in preparation.

Miller, R. G. and Phillips, R. A. (1969). *J. cell. Physiol. 73*, 191-201.

Mishell, R. I., Dutton, R. W. and Raidt, D. J. (1970). *Cellular Immunology 1*, 175-181.

Mosier, D. E. and Coppleson, L. W. (1968). *Proc. natn. Acad. Sci. U.S.A. 61*, 542-547.

Nossal, G. J. V., Shortman, K. D., Miller, J. F. A. P., Mitchell, G. F. and Haskill, J. S. (1967). *Cold Spring Harb. Symp. quant. Biol. 32*, 369-379.

Osoba, D. (1970). *J. exp. Med. 132*, 368-383.

Shortman, K. D., Diener, E., Russell, P. and Armstrong, W. D. (1970). *J. exp. Med. 131*, 461-482.

Syeklocha, D., Siminovitch, L., Till, J. E. and McCulloch, E. A. (1966). *J. Immun. 96*, 472-477.

Towards a Developmental Theory of Immunity: Cell Interactions

Department of Zoology
University of Wisconsin
Madison, Wisconsin, U.S.A.

While the concepts of selective cell aggregation and of inductive interactions at the cell and tissue level have long dominated thinking in the field of developmental biology, only recently have these concepts become central in the area of cellular immunology. If I attempt to transmute immunobiological questions into the language of the developmental biologist, therefore, it is with the belief that such transmutation may be useful in clarifying some aspects of the immune mechanism and leading to application of the discoveries in cellular immunology to problems of embryogeny and cellular differentiation.

Some eight years ago an effort was made to bring together embryologists and immunologists in order to encourage exchange of ideas between these two disciplines. I then attempted to place immune maturation into developmental context by suggesting that the embryogeny of immunity includes a number of distinct phases:

> "The initial phase of thymus development involves an inductive system in which the mesenchyme and epithelium interact to produce the lymphoid organ.
>
> The second step involves the formation of the lymphoid cell series which is characterized, among other things, by its nonadhesiveness.
>
> The third phase involves a migration of cells. In terms of selection, influencing agents, and specificity this phase might be a critical one, (for) subtle influences may be operative during migration or outgrowth of cells.
>
> Finally, when the cells reach the spleen, they form characteristic aggregates (subject to) the concepts of aggregation and selective settling" (Auerbach, 1962).

In the past few years we have had to add to these notions the concept of initial stem cell migration from the yolk sac (Moore and Owen, 1967), and the more complex migration histories of precursor cells of the bone marrow and thymus (Micklem *et al.*, 1966; Ford, 1966). However it still seems clear that the fundamental concepts of induction and aggregation, even if still obscure as regards mechanism, are probably the major elements leading to immunological maturation.

I recently tried to lay the foundations for a developmental theory of antibody formation by drawing an analogy between the developing lymphoid system and the developing germ line—reproductive system (Auerbach, 1970). In doing this I reviewed the inductive, proliferative and migratory events leading to the formation of the cell types involved in the immune system. Continuing this attempt, I would like to examine the post-generative phase of the developing immune system, the cellular interactions that appear to occur during the immune response, again placing these interactions into developmental context.

When we were attempting to miniaturize the *in vitro* spleen response to sheep red blood cells (Auerbach, 1970) we began by modifying the original organ culture procedures of Globerson and Auerbach (1965) by simply reducing the size of explanted fragments. We found that the minimum size required to obtain a regular response consisted of from $1\text{-}2 \times 10^5$ cells per explant. Fragments below this size were unable to survive under our conditions of cultivation. We next used a biological dissection method of separating the naturally-occurring follicular areas (white pulp) of the spleen and sorting these into size classes (Table I). Again a minimum size necessary for viability was reached, this being around 10^5 cells. Turning next to cell suspension systems, we attempted to miniaturize the Marbrook (1967) modification of the Mishell-Dutton culture system (Mishell and Dutton, 1966). Using minimum dilution techniques we found that while the original Marbrook system became limiting at 5×10^6 cells (see also Osoba, 1969) our miniaturization of that system using our standard organ culture filter assemblies (Globerson and Auerbach, 1966) led to a limiting cell number of 2.5×10^5. Looking at the morphology of the cultures it could be seen that in both instances the limiting cell number was the number that represented a *density* of cells sufficient to permit reaggregation *in vitro*. Measurements of the surface area of the Marbrook culture tube and of the organ

TABLE I

In vitro *immunization of spleen follicles against sheep red blood cells*

Follicle size	Number positive	Mean PFC	Range
Large (7.5×10^5)	23/26	72.8	0-374
Medium (3×10^5)	10/19	49.4	0-146
Small (1×10^5)	74/126	14.6	0-108
Culture medium + surrounding cells	7/8	241.0	90-514

culture assembly supported this conclusion, in that the surface area of the former was about 20 times that of the latter. This finding thus provided direct confirmation for the observations of Mosier (1969) that cluster formation was an essential prerequisite for antibody induction in *in vitro* suspension systems.

It is not the purpose of this paper to elaborate on the question of minimum numbers of antigen-sensitive or antigen-reactive cells. Clearly, where these cells are in excess, as in all standard *in vitro* culture systems (Globerson and Auerbach, 1965; Marbrook, 1967; Mishell and Dutton, 1966) limits are placed on the system by a *requirement for organization.*

Many years ago cell density was shown to be a critical factor in optimal aggregation and differentiation of slime moulds (Sussman, 1958). Sorting-out processes in vertebrate cells were described by Holtfreter (1939), and since then the exquisite sensitivity of sorting out-processes has been described in detail by numerous workers, notably Moscona (1965) and Steinberg (1964). Sorting-out can occur in most embryonic reaggregating systems, but appears to be subject to the limits imposed by optimal cell densities (Ball, 1965).

Both in the earlier amphibian work (Holtfreter, 1939; Townes and Holtfreter, 1955) and the subsequent chick and mouse studies (Moscona, 1965; Steinberg, 1964) sorting-out of embryonic cells occurred in a highly predictable, tissue-specific manner. Even in complex masses of three tissue components reorganization of the cells to give a characteristic, organized, aggregate occurred. Of special interest is the fact that reconstruction of tissues appears to be identical whether the starting point for such reconstruction is dispersed cells, or simply involves fusion of tissue masses. The parallel to reconstruction of thymus and marrow-dependent cell populations *in vitro* in organ cultures (Globerson and Auerbach, 1967; Globerson and Feldman, 1969) or cell cultures (Hartmann, this symposium) is obvious.

One may ask just how much specificity there is in cell surface recognition processes leading to sorting-out and reorganization *in vitro*. Not only does cartilage recognize other cartilage, but mouse cartilage recognizes chick cartilage as similar but not identical (Burdick and Steinberg, 1969), and limb cartilage recognizes itself as different from the limb mesenchyme from which it is derived (Zwilling, 1968).

It has been suggested that sorting-out processes following initial aggregation of cells involve a sequence of exchanges, in which cells bonded with higher affinity take precedence over cells bonded with lower affinity. Thus sorting-out represents a series of exchanges of low-strength adhesions with progressively higher-strength ones (Steinberg, 1964). The materials involved in cell adhesions, termed ligators, appear to have considerable specificity (see Lilien, 1969). Specific cell types produce ligators with ability to bind selectively to specific cell surfaces. For embryonic retina such material appears to be glycoprotein in nature (Lilien, personal communication).

Applied to the immunological system involving "marrow-derived" and "thymus-dependent" non-adherent cells and macrophage-like "adherent" cells, several points can be suggested:

1. To permit adhesion (aggregation), since bone marrow and thymus-dependent cells are both non-adherent there must be induction or mediation of adhesion. This might be mediated by the adherent cell type, by a secretory product from such a cell, or by some form of activation where activation is defined as (antigen?-) induced alteration of adhesive properties. All three of these alternatives may be operative. A number of three-cell interaction systems have been described, involving an adherent cell population that functions in some as yet unexplained manner to permit thymus-dependent and marrow-derived cell interactions (Haskill *et al.*, 1970; Mosier and Coppleson, 1968; Talmage *et al.*, 1969; Alter and Bach, 1970). In other developing systems, mesenchyme appears to serve a similar function in permitting non-coaggregating ectoderm and endoderm cells (pituitary gland) (Auerbach, unpublished observation) or ectodermally-derived lens and optic vesicle cells (Muthukkaruppan, 1965) to interact. Exudates from adherent cell cultures may to some extent mimic the adherent cell functions in immune systems (see Bach *et al.*, 1970), just as exudates from mesenchymal cells may be functional in certain inductive systems (Konigsberg and Hauschka, 1965). Finally, activation by phytohaem-agglutinin may provide non-adherent cells with a possibility of interacting, by making them sticky and causing their agglutination (Lilien, 1969).

2. Drawing an analogy between ligator models of cell receptor sites and the marrow-thymus interacting system, it might be suggested that the strongest adhesion, hence functional integrity, would be formed between two cells sharing common (complementary?) sites. If specific antigen receptor sites do exist on the surface of the two populations of non-adherent cells, one might expect that the tightest connections, hence the most stable ones, would be formed between cells sharing these receptor identities.

3. If any type of information transfer is required during the immune response, one might predict that it would require the tight cellular connections produced only in strong type-specific ligations. This type of connection has been shown to be established only as a result of type-specific cellular associations in embryonic reaggregation studies (Loewenstein, 1967).

4. Explanations for non-specific effects of surface-active agents, of poly-amino sugars, and of cyclic AMP become self-evident.

A number of logical consequences of accepting the analogy of cell aggregation systems can be subjected to test. For example, non-specific binding, binding to weakly cross reactive cells, or coating with antibody should lead to tolerance if the timing of such binding or antibody treatment were to preclude exchange of "low-affinity" for "high-affinity" adhesion. For the mouse this would most readily be accomplished in the prenatal stage of development, when the thymus type cells are already in existence, and the bone marrow type cell is yet to be liberated. Activation of cells of both of the non-adherent classes in the perinatal period should still be quite effective since the third cell type (adherent) appears to be lacking at this time (Argyris, 1969).

While antibody fragments on cell surfaces are attractive candidates for specific receptor sites for antigen binding, it would seem likely that for the thymus-marrow type interaction the binding sites include a carbohydrate or glycoprotein component as initiator, i.e. the active site for this interaction must be more than an antigen receptor. Perhaps the theta antigen may turn out to be an essential component in cell interaction.

More specifically, one might predict that activated thymus-dependent cells could produce ligator-type molecules, which could select appropriate bone marrow-derived cells from a general population of cells. One might suspect that trypsin treatment of cells before transfer might abrogate low-zone tolerance in a passive transfer system. Finally one would think that low responding animals could become high responding ones by means of aggregation-enhancing agents or by the use of cellular exudates obtained from adherent cells of a high-responding strain.

ACKNOWLEDGEMENTS

Supported by grants CA 5281 from the National Institutes of Health and GB 6637X from the National Cancer Institute.

The technical assistance of Heidi Anderson and Louis Kubai is gratefully acknowledged.

REFERENCES

Alter, B. J. and Bach, F. H. (1970). *Cellular Immun. 1*, 207
Argyris, B. F. (1969). *Transplantation 8*, 241-248.
Auerbach, R. (1962). *J. cell. comp. Physiol. 60*, Suppl. 1, 159.
Auerbach, R. (1970). *In* "Developmental Aspects of Antibody Formation and Structure" (J. Šterzl and I. Říha, eds), p. 23.
Bach, F. H., Alter, B. J., Solliday, S., Zoschke, D. C. and Janis, M. (1970). *Cellular Immun. 1*, 219.
Ball, W. D. (1965). Ph.D. Thesis, University of Chicago.
Burdick, M. L. and Steinberg, M. S. (1969). *Proc. natn. Acad. Sci. U.S.A. 63*, 1169.
Ford, C. E. (1966). *In* "Thymus" (G. E. W. Wolstenholme and R. Porter, eds), Ciba Fdn Symp., p. 131. Churchill, London.
Globerson, A. and Auerbach, R. (1965). *Science, N.Y. 149*, 991-993.
Globerson, A. and Auerbach, R. (1966). *J. exp. Med. 124*, 1001-1016.
Globerson, A. and Auerbach, R. (1967). *J. exp. Med. 126*, 223-234.
Globerson, A. and Feldman, M. (1969). *In* "Lymphatic Tissue and Germinal Centers in Immune Responses", p. 407. Plenum Press.
Haskill, J. S., Byrt, P. and Marbrook, J. (1970). *J. exp. Med. 131*, 57-76.

Holtfreter, J. (1939). *Arch. exp. Zellforsch. 23,* 169.
Konigsberg, I. R. and Hauschka, S. D. (1965). *In* "Reproduction: Molecular, Subcellular and Cellular" (M. Locke, ed.), p. 243. Academic Press, New York.
Lilien, J. E. (1969). Current Topics in *Devl. Biol. 4,* 169.
Loewenstein, W. R. (1967). *Devl. Biol. 15,* 503.
Marbrook, J. (1967). *Lancet ii,* 1279-1281.
Micklem, H. S., Ford, C. E., Evans, E. P. and Gray, J. G. (1966). *Proc. R. Soc. B 165,* 78.
Mishell, R. I. and Dutton, R. W. (1966). *Science, N.Y. 153,* 1004-1006.
Moore, M. A. S. and Owen, J. J. T. (1967). *J. exp. Med. 126,* 715-726.
Moscona, A. A. (1965). *In* "Cells and Tissues in Culture" (E. N. Willmer, ed.), Vol. 1, p. 489. Academic Press, New York.
Mosier, D. E. (1969). *J. exp. Med. 129,* 351-362.
Mosier, D. E. and Coppleson, L. W. (1968). *Proc. natn. Acad. Sci. U.S.A. 61,* 542-547.
Muthukkaruppan, V. (1965). *J. exp. Zool. 159,* 269.
Osoba, D. (1969). *J. exp. Med. 129,* 141-152.
Steinberg, M. S. (1964). *In* "Cellular Membranes in Development" (M. Locke, ed.), pp. 321-366. Academic Press, New York.
Sussman, M. (1958). *In* "The Chemical Basis of Development" (W. D. McElroy and B. Glass, eds), p. 264. John Hopkins Press, Baltimore.
Talmage, D. W., Radovich, J. and Hemmingsen, H. (1969). *Wistar Inst. Symp. Monograph 9,* 151.
Townes, P. L. and Holtfreter, J. (1955). *J. exp. Zool. 128,* 53.
Zwilling, E. (1968). *Devl. Biol. Suppl. 2,* 184.

Studies of Cellular Recognition *in vitro*: Role of T Lymphocytes and Some Effects of a Lymphoblast-derived Inhibitor of Cell Proliferation

RICHARD T. SMITH, WILLIAM H. ADLER, TOMOO TAKIGUCHI, DUANE PEAVY and JUDE BAUSHER

Tumor Biology Unit, Department of Pathology, University of Florida College of Medicine, Gainesville, Florida, U.S.A.

We have recently examined in some detail the role of populations of thymus cells and thymus-dependent spleen cells and their fractionated subpopulations, in terms of cellular recognition *in vitro*. The methods employed, with the exception of those described herein, have been recently reported (Adler *et al.*, 1970a, b, c, d; Smith and Landy, 1970; Smith *et al.*, 1969, 1970). Each depends on the addition of various antigens, mitogens, nondividing cells, or cell membranes, *in vitro*, to reactive whole spleen, lymph node or thymus cell populations, or albumin gradient-fractionated subpopulations of these cells, and then measuring the incorporation of tritiated thymidine after an appropriate culture period.

Normal spleen, lymph node and thymus cell populations were compared in their responsiveness to a panel of cell stimulants. Although some strain variation was encountered, the data summarized in Table I are representative of the results obtained.

The most striking observation was the absence of LPS responsiveness in thymus cell populations as contrasted to lymph node or spleen cells. In contrast all populations were highly reactive to SEB-stimulation. The insusceptibility of thymus cells to LPS is as yet unexplained. The very weak responses of thymus cell populations to PHA and alloantigens will be shown to be due to a minor sub-population of thymus cells which are actually quite reactive.

Evidence was next sought for early binding of mitogens to the cell membrane of lymphoid cells. A technique of sandwich labeling was developed in which antibody against the mitogen was used in conjunction with fluorescent anti-gamma globulin as an indicator. These studies indicated that within a very few minutes mitogens, such as PHA, were engulfed rapidly by spleen cells, presumably by micropinocytosis. Under normal circumstances no surface

TABLE I

Relative responses of normal mouse thymus, spleen or lymph node cell suspensions in vitro *to various stimuli or mitogens*

Antigen or mitogen	Relative degree of response evoked[a]		
	Thymus cells	Spleen cells	L.N. cells
PHA	+	++++	++++
Staphylococcal enterotoxin B (SEB)	++++	++++	++++
Lipopolysaccharide (LPS) *(E. coli)*	0	++++	++++
Alloantigens (H-2 different)	±	+++	+++
Sheep RBC membrane preparations	0	+	+
PPD	0	0	0

[a] Degrees of response are graded arbitrarily on a 0-4+ scale, reflecting the degree of thymidine incorporation stimulated by each of the mitogens.

membrane binding could be visualized. Uptake was easily detected within a few minutes and was far advanced by 30 min. Uptake was blocked by neuraminidase treatment or by incubation at $4°$ C; under these conditions mitogen bound to the living cell surface was easily seen. Apparently uptake is so fast that insufficient surface mitogen accumulates to permit detection unless membrane turnover is reduced. Surface membrane binding like that seen in blocked cells was observed in lymphoblasts, of both mouse and human origin. Membrane binding was confirmed by absorption studies, showing that mitogens are actually removed from the media by reactive cells. This approach is now being applied to alloantigens and specific antigens. If mitogens and specific antigens turn out to behave in a similar fashion, these data would indicate that antigen recognition first involves binding to the cell surface membrane, and that this initiates a process of rapid uptake. Whether uptake is actually necessary for the processes of cell differentiation and cell division which follow, is not settled by these observations.

Next, further information was sought on whether antigen and mitogen stimulation of spleen cells, depends upon the presence of T lymphocytes. Accordingly, primary responses of spleen cells taken from neonatally thymectomized and littermate control mice of various strains, were compared with respect to a panel of mitogens and specific stimulants. Of course, responsiveness to antigens such as PPD after BCG stimulation was absent in neonatally thymectomized mice. Spleen cell populations derived from such thymectomized animals were markedly deficient in primary responses to PHA and alloantigens. In contrast, cells from neonatally thymectomized mice were fully responsive to LPS or SEB stimulation. In dose-response characteristics, LPS and SEB were similar in control and thymectomized animals. From this study, we conclude that *in vitro* responsiveness to specific antigens induced by immunization, to PHA, and to alloantigens depends upon a T-lymphocyte function, whereas responses to LPS and SEB are thymus-independent.

It was next of interest to determine whether thymic function of neonatally thymectomized mice could be restored to full responsiveness *in vitro* in terms of alloantigens and PHA. Small grafts of thymus introduced under the renal capsule in neonatally thymectomized mice, gave nearly complete functional restitution, providing the donor thymus was grafted when the animal was young, and that a sufficient interval of time had elapsed for the graft to establish a normal structure at the ectopic site. Table II gives the results of some typical experiments of this type.

TABLE II

Restoration of susceptibility to PHA stimulation in spleen cells from neonatally thymectomized and thymus graft-reconstituted mice[a]

Strain of host	Strain of graft	Littermate control	Thymectomy at birth	Thymectomy birth-thymus graft
		PHA-stimulation of ^{3}H-thymidine incorporation $\left(\text{ratio of incorporation} \dfrac{\text{mean cpm/tube PHA}}{\text{mean cpm/tube control}}\right)$		
C57B1/6	C57B1/6	28.1	2.6	11.5
C57B1/6	CBA	35.6	2.2	5.0
C57B1/6	[DBA x C57B1/6] F_1	27.0	3.2	2.8
[A/J x C57B1/6] F_1	C57B1/6	18.0	7.8	6.5
[CBA x C57B1/6] F_1	C57B1/6	12.0	7.7	9.2
[DBA x C57B1/6] F_1	DBA	15.5	5.9	12.3[b]
		A/J-mito stimulation of ^{3}H-thymidine incorporation in MLC $\left(\text{ratio of incorporation} \dfrac{\text{mean cpm/tube allogeneic}}{\text{mean cpm/tube syngeneic}}\right)$		
C57B1/6	C57B1/6	9.3	2.4	7.0
C57B1/6	[DBA x C57B1/6] F_1	9.6	2.6	2.1[c]

a C57B1/6 mice, or the indicated F_1 hybrids, were thymectomized at birth. At 12-13 weeks of age, seven weeks after receiving a thymus graft of the strain indicated placed beneath the renal capsule, the response of their spleen cells to 10 λ PHA, or A/J (mitomycin-treated) target cells in the MLC was compared with that of littermate controls or thymectomized but not grafted, littermates. The results are expressed as a ratio of incorporation of mean value of duplicate or triplicate cultures.

b This degree of restoration was obtained in about two out of three experiments; in the others the values were control levels.

c Partial restoration was obtained in one-third of such experiments.

These studies complement and extend those reported by Stutman (personal communication) for restoration of immune capacity in terms of graft-*versus*-host reactions, and those of Mosier and colleagues (1970) in which plaque-forming capacity towards SRBC was restored in thymectomized and irradiated mature mice.

However, in some contrast with Stutman's results, restoration of thymic function as reflected by *in vitro* recognition of PHA or alloantigen never occurred when the donor of the restoring thymus was allogeneic, and was never more than slight when the donor was of F_1 hybrid origin. Occasional restoration of PHA responses in F_1 by parental strain grafts also apparently occurred. At least one explanation of apparent restoration of alloantigenic responsiveness is conceivable. Since neonatally thymectomized mice are never totally incompetent immunologically, it is possible that the foreign alloantigens of the F_1 thymus have immunized the host against antigenic determinants shared by the target cell strain used for testing primary responsiveness.

Having demonstrated the thymus dependence of cellular recognition of certain antigens and mitogens, and that syngeneic thymus grafts could restore the responsiveness of spleen cell populations in thymectomized animals, it was next of interest to determine if any particular sub-population of the spleen might be responsible for thymus-dependent functions, and whether such a subpopulation was present in the grafts that restored thymus function. For this purpose discontinuous albumin density gradients were prepared according to the method of Raidt *et al.* (1968), and cells from spleens of thymectomized and littermate control animals, and from aliquots of thymus cells, used for restoration of function, were fractionated and examined for reactivity to PHA and alloantigens *in vitro*.

The proportional distribution of cells in these gradients is shown in Table III.

The largest proportion of spleen cells in such gradients had average densities that placed them in the C and D fractions. These consisted chiefly of small lymphocytes with less than 1% larger lymphoid cells, and no reticular cells, blasts or macrophages. It was also evident that thymectomy had no detectable effect upon the distribution of these density-dependent subpopulations, despite the clear cut deficiency of *in vitro* function in the spleen populations from which they were derived.

Donor thymus cell populations on the other hand, consisted mostly of the densest small lymphoid cells. But a significant small subpopulation of larger cells and blasts was present. The morphologic attributes of thymus cells in these various catagories are shown in Table IV. It was noteworthy that cells having the appearance of small lymphocytes were distributed throughout the gradient, suggesting that there is a density-related heterogeneity of identically-sized thymus small lymphocytes.

The reactivities of both spleen and thymus gradient-separated subpopulations

TABLE III

Density distribution of gradient-fractionated subpopulations of spleen or thymus cells from neonatally thymectomized and control C57B1/6 mice

Cell source and treatment of mice	Fraction of the gradient	Cell distributions	
		Number of cells	Percent in group
Spleen cells from	A[a]	10×10^6	1.9
neonatally thymectomized	B	130×10^6	23.61
mice:	C	315×10^6	57.8
	D	90×10^6	16.5
		$545 \times 10^{6\,b}$	
Spleen cells from controls:	A	15×10^6	5.1
	B	100×10^6	30.0
	C	165×10^6	48.8
	D	54×10^6	16.1
		334×10^6	
Thymus cells from control	A	5×10^6	0.8
animals:	B	90×10^6	14.4
	C	345×10^6	55.2
	D	185×10^6	29.6
		625×10^6	

300×10^6 spleen cells from neonatally thymectomized or paired control littermate or thymus cells from C57B1/6 mice were suspended in 35% BSA, and introduced into the lowest portion of a discontinuous albumin density gradient. The tubes were then centrifuged at 13,500 r.p.m. for 30 min in a SW 39 swinging bucket rotor. The cells were carefully harvested by aspiration, then enumerated as indicated.

[a] Fraction A consisted of cells above the 23% albumin layers, layer B above 26%, layer C above the 29% and layer D above the 35%.

[b] Total cells counted in all fractions.

to PHA is shown in Table V, and of spleen cells to alloantigens is shown in Table VI.

PHA-stimulated incorporation of tritiated thymidine was maximal in the dense cell populations of layers C and D of spleen cells from normal animals, and this responsiveness was markedly decreased in those taken from thymectomized animals. These data are given as ratios of PHA-stimulated to control, and thus represent PHA-stimulated incremental incorporation. The A and B layers on the other hand, as has been reported by others, showed the highest "spontaneous" levels of incorporation, demonstrating that these cells synthesize DNA without identified stimulation. In contrast, thymus cell populations show the greatest incremental PHA reactivity in the B layer, which contains the largest proportion of blast cells, but also 30% small lymphocytes. Background incorporation was

TABLE IV

Distribution of CBA thymus cells in albumin density gradients

Layer	Approximate % of cells in layer	% of cells in category			
		Macrophage-like	Mitotic cells	Large lymphocytes and blasts	Small lymphocytes
A	<1	4	28	30	38
B	3	–	–	65	35
C	30	–	–	7	93
D	67	–	–	3	97

Gradient-separated thymus cells were fixed, stained with Giemsa stain and assigned to the various morphologic categories. The data represent the averages of two independent counts.

TABLE V

Comparison of PHA stimulation of gradient-fractionated spleen and thymus cells taken from normal C57B1/6 mice, with spleen cells from neonatally thymecto-mized animals

Fraction	PHA-stimulation $\left(\text{ratio of incorporation} \dfrac{\text{mean cpm PHA stimulated cultures}}{\text{mean cpm controls}}\right)$		
	Spleen cells		Thymus cells
	Littermate controls	Neonatal thymectomy	
A	0.8	0.2	1.4
B	1.1	0.5	7.7
C	2.7	1.9	5.0
D	11.3	1.9	1.0

Cells from normal or neonatally thymectomized animals were separated in albumin density gradients, and the responses of the resulting cell preparations to PHA were assayed *in vitro*. The results of ^{3}H-thymidine incorporation on the second day of a two-day culture period are per 10×10^6 cells from each fraction, and expressed here as the ratio of PHA and control cultures.

highest in the A and B layers of thymus as in spleen. Thus, the PHA reactivity of subpopulations of spleen and thymus differed markedly. As in the case of PHA reactivity, alloantigen stimulation was practically limited to the C and D subpopulations, and in this population the deficiency in thymectomized animals was most evident.

Direct primary stimulation of all spleen cell subpopulations by SRBC stroma in thymectomized animals was greatly decreased or undetectable as compared to

TABLE VI

Effect of neonatal thymectomy on allogeneic stimulation of gradient-fractionated spleen cells taken from C57B1/6 mice

Source of reacting cells	Gradient fraction of reacting cells	^{3}H-thymidine incorporation (mean cpm/tube)		Ratio of incoporation: Allogeneic/Syngeneic
		Syngeneic target cells	Allogeneic target cells	
Thymectomy at birth	B	6319	8278	1.3
	C	1502	1662	1.1
	D	1363	1477	1.1
	Unfractionated cells	2401	2624	1.1
Littermate controls:	B	3493	7836	2.2
	C	1075	3463	3.2
	D	448	1573	3.5
	Unfractionated cells	998	2807	2.8

Various fractions resulting from gradient separation of the spleen cells from neonatally thymectomized or control mice, 15×10^6 cells per tube, were incubated with an equal number of C57B1/6 (syngeneic) or A/J (allogeneic) mitomycin-treated unfractionated target cells. Layer A cells were insufficient for assay in this experiment. ^{3}H-thymidine incorporation was assayed during the final 24 hr of a three-day culture period, and the results expressed as the mean cpm/tube for two or three replicate cultures.

normal mice. Primary stimulation was limited to the C-D layers in density gradient-fractionated spleen cells from normal mice. In contrast in SRBC-immunized animals stimulated incorporation of ^{3}H-thymidine, as well as numbers of plaque-forming cells, were greatly increased in the B layer fraction. This suggested an opportunity to extend the model developed by Shearer and Cudkowicz (1969) in such a way as to determine which subpopulation of thymus cells was responsible for restoration of SRBC reactivity *in vitro*. The results of such experiments are exemplified by that shown in Table VII.

The T-lymphocyte component that recognizes SRBC in this model was found in a subpopulation having B layer density characteristics, and not in the other layers. This suggests that restoration of T-lymphocyte function, as indicated by restoration of antigenic recognition and cellular synthesis of anti-SRBC, is a property of a relatively non-dense cell population consisting mostly of large lymphocytes and blasts. However, since at least 30% of these cells are less dense, small lymphocytes, we cannot definitely conclude that the large lymphocyte population is actually responsible for restoration. On the other hand, the small lymphocytes that compose most of the dense layer cells, did not have this

TABLE VII

Restoration of antibody formation at the cellular level, by subpopulations of thymus cells in irradiated CBA mice

	Primary recipient	Secondary recipient	Mean direct PFC/spleen
(a)	Unfractionated thymus cells−SRBC	SPL^a + BM + SRBC	2271
(b)	SRBC alone	SPL^b + BM + SRBC	183
(c)	−	SRBC	150
(d)	Fraction B thymus + SRBC	SPL^d + BM + SRBC	5649
(e)	Fraction C thymus + SRBC	SPL^e + BM + SRBC	561
(f)	Fraction D thymus + SRBC	SPL^f + BM + SRBC	189

CBA mice, irradiated with 900 r ("primary recipients") received 100×10^6 unfractionated or 70×10^6 gradient-separated thymus cells, together with 5×10^6 SRBC. Five days later, their spleens were removed and the cells (SPL) given intravenously to a second 900 r CBA mouse ("secondary recipient") together with 10^7 CBA bone marrow cells (BM) and 5×10^6 SRBC. Numbers of direct PFC per spleen was determined on these animals eight days later. Each figure for PFC represents the mean of three animals in the group.

property. Further attempts are under way to characterize these restorative cells from the thymus by various functional characteristics.

Finally, we shall describe some recent experiments that affect in several ways the interpretation of all *in vitro* assays of cellular responsiveness to mitogenic stimuli, cell-cell interactions, and possibly their *in vivo* counterparts. The basic finding is that lymphoblasts produce in their immediate environment during proliferation, a powerful inhibitor of DNA synthesis, and in addition, although masked *in vitro* by the inhibitor, a nonspecific mitogenic substance or substances.

The problem is best clarified by the experiment shown in Table VIII. Here the culture fluid from a freshly established human lymphoblast cell line was added to a population of normal human peripheral leucocytes. The effect of heating the medium was compared in two situations. In (a), the source of stimulation in a seven-day culture was probably fetal calf serum. In (b) the source in a three-day culture was PHA. Both inhibitor and mitogenic effects are demonstrated.

We call the first factor IDS (Inhibitor of DNA Synthesis) and have now characterized it quite extensively. The potency of the inhibitor is indicated by frequent observation of its activity in dilutions as high as 1 : 50,000. The inhibitor masks the mitogenic factor in the same medium, unless inactivated by heating at $56°$ C for 60 min. It is non-dialyzable, and has marked affinity for

TABLE VIII

Demonstration of IDS and mitogen effects on normal and PHA-stimulated cells

Procedure	^{3}H-Tdr mean CPM per tube	
	56°C 60 min	Unheated
(a) SN-312A culture medium added to human peripheral leucocytes at varying dilution; the only stimulant in the 7-day culture is fetal calf serum of the medium		
1 : 5	64,016	550
1 : 50	31,667	737
1 : 500	12,844	658
Control	5,419	11,408
(b) Three different culture fluids added at 1 : 25 dilution to human peripheral leucocytes plus 10 λ PHA 3-day culture		
SN 315C	318,830	16,002
SN 308A	274,018	12,862
SN 312A	285,915	14,952
Control medium	270,032	252,068

The donor of peripheral leucocytes was the individual from whom line SN-312A was derived.

cellulose. It is effective over a wide range of stimulant to cell ratios. Its effects are nearly completely reversible. Most significantly, it fails to inhibit intracellular differentiation as eventually expressed by blast transformation, despite the fact that nearly complete inhibition of DNA synthesis is achieved (Table IX).

IDS of apparently identical properties, is produced by mouse and human lymphoblasts resulting from antigenic or alloantigenic stimulation of normal lymphocytes, as well as from continuous established lines. For example, the supernatant fluid from mixed lymphocyte cultures of mouse cells markedly inhibits thymidine incorporation in PHA-stimulated human cells. Likewise, human IDS markedly inhibits all types of stimulation of mouse lymphoid cells *in vitro*.

The properties of IDS as currently elucidated are summarized in Table X. Whether IDS is similarly effective *in vivo* is as yet unknown. It does appear to represent a candidate for effecting feedback inhibition of lymphoid cell proliferation produced by a variety of antigens. If it is actually shown to be active *in vivo*, counteractivators or inhibitors of its activity should abound in the

TABLE IX

Effects of IDS on PHA-stimulated incorporation of nucleoside and amino acids, and upon blast transformation

Time culture terminated (hr)	Material added to cultures	^{3}H-thymidine		^{3}H-uridine		^{14}C-leucine		Blasts in culture (percent)
		Mean cpm per tube	Ratio of incorporation	Mean cpm per tube	Ratio of incorporation	Mean cpm per tube	Ratio of incorporation	
18	Control + PHA	538	0.90	33,984	7.96	2,796	1.70	6
	SN-315C + PHA	4,180	0.98	10,628	1.80	3,244	1.78	6
42	Control + PHA	87,961	203.14	170,170	30.82	12,209	9.09	51
	SN-315C + PHA	5,720	1.02	62,889	2.40	11,057	5.95	46
66	Control + PHA	166,935	67.75	160,879	18.20	61,710	32.48	81
	SN-315C + PHA	3,606	2.00	41,956	2.09	20,138	8.89	78

Culture fluid SN-315C, diluted 1 : 25, was added to cultures of allogeneic human peripheral leucocytes, together with 25 λ PHA, and the incorporation of the indicated precursors assayed independently in replicate cultures for the 18-hr period preceding the indicated time of termination of the culture. Ratio of incorporation is the PHA-stimulated incorporation divided by background level of incorporation in identical cultures containing no PHA; the mean cpm are averages of duplicate tubes.

TABLE X

Properties of IDS

---

1. Produced by 20/20 human lymphoblast cell lines examined
2. Produced by alloantigen and PHA-stimulated human and mouse lymphocytes
3. Secreted rapidly to high concentration into any medium by lymphoblasts
4. Inhibits DNA synthesis and mitosis in stimulated lymphocytes nearly completely, but has no apparent effect on differentiation to blast morphology
5. Inhibits DNA synthesis in stimulated mouse lymphocytes and mouse L-cells
6. Effect is reversible
7. Labile to heat (56°C 60 min), and pH changes
8. Molecular size undetermined, but is not dialyzable
9. Marked affinity for cellulose acetate and cellulose nitrate

---

lymphoid organs, or in the fluids that perfuse them. Consequently its activity would probably be local and intimate.

As an *in vitro* phenomenon IDS merits consideration in every situation in which cell-cell interactions are interpreted in terms of proliferation kinetics or DNA synthesis is used as an endpoint, for IDS masks all mitogenic effects both specific and nonspecific and could conceivably be responsible for misinterpretations in a variety of experimental circumstances.

It will be of special interest to learn whether T lymphocytes and B lymphocytes are equally adept at producing IDS, as well as to test whether they are equally susceptible to its effects. Experiments designed to answer these questions are in progress.

ACKNOWLEDGEMENTS

This work was supported in part by the American Cancer Society ACS-T-463, the American Cancer Society Institutional Grant IN-62-G, the American Cancer Society, Florida Division F7OUF, NIH HD-00384, National Institution of General Medical Sciences Training Grant GM-01996, and National Institution of Allergy and Infectious Diseases Training Grant AI-00401-01.

REFERENCES

Adler, W. H., Takiguchi, T., Marsh, B. and Smith, R. T. (1970a). *J. exp. Med.* *131*, 242.

Adler, W. H., Takiguchi, T., Marsh, B. and Smith, R. T. (1970b). *J. Immun. 105,* 984-1000.

Adler, W. H., Peavy, D. and Smith, R. T. (1970c). *Cellular Immunology 1,* 78-91.

Adler, W. H., Takiguchi, T. and Smith, R. T. (1971). *In:* "In Vitro Methods in Cell-Mediated Immunity" (B. Bloom and P. Glade, eds), pp. 433-439. Academic Press, New York.

Mosier, D. E., Fitch, F. W., Rowley, D. A. and Davies, A. J. S. (1970). *Nature, Lond. 225,* 276-277.

Raidt, D. J., Mishell, R. I. and Dutton, R. W. (1968). *J. exp. Med. 128,* 681-698.

Shearer, G. M. and Cudkowicz, G. (1969). *J. exp. Med. 130,* 1243-1261.

Smith, R. T. and Landy, M. (1970). *In* "Immune Surveillance". Academic Press, New York. (In press.)

Smith, R. T., Lawrence, H. S. and Landy, M. (1969). *In* "Mediators of Cellular Immunity" (H. S. Lawrence and M. Landy, eds), pp. 44-50. Academic Press, New York.

Smith, R. T., Bausher, J. A. and Adler, W. H. (1970). *Am. J. Path. 60,* 495-504.

Purification and Characterization of two Mediators in Delayed Hypersensitivity: MIF and Leucotactic Factor for Monocytes *

HEINZ G. REMOLD, PETER A. WARD and JOHN R. DAVID

*Robert B. Brigham Hospital, Harvard Medical School, Boston, Mass.
and Immunology Branch, Armed Forces Institute of Pathology
Washington, D.C., U.S.A.*

Peritoneal exudate cells from guinea pigs exhibiting delayed hypersensitivity are inhibited from migrating out of capillary tubes by specific antigen (George and Vaughan, 1962; David *et al.,* 1964). Later studies have revealed that lymphocytes obtained from sensitized guinea pigs, when stimulated with specific antigen *in vitro,* produce a soluble substance which inhibits the migration of normal peritoneal cells (Bloom and Bennett, 1966; David, 1966). It is referred to as MIF. Recently, a number of soluble mediators from the supernatants of sensitized and stimulated guinea pig lymph node cells have been reported to have various biological activities as detected by various assay methods. These mediators include, in addition to MIF, a factor causing redness and mononuclear infiltration in the skin (Bennett and Bloom, 1968), a leucotactic factor (Ward *et al.,* 1969), factors that are cytotoxic to culture cells or prevent cell growth (Ruddle and Waksman, 1968; Williams and Granger, 1969; Kolb and Granger, 1968), a blastogenic factor that causes normal lymphocytes to divide (Valentine and Lawrence, 1969; Dumonde *et al.,* 1969) and substances with interferon-like activities (Green *et al.,* 1969).

This paper will deal with the characterization and purification of two of these mediators, the migration inhibitory factor and the leucotactic factor for monocytes. The experiments provide some data relating to the question: are the different effects of the lymphocyte factors due to different products of stimulated lymphocytes or are they merely multiple activities of single substances? The results show that MIF activity could be clearly separated from leucotactic activity on disc electrophoresis.

* Supported in part by National Institutes of Health Grants AI-07685, AI-08026, AIA-07291, AM-12051, AM-05577, RR-05669, and by a grant from the John A. Hartford Foundation.

411

METHODS

Assay for MIF activity. Peritoneal exudate cells from normal guinea pigs were induced by oil and collected as previously described (David *et al.,* 1964; David and Schlossmann, 1968). The cells were washed in Hank's balanced salt solution and suspended to 10% by volume in tissue culture medium consisting of MEM-PS containing 15% guinea pig serum. Capillary tubes were filled with the cell suspension, sealed with wax, and centrifuged. The tubes were cut, and the portion containing the cells placed in Mackaness-type chambers, two per chamber. The chambers were filled with fractions to be tested. Each dilution of each fraction was assayed on cells from two different guinea pigs or two different guinea pig cell pools. The fractions from active supernatants were always compared to the same fractions from control supernatants: normal medium (MEM-PS, 15% serum) and medium with antigen were included as additional controls.

The chambers were incubated for 18 hr at $37°C$, and the area of migration was measured by planimetry, as described previously.

$$\% \text{ inhibition} = 100 - \frac{\text{average migration of normal PE cells in fractions from lymphocytes incubated with antigen}}{\text{average migration of normal PE cells in fractions from lymphocytes incubated } \textit{without} \text{ antigen}} \times 100$$

It should be noted that these data are presented in percent of inhibition rather than in percent of migration as in previous papers. Using this assay at least 20% inhibition must be obtained for significant activity.

Chemotactic assay. Peritoneal exudates rich in mononuclear cells were induced in outbred guinea pigs and adult male New Zealand rabbits by the injection of mineral oil four days prior to abdominal paracentesis. Except where stated, rabbit mononuclear cells were used in the chemotactic assay. Details of this method and the chemotactic assay are given elsewhere (Ward, 1968). The chemotactic assay features the micropore filter method in which leukocytes migrate through the filter toward the opposite (lower) compartment containing a chemotactic material (usually in a volume of 0.2 ml, diluted to 1.2 ml in Medium 199 without serum). Chemotactic activity is defined as the number of cells that have migrated completely through the filter in five fields, examined under light microscopy at high power.

The methods used for Sephadex chromatography, disc gel electrophoresis, and ultracentrifugation are described in detail elsewhere (Remold *et al.,* 1970).

RESULTS

Fractionation of MIF on Sephadex G 100. The results of seven experiments show that after supernatants were filtered on Sephadex G 100, MIF activity was recovered only from the region containing molecules the size of albumin (G-III) and in the post-albumin region (G-IV). G-III had about half the activity of G-IV; when 0.2 ml was tested G-III gave a mean inhibition of $27.8 \pm 7.1\%$ and G-IV gave a mean inhibition of $61.33 \pm 2.59\%$. In three out of seven experiments there was only minimal activity detectable in fraction G-III. In contrast fraction G-IV always gave good inhibition of macrophages. No activity was found in G-I plus II, the fraction containing immunoglobulins and the stimulating antigen OCB-BGG, nor was any MIF activity recoverable from fraction G-V containing the low molecular weight materials.

Molecular weight can be estimated from the mobility of macromolecules in gel filtration (Crestfield *et al.,* 1962). Calculated from the mobility of ^{125}I-rabbit serum albumin and ^{125}I-lysozyme, G-III contains material of a molecular weight between 35,000 and 55,000, assuming that MIF is a globular molecule.

It is of note that both the MIF-rich and the control supernatants contain the same amounts of protein and give identical extinction patterns at 280 nm following filtration on Sephadex G 100. It should be noted that most of the antigen OCB-BGG is removed by precipitation when the supernatants are dialyzed in distilled water, and by aggregation after lyophilization prior to gel filtration and, thus, does not contribute to the extinction pattern at 280 nm of the effluents.

Fractionation of Chemotactic Factor on Sephadex G 100. Similar experiments were carried out in which supernatants were fractionated by gel filtration on Sephadex G 100, and the samples tested for chemotactic activity. Chemotactic activity was consistently found in G-III (chemotactic activity of a typical experiment was 80), the fraction containing molecules the size of albumin. More activity was in G-IV (chemotactic activity of a typical experiment was 144), the fraction containing materials slightly smaller than albumin.

Fractionation of MIF on sucrose gradients. Fractionation of MIF-containing supernatants using sucrose density gradient ultracentrifugation confirmed the results from gel filtration on Sephadex G 100. When 40 ml equivalents of active and control supernatants were centrifuged on sucrose gradients, the fractions containing the albumin tracer and fractions containing slightly lighter molecules were active (respectively 20%, and 48% inhibition).

The fractions that contained substances larger than albumin or very small molecules were not active in any of the concentrations used (both 0% inhibition). About 75% of the initial activity placed on sucrose gradients was recovered after centrifugation. The sedimentation constant computed from the peak of activity is approximately 3S. This corresponds to a molecular weight of approximately 40,000, assuming a globular molecule. The broad area of activity obtained from the sucrose gradient as well as the limited number of fractions examined did not allow more precise calculations of molecular weight.

Polyacrylamide gel electrophoresis of MIF. Active fractions G-III or G-IV and equivalent control fractions were subjected to electrophoresis on acrylamide gel. After the run, the gels were removed from the glass columns and placed on a transversal gel slicer, then cut with a razor at every fifth slot starting from the buffer front. Identical fractions from five gels were pooled and placed in glass columns and the material eluted by electrophoresis. In eight experiments in which fractions G-III were fractionated on gels, activity was consistently found in A-3, the fraction immediately anodal to the albumin fraction. In six out of nine experiments with G-III, no activity was detected in A-4, the albumin-containing fraction. However, activity in A-4, as well as in A-3, was present in three of the experiments.

Fraction G-IV was separated by disc electrophoresis in five experiments. In four of them activity was found in A-3; in two there was activity in A-4, and in one there was activity in A-5, probably due to overloading.

The average inhibition produced by A-3 was 41% and by A-4, 23%. There was no significant difference between values obtained from Sephadex fractions G-III and G-IV.

The amount of MIF activity recovered from A-3 was approximately 20% of that placed on these gels (after elution with 1.5 mA overnight).

In fraction G-III there were three strong and 4-5 faint bands in the prealbumin and albumin regions of the gel that were not present in the patterns of guinea pig serum. Separate gel electrophoresis of G-IV revealed a pronounced increased intensity of some bands in the prealbumin region, whereas the strong band in the albumin region was much fainter. There was a 3.8- and 22.8-fold purification in G-III and G-IV respectively. The degree of purification of the disc electrophoresis fractions A-3 and A-4, however, was difficult to estimate because the amount of protein present in the active gel fractions was below the sensitivity of the Lowry method. Since the protein content in A-3, which has three faint bands, is less than 0.1 μg, the degree of purification is at least 1600 in terms of activity per original protein content.

Density gradient analysis of chemotactic factor. The electrophoretic fraction A-4, that was rich in chemotactic activity, was further fractionated by density gradient ultracentrifugation and the sedimentation behavior of the mononuclear cell chemotactic factor assessed. A well-defined zone of chemotactic activity was present in a position slightly above the bovine serum albumin (BSA) marker (chemotactic activity of a typical experiment was 120). In other experiments the chemotactic factor occupied a position coinciding with BSA. In addition to this chemotactic factor, a second, more rapidly sedimenting, zone of chemotactic activity was found, lying in a position similar to that of the rabbit IgG marker (chemotactic activity 50). The less well-defined nature of this zone of activity, together with the fact that molecules larger than BSA should have been excluded from the material placed on the gradient, suggests that the more rapidly sedimenting zone of chemotactic activity may represent aggregated products of the factor present in the upper portion of the gradient. This tendency to aggregate may explain the occasional presence of chemotactic activity in Sephadex G 100 fractions G-I and II, containing molecules larger than albumin.

Separation of mononuclear cell chemotactic factor by disc electrophoresis. In order to determine whether the chemotactic factor and MIF could be separated on the basis of differences in electric charge, five different G-III or G-IV Sephadex G 100 fractions were fractionated by acrylamide disc gel electrophoresis. The chemotactic values obtained from the unstimulated control fractions were subtracted from the corresponding antigen-stimulated fractions to determine the chemotactic activity. In each of five experiments, the peak of chemotactic activity was found in gel fraction A-4 (the fraction that also contains albumin). There was a slight overlap of activity in fraction A-5 in four experiments as reported (Remold *et al.*, 1970). In two experiments MIF activity was also found in A-4. In several instances a fraction contained either MIF or chemotactic activity, but not both. These experiments indicate that the mononuclear cell chemotactic factor can be dissociated from MIF by the slower electrophoretic migration of the chemotactic factor. This suggests that it is less negatively charged than MIF.

MIF was further characterized by digestion with insolubilized chymotrypsin and neuraminidase. Both enzymes destroyed MIF activity. This result, together with the findings of disc electrophoresis studies and those from CsCl gradient centrifugation, suggest that MIF is an acidic α-glycoprotein with a molecular weight of 35,000-55,000 (Remold and David, 1970). Other mediators are being analyzed at present using these methods.

DISCUSSION

The experiments described above demonstrate that migration inhibitory factor and leucotactic factor can be separated by polyacrylamide disc electrophoresis (Ward *et al.*, 1970).

MIF and leucotactic factor were both eluted from Sephadex G 100 in a broad area with the highest activity in the fractions eluting immediately after the albumin peak. These findings were substantiated in experiments using sucrose density gradients. Both methods indicate a molecular weight in the range of 35,000-55,000 for both factors.

Chemotactic activity for mononuclear cells is consistently eluted from acrylamide gel fractions containing albumin (A-4), and is not found in the prealbumin region (A-3). In contrast, MIF activity is regularly recovered from fraction A-3. Occasionally MIF is found in both the albumin and the prealbumin regions. In several experiments it was clearly demonstrated that prealbumin fractions with a high level of MIF activity had no chemotactic activity, and, the reverse, that a fraction containing albumin with no MIF activity had good chemotactic activity. It should be emphasized that in these experiments both biological activities were assayed using aliquots of the same sample.

As the chemotactic and MIF activities were found in adjacent gels, several questions had to be explored to rule out the possibility that the apparent separation was an artifact. Because the peak activity of MIF was mainly in fraction A-3, but partly in fraction A-4, it was possible that MIF might be chemotactic if diluted (as in fraction A-4), but not when concentrated (as in fraction A-3). In four experiments fraction A-3 was diluted and was never found to be chemotactic. In contrast, fraction A-4 remained chemotactic when diluted, but was more active when concentrated; it consistently showed a dose response effect. The original supernatants only lost activity when diluted—never gained.

It was necessary to determine whether MIF in fraction A-3 could become chemotactic if bound to one of the proteins in fraction A-4; for example, albumin. If this were the case, it would result in an apparent separation of activities, that were, however, still due to the same molecules, some bound, the others unbound. In order to test this hypothesis, eluates from the control fraction A-4 (from unstimulated cells and with no chemotactic activity) were combined with MIF-rich fraction A-3, to determine whether MIF would attach to the proteins in fraction A-4 and become chemotactic. However such mixtures had no chemotactic activity.

Further studies support the presumption of two separate molecules. When fraction A-4 containing chemotactic acitivity was placed with the cells in the upper chamber it inhibited, as expected, the chemotactic effect of fraction A-4 in the lower chamber, by preventing the formation of a chemotactic gradient. In contrast, the presence of fraction A-3 (with MIF) in the upper chamber did not prevent the effect of chemotactic fraction A-4 in the lower chamber. These findings, taken together, strongly support the existence of two separate substances—one, a chemotactic factor, the other, MIF.

These studies demonstrate that antigen-stimulated lymphocytes produce a complex mixture of mediators, some of which appear to be distinct molecules. It will be of great interest to determine how many of the other biologic activities, such as cytotoxic factor, cloning inhibition factors, blastogenic factors, interferon activity, can be distinguished by similar means.

REFERENCES

Bennett, B. and Bloom, B. R. (1968). *Proc. natn. Acad. Sci. U.S.A. 59,* 756-762.
Bloom, B. R. and Bennett, B. (1966). *Science N.Y. 153,* 80-82.
Crestfield, A. M., Stein, W. H. and Moore, S. (1962). *Archs. Biochem. Biophys. Suppl. 1,* 217-222.
David, J. R. (1966). *Proc. natn. Acad. Sci. U.S.A. 56,* 72-77.

David, J. R. and Schlossmann, S. F. (1968). *J. exp. Med. 128*, 1451-1459.
David, J. R., Al-Askari, S., Lawrence, H. S. and Thomas, L. (1964). *J. Immun. 93*, 264-273.
Dumonde, D. C., Wolstencroft, R. A., Panay, G. S., Matthew, M., Morley, J. and Howson, W. T. (1969). *Nature, Lond. 224*, 38-42.
George, M. and Vaughan, J. H. (1962). *Proc. Soc. exp. Biol. Med. 111*, 514-521.
Green, J. A., Cooperband, S. R. and Kibrick, S. (1969). *Science N.Y. 164*, 1415-1417.
Kolb, W. P. and Granger, G. A. (1968). *Proc. natn. Acad. Sci. U.S.A. 61*, 1250-1255.
Remold, H. G. and David, J. R. (1970). *Fedn. Proc. 29*, 339.
Remold, H. G., Katz, A. B., Haber, E. and David, J. R. (1970). *Cellular Immunology 1*, 133-145.
Ruddle, N. H. and Waksman, B. H. (1968). *J. exp. Med. 128*, 1267-1280.
Valentine, F. T. and Lawrence, H. S. (1969). *Science N.Y. 165*, 1014-1016.
Ward, P. A. (1968). *J. exp. Med. 128*, 1201-1222.
Ward, P. A., Remold, H. G. and David, J. R. (1969). *Science N.Y. 163*, 1079.
Ward, P. A., Remold, H. G. and David, J. R. (1970). *Cellular Immunology 1*, 162-174.
Williams, T. W. and Granger, G. A. (1969). *J. Immun. 103*, 170-178.

Studies on the Mechanism of Antigenic Competition

GÖRAN MÖLLER and OLOF SJÖBERG

Division of Immunobiology
Karolinska Institutet
Wallenberg Laboratory
5-10705 Stockholm 50, Sweden

Antigenic competition may occur when two antigens are used to immunize the same animal, and causes a decreased antibody response to one or both of the antigens as compared to the response when each antigen is injected separately (for review see Adler, 1964). The mechanism of antigenic competition remains unknown. The various hypotheses advanced to explain it can be divided into two main categories. (1) It represents an immunologically specific effect. Thus, the antigens may compete for antigen sensitive cells, or the antigens may cross react and the phenomenon be due to induction of tolerance or antibody suppression. (2) Antigenic competition is an immunologically non-specific phenomenon, such as competition for some limiting factors (space, nutrients) or for some of the steps of antigen processing or localization. It could also be caused by the production of inhibitors of antibody synthesis.

Certain recent findings are particularly relevant to an understanding of the mechanism of antigenic competition. Thus, it has been demonstrated (Adler, 1964; Radovich and Talmage, 1967) that antigenic competition was most pronounced when there was a time interval between the two antigens injected. Furthermore, Radovich and Talmage (1967) showed that competition between antigens given to irradiated animals repopulated with spleen cells increased when the animals received a larger number of spleen cells. This suggested that competition was caused by non-specific effects rather than by specific competition for antigen sensitive cells. Brody and Siskind (1969), in contrast to Schechter (1968), demonstrated that antigenic competition between two haptens was equally pronounced whether the haptens were attached to the same or different carriers. These findings argue against competition for common antigen focusing cells, of the type outlined in the two-cell concept of Jerne (1967) and Mitchison (1967). In addition, Brody and Siskind (1969) found that the affinity of the antibodies produced during competition was the same as that in animals immunized against only one antigen, suggesting that neither tolerance nor antibody suppression operated in antigenic competition.

The present experiments represent an attempt to analyse the mechanism of antigenic competition. Particular interest was directed towards the effect of antigenic competition on the number of antigen sensitive cells in the hosts. Furthermore, the ability of spleen cells to produce antibody after transfer into animals exhibiting antigenic competition was investigated in order to detect a possible non-specific influence of antigenic competition on cellular antibody synthesis.

Finally attempts were made to study the phenomenon of antigenic competition by an *in vitro* system, utilizing induction of DNA synthesis in sensitized human lymphocytes confronted with different antigens.

MATERIALS AND METHODS

In vivo studies. The mice were of the inbred strains A, CBA, B10.5M(5M) and F_1 hybrids between them, and of the non-inbred NMRI strain.

The antigens used were sheep red blood cells (SRBC) and horse red blood cells (HRBC). For routine tests 4×10^8 cells were injected intravenously. Other doses and the time intervals between the two antigens will be specified in the text.

Antibody-producing cells were detected by the agar-plaque technique of Jerne and Nordin (1963) for direct plaque-forming cells (PFC) and by the modification described by Dresser and Wortis (1965) for indirect PFC. In the latter test anti-mouse gamma globulin antiserum was raised by immunizing rabbits with complexes of *Salmonella adelaide* and specific mouse antibodies as described before (Möller, 1968). The developing antiserum was used in a dilution of 1 : 100. Although not strictly correct, the direct PFC will be referred to as 19S PFC and the indirect as 7S PFC in the text. The developing serum suppresses more than 90% of the direct PFC and detects most indirect PFC (Möller, 1968).

A transfer system was used to study antigen sensitive cells. Spleen cells from normal or immune animals were made up to 50×10^6 cells/ml and mixed with SRBC. 0.25 ml of this mixture was given intravenously to 350 r irradiated syngeneic recipients, each of them receiving 10^7 spleen cells mixed with 4×10^8 SRBC. Mice were irradiated with 350 r at a rate of 425 r/min. X-Rays were generated in a Siemens X-ray machine at 185 kV and 15 mA, filtration 0.5 mm Cu. Seven days later the number of 19S and 7S PFC was determined in the recipient's spleen. The cell dose used for transfer is such that the number of PFC appearing in the recipient is proportional to the number of antigen sensitive cells in the inoculum (Shearer *et al.*, 1968; Kennedy *et al.*, 1966).

In vitro studies. Lymphocytes were prepared from human venous blood as previously described (Möller, 1969). 10^6 cells were cultivated in Eagle's medium in Earle's solution containing 10% human AB serum. ^{14}C-thymidine (0.2 µc-25 µc/µg) was added for 24 hr at time periods specified in the text. Determination of radioactivity was performed as described before (Möller, 1969). During culture lymphocytes were exposed to different but low concentrations of one antigen. Two to three days after, an optimally stimulating concentration of the same or a different antigen was added, and the degree of DNA synthesis was determined. This was compared to that obtained when the two antigens were added individually in various concentrations, or to that found when they were admixed simultaneously with the lymphocytes.

PPD was obtained from the State Serum Institute, Copenhagen and tetanus toxoid from the Wellcome Company, England. The antigen concentrations and time of addition will be specified in the text.

Results

Kinetics of antigen competition. As shown by Adler (1964), Radovich and Talmage (1967), and Eidinger *et al.* (1968) optimal antigenic competition between two different erythrocytes occurs when they are spaced in time. These findings were substantiated in the present studies (Table I). Simultaneous injection of HRBC and SRBC did not lead to antigenic competition. On the contrary, a slight degree of stimulation was obtained (experiments 1 and 2). However, if one antigen was given 3-10 days before the second, a marked degree of competition was obtained. Furthermore, antigenic competition was pronounced when the second antigen was administered 3-11 days after a secondary immunization to the first antigen (experiment 3). Competition was observed when either SRBC or HRBC were used to pretreat the animals.

Effect of antigenic competition on antigen sensitive cells. Previous studies on antigenic competition have been concerned with the synthetic phase of the immune response. In order to investigate whether antigenic competition affects also the number of antigen sensitive cells, a transfer system was employed. The number of antigen sensitive spleen cells to SRBC was studied in normal animals and in animals that had been immunized twice with HRBC. Spleen cells from these animals were mixed with SRBC and transferred into irradiated syngeneic recipients. The transfer was made five days after the second immunization with HRBC, at the time of maximal antigenic competition and the immune response in the recipients tested seven days later. Parallel groups were immunized directly with SRBC and the extent of antigenic competition determined five days later as before.

The 19S PFC response in recipients of spleen cells from HRBC pretreated mice was slightly lower than that in recipients of untreated cells in two out of three experiments (Table II). In the third experiment there was an increased 19S

TABLE I

Kinetics of antigenic competition between HRBC and SRBC

Exp. (mouse strain)	Immunizing antigen and time of injection				Test:		19S anti-SRBC PFC/10⁶ cells (C. of V.)	% inhibition[a]
	1	At day	2	At day	Antigen	At day		
1 (NMRI)	—	—	SRBC	0	SRBC	4	376 (4.7)	
	HRBC	0	SRBC	0	SRBC	4	722 (9.6)	−92.0
	HRBC	0	—	—	SRBC	4	3 (61.5)	
2 (NMRI)	—	—	SRBC	0	SRBC	5	629 (3.2)	
	HRBC	0	SRBC	0	SRBC	5	815 (5.6)	−29.6
	—	—	SRBC	6	SRBC	9	368 (6.6)	
	HRBC	0	SRBC	6	SRBC	9	7 (90.9)	98.1
3 (CBA)	—	—	SRBC	0	SRBC	4	690 (1.0)	
	HRBC	−16, −3	SRBC	0	SRBC	4	85 (9.6)	87.7
	HRBC	0	SRBC	0	SRBC	4	885 (7.3)	−28.3
4 (A × 5 M)F₁	—	—	HRBC	0	HRBC	5	205 (22.1)	
	SRBC	−13, −3	HRBC	0	HRBC	5	13 (22.7)	93.7

a % inhibition = PFC in group given one antigen minus PFC in group given two antigens divided by PFC in first group, ×100. Minus sign indicates stimulation of response.

TABLE II

Effect of antigenic competition on antigen sensitive cells to SRBC

Strain	Donors given HRBC at day	No. of donors	No. of recipients	19S PFC/10^6 cells (C. of V.)	% inhibition[a] After transfer	% inhibition[a] In the donor	7S PFC/10^6 cells (C. of V.)	% inhibition[a] After transfer	% inhibition[a] In the donor
CBA	—	3	6	130 (5.7)		99.0	1267 (5.3)		
	−29, −5	3	6	77 (10.9)	40.8	99.0	1310 (0.7)	−3.4	93.3
CBA	—	4	8	60 (14.2)			41 (23.5)		
	−20, −5	4	8	41 (13.4)	31.7	72.4	60 (14.2)	−46.3	n.t.
(A x 5 M)F$_1$	—	4	8	8 (39.5)			15 (63.6)		
	−20, −5	4	8	28 (19.8)	−250	64.7	24 (59.8)	−60	n.t

10^7 spleen cells from untreated mice or mice previously immunized twice with HRBC were mixed with 4×10^8 SRBC and injected into 350 r irradiated syngeneic recipients. Seven days later the number of anti-SRBC PFC in the recipient spleens was determined. Each donor suspension was injected into two recipients. n.t. = not tested.

[a] % inhibition after transfer = inhibition (see Table I) of PFC response in recipients of spleen cells from donors given HRBC as compared to PFC response in recipients of unprimed cells. % inhibition in donors refers to parallel tests of the inhibition of PFC response to SRBC in the donor groups given SRBC and tested five days later.

PFC response. The 7S response was higher in recipients given spleen cells from HRBC immunized mice. Thus there was little, if any, effect of pretreating mice with HRBC on the number of antigen sensitive cells to SRBC. However, the immune response to SRBC was severely impaired in parallel groups of animals pretreated with HRBC and given SRBC at the same time as the transfer was performed in the above groups (Table II). Thus antigenic competition appears to affect the increase in antibody-producing cells, but does not lead to a decrease in the number of antigen sensitive cells.

Effect of HRBC pretreatment of recipients on their ability to support antibody synthesis against SRBC by transferred cells. The failure to detect an effect of antigenic competition on the number of antigen sensitive cells does not support the hypothesis that antigens compete for antigen sensitive cells. It rather suggests that multiplication or differentiation of antibody-producing cells is impaired in a non-specific way. If this is so, it would be expected that animals pretreated with one antigen under conditions resulting in antigenic competition, would not adequately support the immune response of adoptively transferred spleen cells stimulated with another antigen. The non-specific factors interfering with multiplication of antibody-producing cells in the animals should also be effective against adoptively transferred cells.

Transfer experiments were performed in order to study whether antigenic competition involves a non-specific impairment of antibody production. Spleen cells were mixed with SRBC and thereafter transferred into lethally irradiated syngeneic recipients, that had been previously immunized twice against HRBC. In the controls, the transferred spleen cells were given to previously untreated animals.

Previous HRBC treatment markedly suppressed both 19S and 7S antibody production by adoptively transferred spleen cells (Table III). Analogous results were obtained when the recipients were irradiated with 350 r (Table III).

In these experiments the same number of antigen-sensitive cells mixed with antigen were transferred into two types of recipients—untreated or previously immunized with a non-cross reacting antigen. Marked suppression of antibody synthesis occurred if the recipients had been pretreated with a different antigen. Thus the effect appears to be caused by some impairment of the differentiation or multiplication steps leading to development of antibody-producing cells. This conclusion is further substantiated by the demonstration (Table III) that spleen cells from donors previously immunized with SRBC were suppressed to the same degree after transfer into irradiated recipients previously sensitized against HRBC. The number of antigen-sensitive cells in these donors was greatly increased by the specific pre-immunization, but still the development of PFC was suppressed by from 54% to 98% after transfer into HRBC pretreated mice as compared to the response in non-immunized irradiated recipients. Thus animals exhibiting antigenic competition do not constitute a suitable environment for

TABLE III

Transfer of CBA spleen cells mixed with SRBC into CBA recipients previously immunized against HRBC

Recipients immunized against HRBC at days	X-Irradiation of recipients	Immune status of donors	19S PFC/10⁶ cells (coeff. of variation) against SRBC in recipients	% inhibition	7S PFC/10⁶ cells (coeff. of variation) against SRBC in recipients	% inhibition
—	—	normal	89 (12.5)		3,210 (4.4)	
−32, −6	—	normal	34 (23.1)	61.8	549 (17.3)	82.9
—	—	imm. SRBC	35 (10.4)		2,465 (0.9)	
−32, −6	—	imm. SRBC	26 (37.6)	25.7	1,124 (6.4)	54.4
—	350 r	normal	199 (10.3)		732 (8.8)	
−29, −5	350 r	normal	72 (18.3)	63.8	59 (66.9)	91.9
—	350 r	imm. SRBC	2182 (21.1)		55,940 (3.1)	
−29, −5	350 r	imm. SRBC	21 (119.0)	99.0	3,760 (11.2)	93.3
—	350 r	normal	67 (13.6)			

antibody production against a non-cross reacting antigen by adoptively transferred spleen cells.

Effect of antigenic competition on DNA synthesis in the spleen. Antigenic competition was induced by pretreating mice two or three times with SRBC. Three or four days after the last injection the mice were divided into two groups; one received HRBC and the other remained untreated. Previously untreated animals served as controls and they were also divided into two groups, one was immunized with HRBC and the other remained untreated. Three days after the injection of HRBC the animals were given 20 μc trypsinized ^{3}H-thymidine in 1 ml saline i.v. and 1 hr later they were killed, their spleen cells removed and put in ice cold saline. A cell suspension was prepared and washed three times. The cells were dissolved in formic acid and the incorporation of ^{3}H-thymidine determined in a liquid scintillation counter. Antigenic competition resulted in a marked decrease in DNA synthesis in the spleen cells (Table IV). The counts per minute per 10^6 or per spleen were lower in animals where there was antigenic competition than in untreated animals. Thus antigenic competition appears to have a general influence on the degree of DNA synthesis in the spleen.

TABLE IV

Suppression of DNA synthesis in spleens of mice subjected to antigenic competition

Exp. no.	Pretreatment of mice	Treatment at day 0	cpm ± SE/10^6 spleen cells	cpm ± SE/spleen
1	—	HRBC	725 ± 107	72,654 ± 5872
	SRBC −14, −4	HRBC	255 ± 57	22,072 ± 2064
2	—	—	27.5 ± 2	2,681 ± 259
	SRBC −10, −6, −3	—	8.5 ± 0.5	1,434 ± 106
	—	HRBC	51.7 ± 8	7,461 ± 1115
	SRBC −10, −6, −3	HRBC	7.1 ± 0.6	1,882 ± 110

The various groups were pretreated with SRBC as indicated. At day 0 some groups were given HRBC. Three days later all mice were given 20 μC ^{3}H-thymidine i.v. They were killed 1 hr later and their spleens suspended, the number of cells counted and their isotope content determined.

Effect on DNA synthesis of human lymphocytes pretreated with antigen in vitro. In an attempt to find an *in vitro* model for the previous *in vivo* studies attention was focused on induction of DNA synthesis in sensitized human lymphocytes exposed to the corresponding antigens. It was previously found that a mixed lymphocyte culture that had been initiated three days earlier, markedly suppressed the ability of the cells to respond to non-specific mitogens, such as phytohaemagglutinin and anti-lymphocyte serum (Möller, 1970). This antigen-induced suppression of stimulation of DNA synthesis after contact with non-specific mitogens appeared to be analogous to the suppression of DNA synthesis by antigenic competition in the spleen. In order to study this further, lymphocytes from human beings sensitized to PPD were cultivated in the presence of low concentrations of PPD, which by themselves did not stimulate DNA synthesis to any substantial degree. Two or three days later the medium of the cultures was changed and replaced with fresh medium containing an optimal concentration of PPD or, in other groups, by an optimal concentration of another antigen (tetanus toxoid). In control groups lymphocytes were cultivated in the absence of PPD for 2-3 days and thereafter treated with an optimal concentration of one of the two antigens involved, or they remained untreated. Three days after the change of medium and the addition of the optimal concentrations of the antigens, ^{14}C-thymidine was added for 24 hr and the

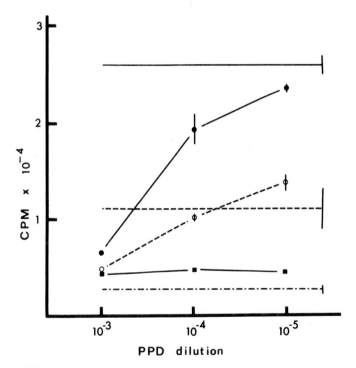

Fig. 1. ^{14}C-thymidine incorporation into untreated human lymphocytes (–·–) or lymphocytes exposed to PPD 1/20 (horizontal line ———) or tetanus toxoid 1/20 (horizontal line – – – –) after two days in culture. In other groups cells were pretreated with the indicated dilutions of PPD for two days and thereafter PPD 1/20 (●——●) or tetanus toxoid 1/20 (O– –O) was added. Cultures exposed to the different low concentrations of PPD alone are also indicated (■——■).

degree of DNA synthesis determined. Pretreatment of cells with low concentrations of PPD caused a pronounced suppression of their ability to respond not only to PPD but also to an unrelated antigen (tetanus toxoid) added three days later (Fig. 1). However, if PPD or tetanus toxoid was added simultaneously with the various low concentrations of PPD there was no suppression of DNA synthesis (Fig. 2).

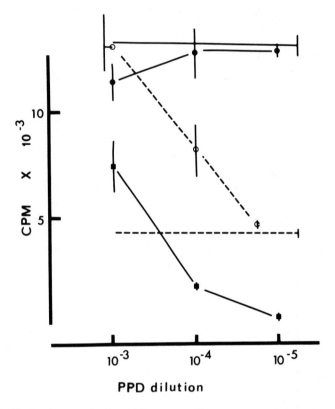

Fig. 2. ^{14}C-thymidine incorporation into human lymphocytes treated with simultaneous addition of PPD and tetanus toxoid. Symbols as in Fig. 1. In this case PPD treatment did not interfere with the response to tetanus toxoid.

Thus it would appear that lymphocytes exposed to one antigen for three days become incompetent to respond to the same antigen and also to another antigen. This is analogous to the antigenic competition phenomenon, but in addition the findings demonstrate that cells treated with low doses of one antigen become incompetent to respond to optimal concentration of the same antigen. Since the response to other antigens is also influenced, the findings strongly suggest that pretreatment changed the culture environment in such a way that cells having other specificities also became unable to respond to their own antigen.

DISCUSSION

It is implied in the clonal selection hypothesis that immunocompetent cells can recognize only one antigen. Thus, in terms of this concept, antigenic competition could not be due to competition for antigen sensitive cells. However, the phenomenon of antigenic competition has been used as an argument for the existence of multipotent antigen sensitive cells (Schechter, 1968; Lawrence and Simonsen, 1967). The demonstration that maximal competition occurs when the two antigens are spaced in time argues against competition for antigen sensitive cells, since that would occur also when two antigens are introduced simultaneously. Furthermore, when a low number of spleen cells was used to repopulate irradiated mice, antigenic competition was less than when a larger number of cells was employed (Radovich and Talmage, 1967). Competition for antigen sensitive cells would be expected to have the opposite result. The present finding, that the number of antigen sensitive cells was not decreased by antigenic competition, also argues against multipotent antigen sensitive cells. The possibility that two antigens compete for helper cells directed against a common carrier determinant is made unlikely by the finding of Brody and Siskind (1969) that competition is equally pronounced whether haptens are attached to the same or to different carrier molecules. Opposite results were reported previously, however (Schechter, 1968), the basis for this discrepancy being unknown. The possibility that antigenic competition is due to tolerance or antibody suppression is based on the assumption that the antigens cross react. Since it was demonstrated (Brody and Siskind, 1969) that the affinity of the antibodies produced in antigenic competition was the same as when only one antigen was used, it is unlikely that tolerance or antibody suppression operates. A lower affinity would be expected in partial tolerance (Theis and Siskind, 1968) and a higher in antibody suppression (Brody et al., 1967). It seems likely, therefore, that antigenic competition is non-specific, and caused by competition for some limiting factors (space, nutrient) or for some aspects of antigen processing or localization, or is due to the production of inhibitors of the immune response. The fact that competition is optimal when the antigens are spaced in time and is greater when a large cell number is used to repopulate irradiated recipients makes it unlikely that "antigenic competition is in fact competition for anything" (Radovich and Talmage, 1967).

In the present experiments, adoptively transferred normal or immune spleen cells mixed with antigen produced less antibody in irradiated recipients previously immunized with a non-cross reacting antigen, than in non-immunized irradiated animals. It is unlikely, but cannot be disproved, that the pretreated animals would have an altered ability to localize the transferred cells and the antigen. The recipients were irradiated with 350 r, which depletes the lymphoid system and causes drastic changes in the architecture of the lymphoid organs. Since the number of antigen sensitive cells is normal in antigen-pretreated mice, it is unlikely that the antigen sensitive cells are killed. It seems more probable that the multiplication or differentiation of the antibody-producing cells is prevented for a short time period, as was actually demonstrated. The *in vitro* studies reported support this conclusion.

In cell-mediated immune reactions, presumed to be carried out mainly by thymus-dependent lymphocytes a variety of non-specific humoral effector molecules are produced with different biological effects. Thus, they can inhibit migration of macrophages and lymphocytes out of capillary tubes *in vitro*, they are competent to transform lymphocytes morphologically and initiate their DNA synthesis and they may transform lymphocytes into a cytotoxic state (David, 1966; Bloom and Bennett, 1966; Dumonde *et al.*, 1969; Švejcar *et al.*, 1968; Falk *et al.*, 1970; Spitler and Lawrence, 1969). It is probable that these factors serve to amplify cell-mediated immunity by acting on non-committed thymus-dependent lymphocytes, thereby transforming them into non-specific effector cells (for discussion see "Mediators of Cellular Immunity", 1969). In the present experiments bone marrow-derived antibody-producing cells have been studied. It seems possible that humoral factors released from antigen sensitive thymus-dependent antigen-stimulated lymphocytes are competent to affect the bone marrow-derived cells responsible for humoral antibody synthesis. This may be either directly, for instance by preventing division of such cells, or indirectly, for instance by making the specific antigen sensitive thymus-dependent cells incompetent to interact with and stimulate the bone marrow-derived cells to differentiate into antibody releasing lymphocytes. In both cases the humoral factors could prevent the appearance of antibody-producing cells, thus causing the phenomenon of antigenic competition. This effect of thymus-dependent lymphocytes would be purposeful in cell-mediated immune reactions, since humoral antibodies may efficiently suppress the contact between the thymus-dependent lymphocytes and the target antigen, because the antibodies and the lymphocytes compete for the same antigen. Therefore, humoral antibody synthesis may be a disadvantage in various tissue-damaging immune processes. It would follow from this concept that antigens that do not stimulate thymus-dependent lymphocytes would not cause antigenic competition.

ACKNOWLEDGEMENTS

This work was supported by the Swedish Medical Research Council, the Swedish Cancer Society and the Damon Runyon Memorial Fund (DRG-1038).

The technical assistance of Mrs Birgitta Frimodig, Mrs Kerstin Andersson, Miss Lena Lundin and Mrs Catharina Teichert is gratefully acknowledged.

SUMMARY

Antigenic competition between the primary immune response to non-cross reacting sheep and horse red blood cells was observed only when the two antigens were spaced in time. A secondary immune response to one antigen markedly suppressed the primary response to the other. The number of antigen sensitive cells was the same in normal mice and in animals expressing antigenic competition, as demonstrated by a transfer system. Adoptive transfer of normal or sensitized spleen cells mixed with antigen, into irradiated recipients exhibiting antigenic competition resulted in marked suppression of cellular antibody production, as compared to the response in non-immunized recipients. DNA synthesis in spleens of animals exhibiting antigenic competition as measured by ^{3}H-thymidine uptake was markedly suppressed, and was lower than that in untreated animals.

By treating human lymphocytes in culture with non-mitogenic doses of PPD for two days their ability to respond to an unrelated antigen (tetanus toxoid) or to an optimal dose of PPD added two days later was markedly suppressed. This may represent an *in vitro* counterpart of the *in vivo* findings on antigenic competition.

REFERENCES

Adler, F. L. (1964). *Progr. in Allergy 8*, 41-57.
Bloom, B. R. and Bennett, B. (1966). *Science, N.Y. 153*, 80-82.
Brody, N. I. and Siskind, G. W. (1969). *J. exp. Med. 130*, 821-832.
Brody, N. I., Walker, J. G. and Siskind, G. W. (1967). *J. exp. Med. 126*, 81-91.
David, J. R. (1966). *Proc. Natn. Acad. Sci. U.S.A. 56*, 72-77.
Dresser, D. W. and Wortis, H. H. (1965). *Nature, Lond. 208*, 859-861.
Dumonde, D. C., Wolstencroft, R. A., Panayi, G. S., Matthew, M., Morley, J. and Howson, W. T. (1969). *Nature, Lond. 224*, 38-42.
Eidinger, D., Khan, S. A. and Millar, K. G. (1968). *J. exp. Med. 128*, 1183-1200.
Falk, R. E., Falk, J. A., Möller, E. and Möller, G. (1970). *Cellular Immunology 1*, 150-161.
Jerne, N. K. (1967). *Cold Spring Harb. Symp. quant. Biol. 32*, 591-603.
Jerne, N. K. and Nordin, A. A. (1963). *Science, N. Y. 140*, 405.
Kennedy, J. C., Till, J. E., Siminovitch, L. and McCulloch, E. A. (1966). *J. Immun. 96*, 973-980.
Lawrence, W., Jr. and Simonsen, M. (1967). *Transplantation 5*, 1304-1322.
"Mediators of Cellular Immunity" (M. Landy and H. S. Lawrence, eds). Academic Press, New York.
Mitchison, N. A. (1967). *Cold Spring Harb. Symp. quant. Biol. 32*, 431-439.
Möller, G. (1968). *J. exp. Med. 127*, 291-306.

Möller, G. (1969). *Clin. exp. Immun. 4*, 65-82.
Möller, G. (1971). *Immunology 20*, 597-609.
Radovich, J. and Talmage, D. W. (1967). *Science, N.Y. 158*, 512-514.
Schechter, I. (1968). *J. exp. Med. 127*, 237-250.
Shearer, G. M., Cudkowicz, G., Connel, M. J. and Priore, R. L. (1968). *J. exp. Med. 128*, 437-457.
Spitler, L. E. and Lawrence, H. S. (1969). *J. Immun. 103*, 1072-1077.
Švejcar, J., Pekárek, J. and Johanovský, J. (1968). *Immunology 15*, 1-11.
Theis, G. A. and Siskind, G. W. (1968). *J. Immun. 100*, 138-141.

EFFECTOR MECHANISMS SECONDARY
TO LYMPHOCYTE ACTIVATION

Studies of Surface-bound Activities in Lymphocyte-mediated Cytotoxic Reactions

P. PERLMANN and H. PERLMANN

Department of Immunology, the Wenner-Gren Institute
University of Stockholm, Sweden

The destruction of tissue culture cells or red blood cells ("target cells") by lymphocytes *in vitro* is believed to reflect lymphocyte effector functions in the immune response *in vivo*. Table I shows three different cytotoxicity models (reviewed in Perlmann and Holm, 1969). These lymphocyte-mediated cytotoxic reactions may be viewed as two step phenomena. Step 1 involves lymphocyte activation. Activation in model 3 of Table I is caused by added stimulants, but in models 1 and 2 it seems to require lymphocyte-target cell contact. Activation may or may not be followed by blast transformation and enhanced DNA synthesis. Step 2 involves the actual lysis of the target cells. This step is considered to be non-specific, implying that it is immunologically independent of the specific immune reaction that caused lymphocyte activation.

TABLE I

Cytotoxicity of lymphocytes in vitro

---

1. Lymphocytes from donors sensitized to target cell antigens.
2. Lymphocytes + antibodies to target cell antigens. Lymphocyte donors not sensitized to target cell antigens.
3. Lymphocytes activated by phytohaemagglutinin, mixed culture conditions, or antigens unrelated to target cell antigens. Lymphocyte donors not sensitized to target cell antigens.

In the following we will discuss some of the features governing reactions in cytotoxicity models 2 and 3. The discussion will mainly deal with experiments with human blood lymphocytes and chicken red blood cells. Highly purified lymphocytes are added to ^{51}Cr-labelled chicken erythrocytes at optimal lymphocyte : erythrocyte ratios (10:1 — 25:1). Release of isotope from red cells to the medium is used as a measure of the lymphocyte-induced damage to the target cells. Under our experimental conditions target cell damage reaches an end point within 1-2 days. All experimental details have been described (Perlmann *et al.*, 1968; Perlmann and Perlmann, 1970).

INDUCTION OF LYMPHOCYTE CYTOTOXICITY BY ANTIBODIES TO TARGET CELL ANTIGENS

Lymphocytes from specifically sensitized animals are cytotoxic to target cells carrying the antigen against which the lymphocyte donors were sensitized (Perlmann and Holm, 1969). This reaction has, in certain instances, been shown to reflect a state of cell mediated immunity which does not involve humoral antibodies. In some of these models antibodies can inhibit the cytotoxic effect of immune cells on target cells (Möller, 1965; Brunner *et al.*, 1968). Therefore it is interesting that even very low concentrations of certain antibodies to target cell antigens may induce strong cytotoxic effects in lymphocytes from normal donors (Table II).

TABLE II

Induction of lymphocyte-mediated cytotoxicity by antibodies

Lymphocytes	Antibodies	Target cells
Human (blood)	Rabbit anti-CRBC	CRBC
Human (blood)	Rabbit anti-Chang	Chang liver cells[a]
Human (blood)	Human anti-?	Chang liver cells
Rat (spleen)	Rat anti-Chang	Chang liver cells
Guinea pig (blood, spleen)	Guinea pig autoantibodies	CRBC coated with guinea pig thyro-globulin
Duck (blood; donor tolerant to target cell alloantigens)	Duck, anti-target cell alloantigens	Duck-RBC or fibroblasts (allogeneic to lymphocyte donor)

For references see Perlmann and Holm (1969).
CRBC = chicken erythrocytes.
[a] Human cell strain.

The available evidence suggests that not all antibodies can induce lymphocytes to become cytotoxic (Figs 1 and 2). Figure 1 shows the conventional complement-mediated lysis of chicken red blood cells (CRBC) by two different rabbit antisera. One was a hyperimmune anti-CRBC serum, containing a high proportion of IgG antibodies. The other, almost entirely IgM, was an anti-sheep haemolysin, containing antibodies against the Forssman

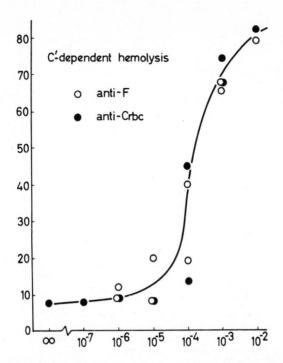

Fig. 1. % Isotope release (ordinate) from chicken erythrocytes (Crbc) treated for 30 min with antiserum and complement (fresh guinea pig serum, final concentration 3.5%). Abscissa: final dilution of antiserum in incubation mixture (∞ = heat-inactivated normal rabbit serum, diluted 10^{-4}, instead of antiserum). Anti-F = rabbit anti-sheep haemolysin (anti-Forssman). Both anti-F and anti-Crbc were heat-inactivated.

antigen, common to sheep and chicken erythrocytes. The haemolytic activity of these two sera was very similar. However, in the absence of added complement, only the anti-CRBC serum was capable of inducing lymphocyte-mediated cytotoxicity (Fig. 2) (Perlmann and Perlmann, 1970). When the antibodies were separated into 7S and 19S fractions on Sephadex G200, the activity was found in the 7S fraction. MacLennan et al. (1969, 1970) also showed that the rat and human antibodies active in the cytotoxic test were IgG. Their results also suggest that the inducing antibodies may belong to a particular subpopulation within the IgG fraction.

Induction of cytotoxicity by antibodies requires contact between target cells and lymphocytes. In these experiments, heat-inactivated antiserum is either added to the target cell-lymphocyte mixture, or the target cells are pretreated with antiserum and washed before the lymphocytes are added. To establish whether the inductive antibodies are cytophilic for lymphocytes, the lymphocytes were treated with antibodies before addition of target cells

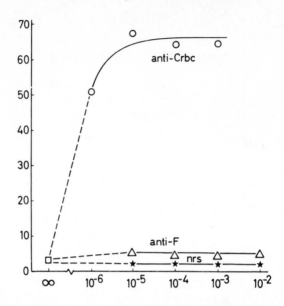

Fig. 2. % Isotope release (ordinate) from chicken erythrocytes (Crbc) incubated for 42 hr with purified lymphocytes (99.6%) in the presence of rabbit anti-Crbc (circles), rabbit anti-Forssman serum (triangles), or normal rabbit serum (asterisks). Rectangle: normal rabbit serum diluted 10^{-4}. Abscissa: final dilution of antiserum in incubation mixture. Lymphocyte: Crbc ratio 25 : 1.

(Perlmann and Perlmann, 1970). Incorporation of antiserum in the incubation mixture resulted in 50% haemolysis at antiserum dilutions of 10^{-7}-10^{-6} (Fig. 3). In contrast, 10^4-10^5 times more antiserum was needed for a corresponding effect if the lymphocytes were pretreated. This suggests that the affinity of the antibodies for the lymphocytes is rather low. In this respect, the inductive antibodies seem to be different from those that confer immunological activity on macrophages or monocytes (Berken and Benacerraf, 1966). This is supported by the fact that induction of lymphocyte cytotoxicity by antibodies is only inhibited by large amounts of normal rabbit IgG, while erythro-phagocytosis by monocytes is easily inhibited by small amounts (Huber and Fudenberg, 1968; Huber et al., 1969). Complement (C3)-mediated adherence of chicken erythrocytes to lymphocytes is not involved in this cytotoxic reaction (Perlmann et al., 1969a; Müller-Eberhard et al., 1969). (It should be noted that lymphocytes with affinity for activated C3 (Nussenzweig, this symposium) have been removed from our effector population by nylon filtration (Perlmann and Perlmann, 1970).) Taken together these data suggest that formation of antigen-antibody complexes on the target cell surface provides the stimulating agent in this cytotoxicity model. Whether this involves a conformational immunoglobulin change which produces affinity for lymphocytes, or whether it

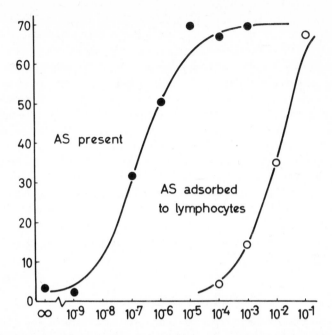

Fig. 3. % Isotope release (ordinate) from chicken erythrocytes (CRBC), incubated for 42 hr with purified lymphocytes (99.2%; lymphocyte : CRBC ratio 50 : 1). AS present: anti-CRBC serum incorporated in incubation mixture at dilution given by abscissa. AS adsorbed to lymphocytes: lymphocytes pretreated for 1 hr at 4°C with antiserum at dilution given by abscissa, washed four times with 10 ml cold Hank's solution. ∞ = antiserum replaced by normal rabbit serum. All sera were heat-inactivated.

simply means that larger complexes rather than single molecules are required for successful interaction is not known.

Interaction between antibody-coated CRBC and lymphocytes does not lead to the formation of stable mixed aggregates (Perlmann and Perlmann, 1970). However increased frequency of uropod formation indicates that the lymphocytes are activated by contact with the target cells (Biberfeld and Perlmann, 1970). Others have shown that contact with antigen-antibody complexes may lead to blast transformation of lymphocytes and enhanced DNA synthesis (Bloch-Shtacher *et al.*, 1968; Möller, 1969). Contact between human peripheral blood lymphocytes and antibody-coated CRBC under our conditions also induces blast transformation and the cytotoxic potential of such populations is strongly enhanced (Stejskal and Perlmann, 1970). While these experiments suggest that the initial contact is followed by lymphocyte activation, they do not imply that blast transformed cells are the actual effector cells in this cytotoxicity model. Blast transformation after activation with antigen-antibody complexes is a late event (4-6 days) while cytotoxicity becomes significant within a few hours (Perlmann and Perlmann, 1970). Moreover these results do not even prove that the cells that will undergo blast transformation are the cytotoxic effector cells during the early phase of interaction.

The origin of the effector lymphocytes in antibody-induced cytotoxicity is unknown. The reaction of lymphocytes from sensitized mice against foreign H-2 mouse tumour cells has recently been shown to involve thymus-dependent lymphocytes, since the reaction can be inhibited by anti-theta serum (Cerottini *et al.*, 1970). This is also believed to be the case in phytohaemagglutinin (PHA)-induced cytotoxicity, since the PHA-responsive cells are assumed to be thymus-dependent. In contrast, recent experiments with rats suggest that antibody-induced cytotoxicity involves lymphocytes of other origin (MacLennan and Harding, 1970). Further experiments are needed to establish the extent to which effector cells of different origin are involved in the different models depicted in Table I. It also remains to be established whether or not co-operation between cells of different origin or function takes place in any of these *in vitro* models.

Lytic Mechanisms

It has recently been reported that cytotoxic factors are released by lymphocytes activated by PHA, antigen, or mixed culture conditions. When media containing such "lymphotoxins" are added to target cells, these are killed in an immunologically non-specific way (Granger and Williams, 1968; Kolb and Granger, 1968; Williams and Granger, 1969). Lymphocyte-derived toxins are assumed to be important mediators of cell-mediated immunity. It should be stressed that most experiments of this type have been performed with target cells growing in monolayers. We have not been able to lyse target cell suspensions by such media, even when they were derived from strongly cytotoxic lymphocytes (Table III). Similar results were obtained with ^{51}Cr-labelled Chang liver cell suspensions (human cell strain). Media that were not lytic for suspended Chang cells sometimes weakly inhibited the growth of monolayers of Chang cells (Holm and Perlmann, 1970). These results indicate that target cells may be destroyed by a number of different pathways. It remains to be established which of these are biologically important, and which represent tissue culture artifacts.

In antibody-induced cytotoxicity the available evidence also supports the notion that cytotoxicity requires close proximity between effector cells and target cells. Thus when mixtures of antibody-coated CRBC and ^{51}Cr-labelled duck erythrocytes were exposed to lymphocytes, no label was released even upon prolonged incubation (three days). The same negative results were obtained in the reverse situation (antibody on duck erythrocytes, label in chicken erythrocytes). In both cases a normal cytotoxic reaction was found

TABLE III

Effect of lymphocytes or various incubation media on ^{51}Cr-labelled chicken erythrocytes (CRBC)

CRBC incubated with	% isotope release from CRBC
Lymphocytes + PHA	59.4
Lymphocytes, no PHA	2.8
Medium from lymphocytes + PHA[a]	1.7
Medium from lymphocytes, no PHA	1.2
Medium from CRBC + PHA[b]	1.5
Medium from CRBC, no PHA	0.8
Fresh medium + PHA[c]	5.7
Fresh medium, no PHA	1.2

2.5 x 10^6 human peripheral blood lymphocytes, purified by nylon fibre filtration, incubated for 44 hr with 1 x 10^5 CRBC in 1.5 ml of Parker's medium 199, containing 20 μg PHA-P (Difco) and 5% fetal calf serum. No PHA in the control. (H. Nilsson, to be published.)

[a] Medium from aliquot of the same lymphocytes, incubated with or without PHA for 44 hr, prepared according to Kolb and Granger (1968). Isotope release from CRBC measured after 24 hr.

[b] Lymphocytes replaced by 1 x 10^6 unlabelled CRBC. Other conditions as in [a].

[c] No cells. Other conditions as in [a].

when antibody and isotope were on the same cells (Perlmann and Perlmann, 1970). Similar results have been obtained with tissue culture cells (Holm and Perlmann, 1969a; MacLennan *et al.*, 1970). They constitute strong evidence against non-specific effects caused by mediators acting on bystander cells not carrying the inducing complex. Corresponding findings have recently also been made when sensitized mouse lymphocytes were added to target cell mixtures of different H-2 type (Brunner, personal communication). Thus the lytic step as well as the inducing step in these cytotoxicity models appears to require close contact between effector cells and target cells. The concept of contactual lysis does not exclude release of cytotoxic mediators acting over a short range in the contact area of the interacting cells.

Extensive experiments have shown that phagocytosis by lymphocytes is of minor importance in killing target cells in the antibody-induced cytotoxic reaction (Perlmann and Perlmann, 1970). Neither does it play a role in the PHA-induced reaction. Some typical micrographs of the latter with CRBC as target cells are shown in Fig. 4. PHA causes formation of mixed aggregates, with lymphocytes frequently being attached to erythrocytes by their uropods. After some hours, progressive destruction of the target cells inside the aggregates can be seen. After 48 hr almost no intact erythrocytes remain in these cultures.

Figure 4(d) shows that the cytotoxic reaction can be completely inhibited in spite of mixed aggregation and blast transformation. In this experiment the lymphocytes were treated with both PHA and the plant mitogen Concanavalin A, which suppresses cytotoxicity but potentiates blast transformation (Perlmann *et al.*, 1970). These findings emphasize again that cytotoxicity and blast transformation are different expressions of lymphocyte activation.

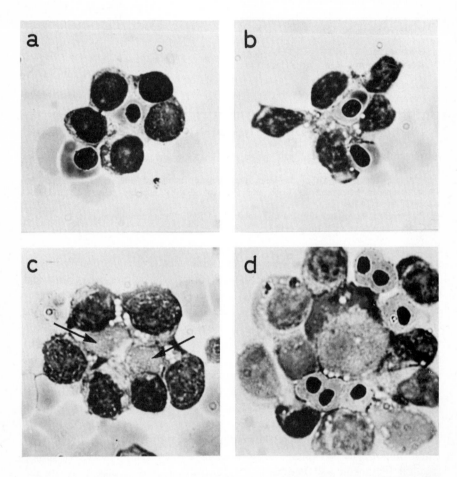

Fig. 4. Interaction of purified lymphocytes with chicken erythrocytes (CRBC) in the presence of PHA (20 μg per tube, see table III). After different times of interaction the cells were sedimented on microscope slides by centrifugation in a cytocentrifuge, and fixed and stained with May-Grünwald Giemsa. Micrographs taken after (a) 3 hr, (b) 15 hr and (c) 24 hr. Arrows show lysed CRBC inside aggregates. (d) Lysis of CRBC was inhibited by Concanavalin A (Perlmann *et al.*, 1970). Micrograph taken after 64 hr of incubation (1000x).

Antibodies to certain lymphocyte-associated immunoglobulin determinants have also been found to suppress cytotoxicity. Table IV shows a typical experiment in which the PHA-induced cytotoxicity of human lymphocytes to CRBC was reduced by addition of rabbit antiserum to human IgG. This treatment did not suppress PHA-induced blast transformation and DNA

TABLE IV

Inhibition of PHA-induced cytotoxicity by rabbit antiserum to human IgG

| Incubation mixture | % isotope release from chicken erythrocytes exposed to human lymphocytes | | | |
| | With PHA | | Without PHA | |
	5 hr	20 hr	5 hr	20 hr
Normal rabbit serum	12.6	54.2	4.1	5.3
Anti-IgG	7.3	24.1	3.6	4.3
Anti-IgG, absorbed[a]	11.6	51.6	4.0	3.9

Conditions of incubation as in Table III. In addition to 5% fetal calf serum, each tube contained 0.3% heat-inactivated normal rabbit serum or rabbit antiserum to human IgG. From Holm and Perlmann (1969b).

[a] Absorbed with human IgG.

synthesis. Figure 5 shows the suppression of the antibody-induced reaction (rabbit anti-CRBC serum) by rabbit antiserum to human kappa or lambda chain determinants. Thus far we have found antibodies to light chain determinants to be the most efficient inhibitors, while antibodies to heavy chain determinants have been only slightly active or inactive (Holm and Perlmann, 1969b). Others have also found anti-L chain antibodies active in suppressing lymphocyte stimulation by antigen or by mixed culture conditions (Greaves *et al.*, 1969). In addition we found an increased surface stainability of stimulated lymphocytes by anti-L chain sera in the indirect immunofluorescence assay (Hellström *et al.*, 1971). These results do not prove that lymphocyte-associated immunoglobulins are directly involved in the cytotoxic reaction. However they do show that immunoglobulin light chain determinants are present on some participating cells in the effector population.

There is some evidence that C8, a late acting complement component, may be of importance for the lymphocyte-mediated cytotoxic reaction. Thus CRBC brought to the C7-stage by addition of small amounts of anti-Forssman antibodies, not producing lysis by themselves, and the first seven complement components in sequence, were readily lysed by lymphocytes (Perlmann *et al.*, 1969a; Müller-Eberhard *et al.*, 1969). Recent experiments, performed in collaboration with Dr P. Lachmann, have also shown that this lymphocyte effect

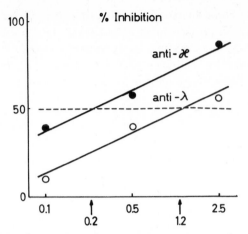

Fig. 5. % Inhibition (ordinate) of isotope release from chicken erythrocytes (CRBC). CRBC were pretreated with heat-inactivated rabbit anti-CRBC serum, diluted 10^{-3}, and incubated for 44 hr with lymphocytes at a lymphocyte : CRBC ratio of 25 : 1. Lymphocytes were purified by nylon filtration and incubation in glass flasks for a day (99.8% lymphocytes). For inhibition they were pretreated for 2 hr at 37°C with heat-inactivated rabbit antiserum to human immunoglobulin light chains at concentrations given by abscissa (= % antiserum, final concentration used for pretreatment). They were then centrifuged and added to target cells without washing. % Inhibition calculated from isotope release obtained with lymphocytes treated with heat-inactivated normal rabbit serum (40% of isotope released at 44 hr. Isotope release in lymphocyte-free control was 3.5%). Arrows indicate antiserum concentration needed for 50% inhibition.

may be obtained in the "reactive lysis" system (Thompson and Rowe, 1968; Thompson and Lachmann, 1970; Lachmann and Thompson, 1970) in which neither antibody nor the first four complement components are needed. In this system, erythrocytes are brought to the C7-stage by mixing the complex $C\overline{56}$ (present in excess in certain sera) with C7 in the presence of the erythrocytes. Such erythrocytes are very stable but are rapidly lysed by the addition of lymphocytes. Addition of human erythrocytes from the same donors has no effect. In the controls, $C\overline{56}$ is mixed with C7 5 min before addition of CRBC. Such cells are not susceptible to lysis. This very efficient lytic system can be completely inhibited by the addition of rabbit antiserum to human C8, but not by antiserum to C3 or C4. Details will be published elsewhere.

A C8-active fraction has also been found in sonicated lymphocytes (Müller-Eberhard *et al.*, 1969). Since both PHA- and antibody-induced cytotoxicity can be inhibited by antiserum to C8 but not by antisera to C1q, C2, C3 or C4 (Perlmann *et al.*, 1969b; unpublished observations), it may be that activation of lymphocyte-associated C8 in connection with lymphocyte activation may constitute a more general effector mechanism for lymphocyte-mediated lysis. This does not invalidate the concept of contactual lysis, since activated C8 in solution may have a very short half life, and would not be able to act as a soluble mediator. Further experiments are needed to establish the significance of complement components for lymphocyte-mediated target cell destruction and other functions.

ACKNOWLEDGEMENTS

This work was supported by grants no. 2034-29 from the Swedish Natural Science Research Council, and no. B70-16X-148 from the Swedish Medical Research Council. The skilled technical assistance of Miss G. Halldén is gratefully acknowledged.

SUMMARY

Chicken erythrocytes are destroyed by human blood lymphocytes activated by PHA. Minute amounts of certain antibodies to chicken erythrocytes may also induce a lymphocyte-mediated haemolysis. The reaction is brought about by highly purified lymphocytes in the absence of added complement. IgM antibodies seem to be inactive, and the inducing antibodies have a very low cytophilic affinity for lymphocytes. The evidence suggests that the reaction is induced by contact between lymphocytes and antigen-antibody complexes on the surface of the target cells. Induction is followed by lymphocyte activation, leading to blast transformation. However blast transformation is a late event, while cytotoxicity occurs early. It is not known whether the lymphocytes that will become blast cells are the cytotoxic effector cells. The origin of the effector lymphocytes and the possible occurrence of cell co-operation in this model remain to be established.

PHA-activated lymphocytes do not lyse the chicken erythrocytes by releasing soluble mediators. There is strong evidence that the lytic step in antibody-induced haemolysis also requires lymphocyte-erythrocyte contact, or close proximity. Antibodies to human immunoglobulin light chains inhibit both PHA- and antibody-induced cytotoxic reaction. While these data do not prove that lymphocyte-associated immunoglobulins are directly involved in the reaction, they indicate that at least some of the participating lymphocytes carry surface bound immunoglobulin. One of the late acting complement components, C8, seems to be lymphocyte-associated and to be an effector in the cytotoxic reactions described. Whether activation of lymphocyte-borne C8 is important in lymphocyte-mediated cytotoxic reaction in general is not known.

REFERENCES

Berken, A. and Benacerraf, B. (1966). *J. exp. Med. 123,* 119-144.
Biberfeld, P. and Perlmann, P. (1970). *Expl Cell Res. 62,* 433-440.
Bloch-Shtacher, N., Hirschhorn, K. and Uhr, J. W. (1968). *Clin. exp. Immun. 3,* 889-899.
Brunner, K. T., Mauel, J., Cerottini, J.-C. and Chapuis, B. (1968). *Immunology 14,* 181-196.
Cerottini, J.-C., Nordin, A. A. and Brunner, K. T. (1970). *Nature, Lond. 228,* 1308-1309.
Granger, G. A. and Williams, T. W. (1968). *Nature, Lond. 218,* 1253-1254.
Greaves, M. F., Torrigiani, G. and Roitt, I. M. (1969). *Nature, Lond. 222,* 885-886.
Hellström, U., Zeromski, J. and Perlmann, P. (1971). *Immunology.* (In press.)
Holm, G. and Perlmann, P. (1969a). *In* "Proceedings of the 4th Annual Leucocyte Culture Conference" (R. McIntire, ed.). (In press.)
Holm, G. and Perlmann, P. (1969b). *In* "Human Anti-human Gamma Globulins" (R. Grubb, ed.), pp. 217-223.
Holm, G. and Perlmann, P. (1970). In preparation.
Huber, H. and Fudenberg, H. H. (1968). *Int. Archs Allergy appl. Immun. 34,* 18-31.
Huber, H., Douglas, S. D. and Fudenberg, H. H. (1969). *Immunology 17,* 7-21.
Kolb, W. P. and Granger, G. A. (1968). *Proc. natn. Acad. Sci. U.S.A. 61,* 1250-1255.
Lachmann, P. J. and Thompson, R. A. (1970). *J. exp. Med. 131,* 643-657.
MacLennan, I. C. M. and Harding, B. (1970). *Nature, Lond. 227,* 1276-1278.
MacLennan, I. C. M., Loewi, G. and Howard, A. (1969). *Immunology 17,* 897-910.
MacLennan, I. C. M., Loewi, G. and Harding, B. (1970). *Immunology 18,* 397-404.
Möller, E. (1965). *J. exp. Med. 122,* 11-23.
Möller, G. (1969). *Clin. exp. Immun. 4,* 65-82.
Müller-Eberhard, H. J., Perlmann, P., Perlmann, H. and Manni, J. A. (1969). *In* "Current Problems in Immunology" (O. Westphal, Bock and Grundmann, eds), p. 5. Springer-Verlag, Berlin.
Perlmann, P. and Holm, G. (1969). *Adv. Immun. 11,* 117.
Perlmann, P. and Perlmann, H. (1970). *Cellular Immun. 1,* 300-315.
Perlmann, P., Perlmann, H. and Holm, G. (1968). *Science, N.Y. 160,* 306-309.
Perlmann, P., Perlmann, H., Müller-Eberhard, H. J. and Manni, J. A. (1969a). *Science, N.Y. 163,* 937-939.
Perlmann, P., Perlmann, H., Holm, G., Müller-Eberhard, H. J. and Manni, J. A. (1969b). *In* "Proceedings of 4th Annual Leucocyte Conference" (R. McIntire, ed.). (In press.)
Perlmann, P., Nilsson, H. and Leon, M. A. (1970). *Science, N.Y. 168,* 1112-1115.
Stejskal, V. and Perlmann, P. (1970). In preparation.
Thompson, R. A. and Lachman, P. J. (1970). *J. exp. Med. 131,* 629-641.
Thompson, R. A. and Rowe, D. S. (1968). *Immunology 14,* 745-762.
Williams, T. W. and Granger, G. A. (1969). *J. Immun. 102,* 911-918.

GENERAL DISCUSSION

General Discussion

Chairman

N. A. MITCHISON

1. Rosette Forming Cells

NUSSENZWEIG. Excuse me for presenting new data in this general discussion, but what I have to say may be of importance to all "rosettologists" and also to all of those looking for interactions between lymphocytes and a potential antigen. A very crucial question is of course whether the binding of an antigen to a lymphocyte means that it is specifically bound (through an Ig receptor) and whether this lymphocyte is a precursor of an antibody forming cell.

I want to report some data obtained from Dr W. Lay in Sao Paulo, in collaboration with Dr N. Mendes, C. Bianco and myself. We found that human lymphocytes (from the peripheral blood or from the thoracic duct) can, under appropriate conditions, form "rosettes" when incubated with SRBC. The remarkable thing is that 30% or more of the lymphocytes form "rosettes", and this is not related to age, sex or race of the lymphocyte donor. The conditions for obtaining these "rosettes" are:

(a) no rosettes are formed when the incubation is carried out at $0°C$ or $37°C$;

(b) "rosettes" are formed when the mixture of lymphocytes and red cells is incubated at $37°C$ for 30 min, followed by centrifugation and a second incubation in the cold for a couple of hours.

This is in some way a specific interaction because (a) only human lymphocytes (but not polymorphs or monocytes) form "rosettes". Lymphocytes from other species (from blood, lymph nodes or thymus) form only a very small number of "rosettes" with SRBC; (b) only SRBC (and to a much smaller extent pig red cells–PRBC) form "rosettes" with human lymphocytes. Other erythrocytes tested with negative results are: mouse, horse, cow, rat, guinea-pig, rabbit. In addition, it appears that the same population of lymphocytes reacts with both PRBC and SRBC, because when the lymphocytes are incubated with a mixture of SRBC and PRBC, most of the "rosettes" formed are mixed, containing both types of red cells.

It is also curious that in some cases in which it has been possible to test thymocytes and circulating lymphocytes from the same patient, a higher percentage of "rosettes" was found with thymocytes. In one patient this number came close to 100%. On the other hand, lymphocytes from two lines of Burkitt lymphoma were entirely negative.

Now it is rather hard to believe that close to 100% of human thymocytes are cells precommitted to produce anti-SRBC antibodies. I think that this kind of result points to the caution that is needed to interpret "rosette" formation. Unfortunately, the inhibitory effect of anti-Ig sera on "rosette" formation again cannot be considered definitive proof of an immunological mechanism, because of possible steric hindrance from adjacent sites.

GREAVES. We have also found high percentages of "rosette"-forming cells in human blood. I suspect that these reactive cells are not binding by virtue of specific immunoglobulin receptors, since they cannot be inhibited by anti-light chain sera. Very few, if any, of such non-specifically binding cells appear to exist in mice, but this clearly is a problem for those wishing to work with human lymphocytes. However, since these cells appear to constitute a distinct sub-population they can presumably be removed by gradient sedimentation following rosette formation as described by Dr Bach.

2. Stability of the Theta Antigen as a Marker of T Cells: Evidence For and Against the Theta Antigen as a Stable Marker on Thymus-derived Cells

E. MÖLLER. Findings to support the idea that the theta isoantigen is a stable marker on thymus-derived cells in the peripheral lymphoid organs of *normal* mice:

(a) The proportion of anti-theta-sensitive cells is 20-40% in the normal spleen, 65-70% in the lymph nodes, 100% in the thymus and 70-80% in peripheral blood (Raff).

(b) Anti-theta serum treatment of spleen cells *in vitro* significantly inhibits: the primary immune response of transferred cells (Raff, Möller-Greaves), the primary immune response *in vitro* (Dutton), the secondary response of adoptively transferred spleen cells from donors primed with low doses of SRBC (Möller-Greaves), the helper function of anti-carrier cells in an adoptive response (Raff) and the cytotoxic activity of sensitized lymphoid cells on allogeneic target cells *in vitro* (Brunner).

Findings to support the idea that theta is a stable antigen on thymus-derived cells in *thymectomized* animals:

(a) The proportion of theta-sensitive lymph node cells in mice thymecto-mized as adults 5-8 weeks previously is 55% (Raff).

Evidence suggesting that the theta antigen is *not* a stable marker on thymus-derived cells in *thymectomized* animals:

(a) Disappearance of theta-sensitive lymph node cells within 21 days after adult or neonatal thymectomy (Schlesinger-Yron).

(b) Higher proportion of theta positive antigen binding spleen cells in adult thymectomized, irradiated, thymus-bone marrow reconstituted mice than in normal mice (Greaves-Möller).

(c) Lack of theta positive antigen binding cells in animals thymectomized five months prior to immunization with SRBC (Möller).

(d) Gradual decrease of theta positive antigen binding cells in mice 25-30 days after immunization with SRBC on day 0 and thymectomy on day 7 (Möller).

MITCHISON. Would you conclude from these experiments that theta is an unstable marker?

E. MÖLLER. No, we must leave this open.

RAFF. In view of the adult thymectomy data I think that theta is not markedly unstable.

NOTE ADDED AFTER MEETING

RAFF and GREAVES. In view of the striking discrepancies in the effect of adult thymectomy on θ-bearing peripheral lymphocytes we have done a further experiment following the Helsinki meeting to try to clarify this issue. Looking at CBA mice, 5½ months post-adult thymectomy (done at 6 weeks of age) we found that 55% of the peripheral lymph node cells were θ-bearing by dye exclusion cytotoxic testing as opposed to 72% in age-matched control mice. The cytotoxic titre was the same for lymph node cells from the thymectomized and normal mice. Seven days after i.p. injection of 4×10^8 SRBC, there were slightly less rosette-forming cells (RFC) in the spleens of the thymectomized mice when compared to controls, but 30% of the RFC could be inhibited by anti-θ in both groups.

We conclude that θ is a stable marker of thymus-derived lymphocytes in CBA mice and propose to exchange antisera and possibly mice with those workers who are obtaining results that differ from ours.

3. TRIGGERING OF CELLS, AND TRIGGERING THRESHOLDS FOR IMMUNITY AND TOLERANCE

HUMPHREY. Dr Mitchison asked me to comment on the problem of why it should apparently be easier to paralyze T cells than B cells. One reason why it is difficult to answer this question is that we still do not know how any cell becomes paralyzed. The hypothesis which is simplest, and is in least conflict with experimental evidence is that specific paralysis depends upon the physical elimination of specifically responsive lymphocytes as a result of combination of their receptors with the antigen. This could occur in one of two ways. The first would be intensive stimulation (by a large excess of antigen) of potential antibody producers (B cells) to differentiate into terminal plasma cells, without simultaneous generation of memory cells (suicidal proliferation) thereby exhausting the stock of responsive precursors of producing cells. The second would be elimination, either by opsonization and ingestion by macrophages or by C' dependent lysis, and would in principle apply to both T and B cells. It implies that interaction of antigen with the cell receptors under certain circumstances activates C' at the cell surface, with disastrous effects for that cell. In a very general way this is what Bob Good and his colleagues have been setting out to prove by testing whether tolerance is more difficult to induce under conditions of C' deficiency (Yunis, Pickering and Good; *Lancet* 1968, *i*, 1279).

Diener's very important experiments (Feldman and Diener, *J. exp. Med.* 1970, *131*, 24) referred to by Gus Nossal point in the same direction, since they indicate that very low levels of antigen may induce tolerance, in the presence of traces of antibody, by a process involving building up an antigen-antibody lattice on top of the cell receptors. In discussing with Dr Paul Plotz a possible way of putting this idea to the test I learned that he was actually doing the experiment! If the hypothesis is correct it should be possible to show that cells with specific receptors treated with antigen and then with antibody against the antigen, or even with the antigen alone, would be selectively removed *in vitro* by macrophages and/or killed by C'; this would be revealed as a specific depletion of such cells from the treated population. I hope that the experiment will be successful, but since the outcome is not known there is nothing to inhibit further speculation on my part. In order to explain greater susceptibility of T cells to paralysis one could postulate that combination of free antigen with their receptors (± antibody on top) results in more effective opsonization than in the case of B cells. This could be either because of the intrinsic nature of T cell receptors (e.g. IgM, permitting multipoint attachment), or because opsonized T cells are more exposed to removal during their recirculation, or both. If B cells include a variety with receptors representing different immunoglobulin classes, it seems to me quite conceivable that interaction of some of these with free antigen would fail to result in opsonization, and that such cells would be particularly difficult to paralyze. (A consequence of this hypothesis would be that a regime which caused only partial paralysis would selectively affect production of one class of antibody more than another—I do not know whether this is what actually occurs.) However, they might also survive better because their exposure to antigen occurred in a different micro-environment. To bring in the micro-environment is a confession of ignorance, but I fear that it cannot be ignored.

G. MÖLLER. It seems likely that lymphocytes have surface receptors for antigen. These receptors would have very different binding affinities for the antigen on different cells. I suggest that lymphocytes become triggered when the binding energy at the surface caused by the reaction between receptor and antigen is sufficiently strong. As an example it may require 100 antigen molecules bound simultaneously for 10 ms in order for the lymphocytes to become triggered. This condition would be more easily met with high affinity cells which would bind the molecules efficiently. However, with low affinity cells the rate of dissociation would be so high that the chance of binding a sufficient number of antigenic antigens for a sufficient time would be too low for triggering to occur. Therefore, a variety of helper mechanisms have been developed, which all act to decrease dissociation between receptors and antigen. Five such helper mechanisms are listed below.

1. T and B cell interaction.

2. Macrophage and B cell interaction.
3. Repeated antigenic determinants (polysaccharides).
4. Antigen bound to antibodies (IgM).
5. Naturally concentrated antigens (cell surface histocompatibility antigens).

The T and B cell interaction, which has been experimentally verified would, according to this concept, act by binding the antigen at two sites. By doing this the chance of dissociation of the whole molecule from both cells would be much lower than if the molecules were only bound to one cell. Even when antigen dissociates from the B cells it would be sticking to the T cells and reassociation to the B lymphocytes would be a likely event. In this interaction it is not necessary that the helper cell is a T lymphocyte. It could equally well be another cell type, for instance a macrophage. The main purpose of the interaction would be to decrease dissociation of antigen from the receptors of the B cells.

Another situation in which increased association between antigen and cell surface receptors would occur is when the antigen is made up of repeated determinants. Polysaccharides can be considered as having repeated antigenic determinants on a common backbone. If such an antigen reacts with antigen-binding receptors, the chance of dissociation of the entire molecule from the cell is much reduced, since it requires that all determinants dissociate from the receptors at the same time. Therefore, repeating antigenic determinants would constitute their own helper function and would be competent to induce an immune response by themselves. It has been shown by us that endotoxin, which is a molecule of this type, is not dependent upon thymocytes for its immunogenicity. Furthermore, such an antigen would also be highly competent to induce tolerance. It may also be argued that immunization with this antigen would induce tolerance in the thymus-dependent lymphocytes, because these cells are likely to have a lower threshold of triggering. It was actually found by Sjöberg that the rosette-forming cells binding endotoxin were theta negative, both in immune and tolerant animals, and therefore not derived from the thymus.

If a divalent antibody binds two antigen molecules and the antigen is presented to the cells in this form, it could react with the lymphocyte receptors. Since the antigen is bound at two sites the chance for dissociation from the cell surface has decreased considerably. If the antigen is bound to an IgM antibody, capable of binding five antigen molecules, the degree of dissociation would be even less likely and in this case triggering would be efficient. It is well known that IgM over wide dose ranges is highly stimulatory of the immune response.

If the antigen is naturally concentrated on a particle—the most relevant example being cell surface histocompatibility antigens of the strong H-2 or HL-A locus—the situation would be analogous to that with repeated antigenic determinants. In this case the antigen molecules would be tightly packed on the surface of a lymphocyte, and in that way a number of receptors on the responding lymphocytes would bind to the antigen, and the chance of dissociation would be much decreased. In this situation, as well as when repeated antigenic determinants are presented to a cell, it would be expected that even low affinity cells would have a high chance of being stimulated by antigen. Therefore, one would also expect that the number of antigen-sensitive cells would be higher. It is well known that histocompatibility antigens differ from other antigenic systems by the large number of antigen-sensitive cells. As shown by Wilson this number is as high as 1-3%, whereas it is 0.0001-0.01% with most other antigens. This high frequency of antigen-sensitive cells constitutes the basis for Jerne's hypothesis on antibody variability.

I would like to suggest that the high frequency of antigen-sensitive cells to histocompatibility antigens can be explained by two pecularities of these antigenic systems. 1. The test systems which have been used for determination of the high frequency of antigen-sensitive cells have been the mixed lymphocyte culture *in vitro* or the graft-*versus*-host reaction *in vivo*. Both test systems most probably detect T lymphocytes specifically, whereas B lymphocytes are not involved. As already stated it appears likely that T lymphocytes have a much lower threshold of triggering than B lymphocytes. The evidence from this comes from various experiments (e.g. E. Möller and M. Greaves in this volume). Therefore, cells having low affinity receptors on their T cells would be triggered in a higher proportion in test systems detecting preferentially T lymphocytes. It follows from this concept that a high number of antigen-reactive cells to histocompatibility antigens would not be found if humoral antibody synthesis was studied, since the B cells require a higher threshold for triggering. 2. The transplantation antigens are naturally concentrated at the surface of the inducer lymphocyte and, according to the above concept, would efficiently bind to the receptors of the responding lymphocytes. By the natural antigen concentration, dissociation between the antigen and the responding cell receptors would be decreased. This would result in stimulation of low affinity cells. The fact that the T lymphocytes are studied in tests for antigen-sensitive cells against histocompatibility antigen, and that the antigens in question are naturally concentrated and therefore highly competent to induce immunocytes, may be sufficient to explain the high number of antigen-reactive cells against histo-compatibility antigens. According to the argument most of the responding cells would be T lymphocytes with low affinity receptors.

There are some facts suggesting that the histocompatibility antigens by themselves are not highly immunogenic, but only when they are presented as a naturally concentrated antigen packet. Thus, a mixed lymphocyte culture only works with intact living lymphocytes, whereas it is very weak or absent if allogeneic fibroblasts are added or if the lymphocytes have been broken up by freezing and thawing or ultrasound treatment even though the antigenicity has been preserved in these extracts.

Author Index

Numbers with an asterisk refer to pages on which references are listed at the end of the paper

Subject Index

A

Activated lymphocytes, 15
Adherent cell population, 9
Adoptive response, 367
Affinity labelling
  BADE, BADL, 182
  BANIT, 177
  NIP-azide,171
  NIP-CAP-NPE, 177
  serum antibodies, 183
Agammaglobulinaemia, 29, 43
Aggregation, 393
Albumin density gradients, 57, 402
Alkaline phosphatase, 71
ALS, 84, 427
Amplification of responses, 157, 166
Antibody combining site, 188
Antibody-forming cell precursors, 251
Antigen-binding cells, 101, 139, 145
  frequency of, 159
  increase after immunization, 164
Antibody-secreting cells, 197, 232
Antigen-coated columns, 231
Antigen concentration, 3
Antigenic competition, 419
Antigen sensitive cells, 32, 111, 421, 454
  thymic origin, 101
Antigen uptake, 4
Anti-immunoglobulin sera, 127, 146, 164, 198, 237, 443, 450
Anti-lymphocyte serum 84
Anti-macrophage serum, 9
Antimetabolites, 118
Anti-μ chain sera, 128
Anti-theta antibody, 83, 105, 119, 130, 141, 146, 450
Arsanilic acid, 189
Autoimmunity, 261, 275
Avidity of antibody, 325
  effect of antigen dose, 326
  effect of thymectomy, 326
Avidity of PFC, 202

B

BADE, BADL, 182
BANIT, 177
Beads, 231
Beta-galactosidase binding cells, 157
Billiard balls, 160
Blast transformation, 439
B lymphocytes, 325
  memory, 359
  tolerance, 253, 264, 269
Bone marrow grafting, 36, 42
Burkitt lymphoma, 91
Bursal system development, 27
Bursa of Fabricius, 28, 41
Bursectomy, 30
Burst size of clones, 384

C

Carrier specificity, 187, 353, 367
  optical isomerism of carrier, 190
Cell co-operation, 31, 66, 130, 157, 240, 149, 261, 270, 293, 323, 325
  corticosteroid effects, 333, 344
  effect on antibody class, 367
  in development, 393
  in vitro, 9, 15, 294, 373, 379
Cell density, 394
Cell-mediated immunity, 27, 436
Cell recognition, 395, 399
Cell surface IgM, 91
  7S IgM, 94
Cell surface immunoglobulin
  quantitation, 92
Cell surface receptors, 123, 152, 231, 367, 449, 451
  avidity, 235, 452
  number per cell, 159, 256
  on T lymphocytes, 61
  replacement, 118, 175
  similarity to cell's product, 243, 357